Fragment-Based Drug Discovery

Fragment-Based Drug Discovery

A Practical Approach

Editors

EDWARD R. ZARTLER

Merck & Co., Inc., USA

MICHAEL J. SHAPIRO

School of Pharmaceutical Sciences, University of Maryland, USA

WILEY

A John Wiley and Sons, Ltd, Publication

This edition first published 2008

Registered office
John Wiley & Sons, Ltd, The Atrium, Southern Gate, Chichester, West Sussex, PO19 8SQ, United Kingdom

For details of our global editorial offices, for customer services and for information about how to apply for permission to reuse the copyright material in this book please see our website at www.wiley.com.

Library of Congress Cataloging-in-Publication Data

Fragment-based drug discovery : a practical approach / [edited by] Edward Zartler and Michael Shapiro.
p. ; cm.
Includes bibliographical references and index.
ISBN 978-0-470-05813-8 (cloth)
1. Drug development. 2. Drugs—Design. 3. Ligands (Biochemistry) I. Zartler, Edward. II. Shapiro, Michael (Michael J.) [DNLM: 1. Drug Design. 2. Ligands. 3. Magnetic Resonance Spectroscopy—methods. QV 744 F8115 2008]
RM301.25.F73 2008
615′.19—dc22

2008027930

British Library Cataloguing in Publication Data

A catalogue record for this book is available from the British Library.

978-0-470-05813-8 (H/B)

Set in 10/12pt Times by Integra Software Services Pvt. Ltd, Pondicherry, India
Printed in Great Britain by CPI Antony Rowe, Chippenham, Wiltshire

Contents

List of Contributors

Jeffrey S. Albert CNS Discovery Research, AstraZeneca Pharmaceuticals, 1800 Concord Pike, P.O. Box 15437, Wilmington, DE 19850-5437, USA

Mark Brewer Evotec (UK) Ltd, 114 Milton Park, Abingdon, Oxfordshire OX14 4SA, UK

Mike Cherry Accelrys Ltd, 334 Cambridge Science Park, Cambridge CB4 OWN, UK

Brian L. Claus Bristol Myers Squibb, Research and Development, P.O. Box 4000, Princeton, NJ 08543, USA

Keith L. Constantine Bristol Myers Squibb, Research and Development, P.O. Box 4000, Princeton, NJ 08543, USA

J. Christopher Culberson Molecular Systems, Merck Research Laboratories, WP53F-301, P.O. Box 4, West Point, PA 19846, USA

Malcolm E. Davis Bristol Myers Squibb, Research and Development, P.O. Box 4000, Princeton, NJ 08543, USA

Philip D. Edwards CNS Discovery Research, AstraZeneca Pharmaceuticals, 1800 Concord Pike, P.O. Box 15437, Wilmington, DE 19850-5437, USA

Daniel A. Erlanson Sunesis Pharmaceuticals, Inc., 341 Oyster Point Boulevard, South San Francisco, CA 94080, USA

Bradley Feuston Molecular Systems, Merck Research Laboratories, WP53F-301, P.O. Box 4, West Point, PA 19846, USA

Virginie Früh Leiden Institute of Chemistry, Leiden University, Leiden, The Netherlands

Valentina Goldfarb Bristol Myers Squibb, Research and Development, P.O. Box 4000, Princeton, NJ 08543, USA

Stig K. Hansen Sunesis Pharmaceuticals, Inc., 341 Oyster Point Boulevard, South San Francisco, CA 94080, USA

Robert J. Heetebrij ZoBio, Leiden, The Netherlands

M. Katharine Holloway Molecular Systems, Merck Research Laboratories, WP53F-301, P.O. Box 4, West Point, PA 19846, USA

Osamu Ichihara Evotec (UK) Ltd, 114 Milton Park, Abingdon, Oxfordshire OX14 4SA, UK

Adriaan P. IJzerman Leiden/Amsterdam Center for Drug Research, Division of Medicinal Chemistry, P.O. Box 9502, 2300 RA Leiden, The Netherlands

Stephen R. Johnson Bristol Myers Squibb, Research and Development, P.O. Box 4000, Princeton, NJ 08543, USA

Christian Kirchhoff Evotec AG, Schnackenburgallee 114, 22525 Hamburg, Germany

Gary H. Kruppa Bruker Daltonics Inc., 2859 Bayview Drive, Fremont, CA 94538, USA

Patricia A. McDonnell Bristol Myers Squibb, Research and Development, P.O. Box 4000, Princeton, NJ 08543, USA

Georgia McGaughey Molecular Systems, Merck Research Laboratories, WP53F-301, P.O. Box 4, West Point, PA 19846, USA

William J. Metzler Bristol Myers Squibb, Research and Development, P.O. Box 4000, Princeton, NJ 08543, USA

Tim Mitchell Sareum Ltd, 2 Pampisford Park, Cambridge CB2 4EE, UK

Luciano Mueller Bristol Myers Squibb, Research and Development, P.O. Box 4000, Princeton, NJ 08543, USA

Sally-Ann Poulsen Eskitis Institute, Griffith University, Brisbane, Queensland 4111, Australia

Markus Schade Pfizer Ltd, PGRD, Sandwich, Kent, CT13 9NJ, UK

Johan Schultz iNovacia, Lindhagensgatan 133, 11251 Stockholm, Sweden

Michael J. Shapiro Department of Pharmaceutical Chemistry, School of Pharmacy, University of Maryland, Baltimore, MD 21201, USA

Gregg Siegal Leiden Institute of Chemistry, Leiden University and ZoBio, Leiden, The Netherlands

Eelke van der Horst Leiden/Amsterdam Center for Drug Research, Division of Medicinal Chemistry, P.O. Box 9502, 2300 RA Leiden, The Netherlands

Mark Whittaker Evotec (UK) Ltd, 114 Milton Park, Abingdon, Oxfordshire OX14 4SA, UK

Edward R. Zartler Biologics Analytical and Formulation Sciences, Merck & Co., West Point, PA 19486, USA

1

Introduction to Fragment-based Drug Discovery

Mike Cherry and Tim Mitchell

1.1 Introduction

Fragment screening is the process of identifying relatively simple, often weakly potent, bioactive molecules. It is gaining wide acceptance as a successful hit-finding technique both in its own right and as a method of finding hit molecules when traditional high-throughput screening (HTS) methods fail. Fragment hits are typically highly 'ligand efficient', i.e. possess a high binding affinity per heavy atom, and thus are ideal for optimisation into clinical candidates with good drug-like properties. Fragment screening is being increasingly proven as a successful means of generating novel chemical starting material for drug discovery programmes. It has been the subject of numerous publications and reviews in the last few years.[1–5]

Fragment screening was initially developed to generate hit compounds against targets for which other methods, such as HTS, had been unsuccessful.[6] At the same time, many shortcomings of HTS were becoming increasingly apparent:

1. The hits being generated from high-throughput screening of combinatorial chemistry-derived libraries are not particularly suitable for lead optimisation programmes: These compounds tended to be large and hydrophobic and thus had limited potential for development before becoming in violation of 'drug-like' parameters as described by Lipinski *et al.*[7] 'Garbage in, garbage out really applies to drug screening', as Lipinski *et al.* quote.[8]

Fragment-Based Drug Discovery: A Practical Approach Edited by Edward R. Zartler and Michael J. Shapiro

2. Hit rates from HTS are often low and the hits obtained fail to progress into lead optimisation: HTS is the predominant technique for hit finding employed by the majority of pharmaceutical companies and is central to modern drug discovery. However, many scientists are now regarding HTS as a costly necessity rather than a method of choice.[9]
3. HTS only samples a minute fraction of drug-like chemical space: A widely quoted estimate[10] of the number of molecules containing up to 30 C, N, O or S atoms exceeds 10^{60} (and the mass of a 1 pmol sample of each would of the same order as that of the observable universe). An HTS screen of, say, 10^6 molecules would only sample a minute portion of this space. By contrast, the number of synthetically feasible small molecule fragments with masses up to 160 Da has been estimated[11] at 10^7, hence a typical fragment screen of 10^3–10^4 molecules is sampling a much higher proportion of this space.
4. Many companies have realised, to their significant cost, that screening non-proprietary vendor compounds against non-proprietary targets can lead to difficulties in securing a good intellectual property (IP) position: Increasingly, there is the possibility that another company has screened similar compounds against a similar target and therefore it is difficult to obtain strong patent protection against the chemical series. If competitor patents do not appear for some time after the initiation of a drug discovery programme, significant resources can be wasted on attempting to optimise compounds that another organisation had already discovered and patented[12].

Hence there is increased pressure to discover more suitable chemical starting points and the term 'lead-like' has been used to describe these less complex molecules.[13] By its very nature, fragment screening is ideally suited to finding low molecular weight bioactive compounds. Because these compounds tend to be relatively low in potency (typically in the 100–1000 μM range) they are not identifiable by an HTS run at a typical compound concentration of 10–30 μM. Fragment screening collections, even those formed from non-proprietary vendor compound collections, lie beyond the scope of HTS compound collections, thus increasing the chances of identifying novel molecules that can be optimised into patentable series.

Subsequent chapters will discuss in more detail and provide case studies of best practices of using fragments in drug discovery programmes. It is useful, though, to start with an understanding of what is meant by a fragment and how the use of fragments has reached the current level of popularity.

1.2 What is a Fragment?

A quick search of the literature using the term 'drug-likeness' throws up thousands of articles, reflecting the aspiration of the industry to be able to classify compounds readily as drug-like or not. A wide variety of methodologies has been employed in the process of classification, from simple filters based on physicochemical descriptors through to more complex QSAR models. Although the concept of fragment-based drug discovery has been around since the early 1980s, the application is relatively new and as such the volume of literature around the subject can be measured in hundreds, not thousands. There are a rapidly increasing number of studies investigating what makes a good starting point in fragment-based drug discovery and how to formulate libraries to maximize success in the screening process.

The exact nature of a fragment library is very much dependent on the screening protocol; however, the methods employed to construct fragment libraries borrow heavily from experiences in drug-like classification. A simple analogy is that of the Astex 'rule-of-three'[14] as compared with Lipinski *et al.*'s 'rule-of-five'[15]. Here the same physicochemical descriptors are used but compounds are limited to a molecular weight (MW) of 300 Da, three or less hydrogen bond donors and acceptors and a calculated log$P$ of ≤ 3. The Astex library is comprised of several hundred small organic ring systems, is a mixture of target-specific and general-purpose fragments and is used to probe a target site using high-throughput X-ray crystallographic screening.

The application of a filter such as the 'rule-of-three' and subsequent identification of target-specific compounds are in general the penultimate steps in the virtual selection of fragments from a significantly larger chemical space. Additional steps common to most, if not all, selection criteria for fragment libraries, particularly if starting from commercial vendor space, include the removal of undesirable chemical functionality, elimination of poorly soluble compounds, a selection based on synthetic tractability and consideration of scaffold diversity. It is also probably safe to say that the final step in most virtual screening campaigns involves scientists eyeballing compounds, predominantly to make the final selection but also to ensure the baby is not being thrown out with the bath water. The virtual screening process is commutative with respect to the final result, excepting the manual selection; however, the order of operation will impact on efficiency. 1D and 2D filters, such as the exclusion of undesirable chemical functionality using substructure searching, can eliminate a high percentage of compounds,[16] thus reducing the resources needed for computationally more expensive procedures such as pharmacophore searching and high-throughput docking.

SGX[17] outline a series of such criteria in the selection of a $\sim$1000 member diverse fragment library that includes filters for MW, *C*log*P*, compound complexity, exclusion of undesirable chemical functionality, solubility, ring system diversity, synthetic accessibility and, interestingly, a selection based on bromination. All the compounds in the library have $\leq$16 heavy atoms, $\geq$1 ring and at least two synthetic handles. Most of the compounds obey the 'rule-of-three' and, importantly for screening using X-ray crystallography, 60 % have high solubility. Whereas Astex have target-specific fragments, SGX make no such distinction on the basis that hit rates are in general higher in fragment screening and a library of 1000 diverse fragments is deemed large enough to yield sufficient chemical matter to initiate a discovery programme. The size of a fragment library, the complexity of the molecules within the library and the optimisability are all characteristics important to fragment screening and are discussed in more detail below.

Approximately half of the compounds in the SGX library contain bromine, a feature included to enhance synthetic elaboration but also aid in the detection and validation of crystallographic screening data. Compounds in the original 'SHAPES' library developed at Vertex,[18] in addition to many of the aforementioned filters, had to yield simple ^{1}H NMR spectra and contain at least two protons within 5 Å of one another, both aids to the screening of mixtures using nuclear magnetic resonance (NMR) techniques. The library stemmed from previous work investigating the properties of known drugs,[19, 20] where computational methods were used to break down and analyse the constituent components of a database of commercially available drug molecules. Molecules are split into ring systems, linkers, side-chains and frameworks, where a framework is defined as the union of ring systems and linkers in a molecule. A surprisingly small number of frameworks (41), taking into consideration atom types and bond orders, describe the framework of ~24 % of the molecules in the database. These frameworks, along with 30 of the most common side-chains, were used in the process of selecting compounds from the ACD (MDL Available Chemicals Directory) for inclusion in the SHAPES library. The final library contained commercially available compounds that are water soluble at 1 mM, have MW in the range 68–341 Da (average 194 Da), 6–22 heavy atoms and a *C*log*P* of –2.2 to 5.5. The library profile reflects the fact that the design was dominated by the selection of suitable frameworks and side-chains and the requirement for high aqueous solubility. More recent design strategies incorporate physicochemical property filters where it would be unusual to pass compounds with such high log*P* values.

Breaking down drug-like molecules into fragments would at face value seem an obvious starting point for a fragment library. One of the issues, however, associated with earlier attempts to use fragments was that chemistry featured poorly in the fragmentation process, leading to synthetic difficulties in subsequent application. Consequently, the breakdown of drug-like compounds into fragments has been automated in computational methods, such as RECAP,[21] that are chemically intuitive. Originally cited as a means of identifying privileged molecular building blocks for constructing combinatorial libraries rich in biological motifs, these methods are equally suited to generating libraries of fragments for screening and ensuing optimisation studies. Synthetic optimisability features heavily in Novartis's design for a fragment library. First an analysis was undertaken looking at results from NMR screening as compared with HTS results in relation to the Hann complexity model.[22] The analysis, which is discussed in more detail below, supports the basic principles of fragment-based discovery[23] and provided the framework from which to design a next-generation fragment library. Emphasis is placed on the ability to optimise low-affinity hits through incorporating one or more synthetic handles. Investigation of previous attempts to utilize synthetically tractable functionality in fragments highlighted the fact that in these smaller, less complex molecules the functional group is more often than not an integral component in binding to the target protein. Strategies for increasing the likelihood that a synthetic handle is available for modification include masking the functionality, selection of functionality that is normally not recognised by a protein or simply incorporating multiple groups. Novartis employed what they term a fragment pair strategy, where a fragment building block with an exposed synthetic handle is transformed into a screening fragment by masking the functional group but having minimal effect on a range of computed properties. The fragment building block can then be used in subsequent elaboration when a hit is observed for the corresponding fragment screening compound. In addition to library profiles akin to those outlined above, Similog keys are used as a measure of complexity. Similog keys represent pharmacophoric triplets and it was found that, as expected, fragments with micromolar to millimolar potency are significantly lower in complexity than the more drug-like molecules used in HTS.

Vernalis also use NMR screening in their strategy for fragment-based drug discovery called SeeDs[24] (Selection of Experimentally Exploitable Drug start points). A pharmacophore fingerprint is used as a measure of chemical complexity and diversity amongst compounds as part of the virtual selection, to give a fragment library that has evolved over several generations. The fingerprints, which are essentially identical in nature to Similog keys noted above, encode the presence of pharmacophoric triangles comprised of standard features such as hydrogen bond donors/acceptors, hydrophobes and aromatics. The more pharmacophoric triangles in a molecule, the longer is the fingerprint, the length of which is taken to represent the complexity of the molecule. Within the context of the library as a whole, the fragment fingerprints are compared with fingerprints calculated for a drug-like and a protein binding set of reference compounds. The comparison is made at increasing limits in the distance between features in a pharmacophoric triangle, thus functioning as a rough guide to the diversity of the library with increasing molecular size. It also allows the selection of compounds that are novel to either of the reference sets. Novelty, as previously mentioned, is another advantage of fragment-based discovery as it operates in chemical space not normally identified by HTS. The physicochemical profile of the library is very similar to those outlined above; most (99 %) of the fragments in the ~1300 library have MW $\leq$300, *S*log*P* $\leq$3 and $\leq$3 hydrogen bond donors, 90 % $\leq$3 hydrogen bond acceptors and ~80 % $\leq$3 rotatable bonds and polar surface area (PSA) $\leq$60 Å^2.

The SeeDs strategy uses NMR experiments to identify fragments that bind competitively to a specific site of the target protein and then X-ray crystallography to determine the exact pose of a fragment hit. SGX also advocates the combination of methodologies using a high-concentration biochemical assay in conjunction with X-ray crystallography. The combination of approaches helps to circumvent the individual shortcomings of each method if used on an individual basis. For instance, false positives are inevitable when screening at high concentrations in a biochemical assay. X-ray crystallography therefore provides validation as to the mechanism of action of the compound in the assay. Knowledge of the binding mode of fragments is also key to the rapid development of hits that are typically in the micromolar to millimolar potency range. Structure alone, however, tells us nothing about the binding affinity, making it difficult to assess and rank the effectiveness of each fragment hit for further modification. The order in which the methodologies are applied greatly impacts on the size and nature of the fragment library. Fragment libraries screened using X-ray crystallographic methods are at the lower limits, whereas libraries for biochemical screening can be significantly larger in terms of both molecular size and library numbers. Indeed, at the upper limits of what can be construed as fragment screening, Plexxikon[25] use a high-concentration biochemical assay to identify compounds from a library of ~20 000 scaffolds. Scaffolds are noted to be smaller, less potent and less complex than traditional HTS compounds but with MWs up to 350 Da they are obviously larger than the aforementioned library compounds. Hits from the high-concentration biochemical assay are validated using X-ray crystallography, an approach that has found favour in many companies, especially those familiar with crystallography as a tool in drug development.

The definitions described above, although more restrictive than drug-like criteria, still encompass a broad range of molecules, for example optimisation of a 1 mM inhibitor with MW 150 Da is a far better prospect than optimising a 1 mM inhibitor with MW 300 Da. Can you quantify the quality of a fragment hit in terms of its potential for transformation into a drug-like molecule? Hadjuk[26] attempted to rationalise the selection of fragments for initiating discovery programmes through a retrospective analysis of the development of a number of highly potent inhibitors. By tracing back the end compounds through to constituent fragments and analysing the change in physicochemical properties with potency, he formulated a measure of the likelihood of developing a drug-like molecule from a particular hit fragment based on the size and potency of that molecule. Essentially, potency was observed to increase proportionately with mass along an ideal optimisation path, suggesting that ligand efficiency, discussed further below, should be used in the process of both selecting the most desirable hit fragments and also evaluating the effectiveness of each modification in the development phase. Hadjuk's analysis looks at requirements for individual fragment hits; Makara[27] investigated another aspect of fragment screening, library size, in particular in relation to sampling available fragment space. Key conclusions are that a relatively small number of fragments yield hits against many targets with libraries of 10^3 compounds providing more than sufficient chemical matter to follow up. Diversity, in this instance across the reagents used to construct the library, is vital in formulating a fragment screening deck.

What, then, is a fragment? As for drug-like classification, there is no single, unifying definition that can categorically distinguish between fragments and non-fragments, but there are some general rules of thumb:

- Fragments are smaller than drug-like molecules, whether that is defined by MW or the number of heavy atoms.
- Fragments, in addition to being smaller, are less complex than drug-like molecules.
- Fragments, due to the manner in which they are used, should be highly soluble.
- Fragments should be devoid of undesirable chemical functionality while at the same time facilitating rapid development to more potent compounds.

1.3 Why Use Fragments?

There are a multitude of reasons for the current popularity of using fragments in drug discovery stemming from both the conceptual advantages of using fragments and the observed failures of alternative methodologies. As stated above, the concept of using fragments arose in the early 1980s when it was theorized that the total affinity of a molecule could be taken as a function of the affinity of constituent components. However, it was only in the mid-1990s, with technological advances in the ability to detect weakly binding fragments, that theory was put into practice with what is often cited as the first demonstration of fragment-based drug discovery.[28] In the meantime, drug discovery became a game of numbers as it was perceived that quantity could compensate for a lack of understanding with the development of HTS and combinatorial chemistry. HTS has had successes and is still the predominant means of hit identification, but it was quickly acknowledged that screening large compound libraries has many shortcomings and if anything led to higher attrition rates. Subsequently, screening libraries were overhauled based on drug/lead-like characteristics, high-content screening was introduced and ultimately companies looked to alternative means of initiating discovery programmes including ever more sophisticated fragment based approaches.

One of the postulates of fragment-based drug design is that considerable gains in potency can be achieved through linking together fragments binding in different regions of the protein site due to free energy considerations in the shift to a single entity. Indeed, initial attempts to exploit fragment-based drug design centred on the strategy of linking fragments; the ideas behind the strategy have since been explored further. Murray and Verdonk[29] analysed the change in the free energy of binding for a molecule formed through linking two separate fragments. Essentially the total free energy of binding for a fragment is divided into two components, the free energy associated with the loss of rigid body entropy and the remaining free energy contribution, termed the intrinsic free energy, which incorporates factors such as the protein–ligand interactions and intramolecular conformational restriction of the fragment. The free energy for the fragment-linked molecule is then a summation of both fragment intrinsic free energies, additional free energy terms associated with the linking group and only a single rigid body free energy contribution. The magnitude of the rigid body term is estimated to be 15–20 kJ mol^{-1} and independent of the size of the molecules under consideration. The larger fragment linked molecule only then incurs the same entropic penalty as for a smaller fragment, thus leading to a saving in the total entropic penalty as compared with the individual fragments. Fragment linking has shown limited success as it is very much dependent on the ability to link chemically the individual fragments without significantly perturbing the effectiveness of the fragment receptor interactions. As discussed previously, this is by no means easy and careful consideration must be taken in the design of the fragments and the ability to link the fragments synthetically in the context of the receptor. It is also the case that it is less common to identify fragments that could bind simultaneously in adjacent regions of the target binding site. It may be simply because there is insufficient space in the binding site to recognize multiple fragments simultaneously or adjacent sites, if available, may not provide a suitable environment for fragment recognition. As discussed below, features in a binding site are unlikely to be distributed evenly, meaning that there is a greater probability of identifying fragments that match a specific region in a site. Alternatively, what has proved more amenable is the merger of two fragments into a larger, more potent, compound or more simply the use of the structural information to rationally improve upon a fragment hit (Figure 1.1). There is an increased probability of identifying the separate, less complex fragments compared with the larger, more potent, compound and also a greater chance of achieving maximum affinity.

The effectiveness of the interactions between a receptor and a small molecule is nowadays discussed in terms of ligand efficiency. There are many different ligand efficiency indices, but essentially they are all a means of normalising the affinity of a molecule with respect to the size of that molecule,[30] thereby providing a measure of the quality of fit for that molecule to the receptor. Quality of fit is an important consideration in both the selection of compounds from a screen, whether that is a screen of drug-like molecules or a fragment-based screen, and in the development of molecules through to preclinical candidates. The origins of ligand efficiency can be seen in earlier work investigating ligand–receptor interactions. Our understanding of small molecule–protein interactions is as yet insufficient to be able to predict binding affinities accurately, however, from an analysis of existing data Andrews *et al.*[31] formulated a means of ranking drug–receptor interactions based on known functional group contributions. Small charged groups were found to contribute significantly, followed by polar groups and finally nonpolar groups. The experimentally observed binding affinity of a molecule can then be compared with an estimated value obtained by summing the intrinsic binding energies for these constituent groups taking into account entropic penalties. If the affinity is greater than average, then the fit to the receptor

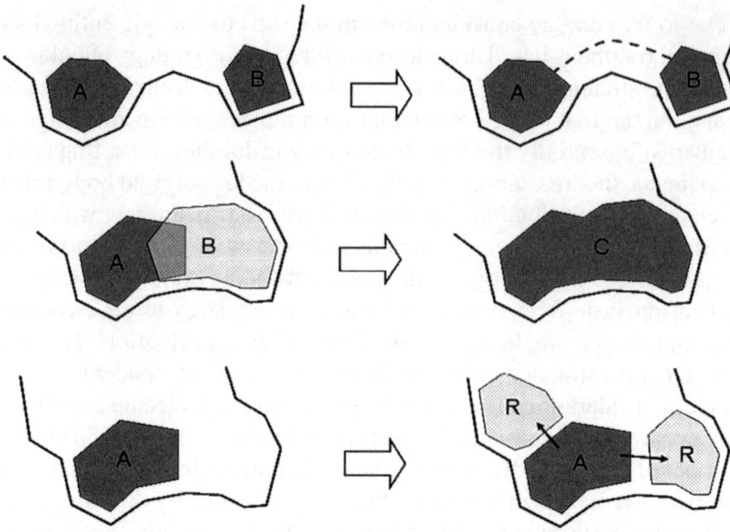

Figure 1.1 Fragment development strategies. Top: fragment linking, where fragments found to bind in adjacent regions of the binding site are linked to create a larger, more potent compound. Middle: fragment fusion, where fragments in overlapping space are amalgamated to form a larger more potent compound. Bottom: fragment growth, where rational design is used to grow the core fragment into adjacent regions of the binding site.

is good, and if not, then it is suboptimal, giving the medicinal chemist a qualitative guide to designing the next round of molecules. Kuntz *et al.*[32] extended the analysis, looking at the maximum affinity of ligands. Strong binding ligands were taken as references for understanding potential free energy gains as the number of heavy atoms in a molecule is increased. They came to the conclusion that increasing the number of non-hydrogen atoms up to 15 heavy atoms can increase the affinity by $\sim$1.5 kcal mol^{-1} per atom; beyond that, the free energy was found not to increase linearly with increasing molecular size. Interestingly, van der Waals interactions and hydrophobic effects are now able to explain affinities for most ligands and only in particular cases do atoms such as metal ions dominate binding. The work of Kuntz *et al.* was instrumental in the use of ligand efficiencies to assess the binding affinity of compounds. With access to superior datasets, subsequent studies suggested that ligand efficiency is dependent on molecular size and does not exhibit the linear relationship below 16 atoms as proposed by Kuntz *et al.* Reynolds *et al.*[33] noted that smaller molecules can demonstrate considerably higher ligand efficiencies than observed for larger drug-like molecules. This has important implications when comparing hits from a wide range of molecular sizes; indeed Reynolds *et al.* propose a size-normalized efficiency scale termed 'fit quality' as a metric for assessing the goodness of fit between ligand and receptor. Essentially, a maximum ligand efficiency is calculated, based on existing data, for each heavy atom count and the ligand efficiency for a particular molecule is scaled according to the optimum curve. Murray and Verdonk[29] also stated that smaller molecules necessitate more optimal binding interactions in order for the intrinsic free energy to surmount the entropic free energy penalty.

Ligand efficiency provides a means, as discussed by Hadjuk,[26] of extrapolating from a screening hit to determine if a potency objective can be achieved through additions to the molecule while maintaining a desired physicochemical property profile. One of the advantages of using fragments is that they leave more scope for improvement based on a typical medicinal chemistry approach (Figure 1.2), where studies[34] have shown that lead development increases both the size and lipophilicity of the original hits. The fact that existing data suggest fragments exhibit higher ligand efficiencies may also be a consequence of traditional development strategies where suboptimal hits were taken as starting points and/or suboptimal modifications were made in the development process. A clearer picture may well develop as current practices of using ligand efficiency in both the selection and development of compounds feeds into our knowledge base. It is also the case that the different features around a receptor binding site lead to an uneven distribution in the maximum binding affinity across the site. The same can be said for screening hits where different components of a ligand will contribute differently to the total binding affinity. Without additional SAR information, it can be very difficult to establish, even with knowledge of the binding pose, the contribution of each component to the total binding affinity, thereby necessitating the deconstruction of that hit and subsequent testing of the individual components, i.e. fragment-based screening.

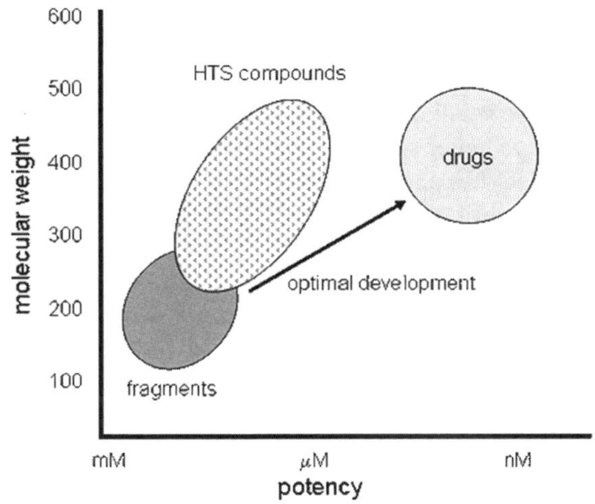

Figure 1.2 *Fragments provide greater scope for development into drug-like molecules as compared with HTS molecules that exhibit suboptimal binding to the target protein.*

The dynamic, complex nature of receptor binding sites also highlights another very important benefit of the fragment-based approach, that of the probability of identifying an optimal match to the receptor. Hann *et al.*[22] calculated the probability of a binding event with varying complexity of ligand–receptor interactions and noted that the chance of observing a useful interaction falls rapidly with increasing complexity. The process of binding was reduced to a simple functional model of molecular recognition between the features of the receptor and a ligand. All the standard pharmacophoric features, such as donors and acceptors, were represented as +s and –s with a positive interaction occurring between a + on the receptor and a – on the ligand or vice versa. An exhaustive calculation was then performed, computing the probability of finding an exact match between a receptor of given complexity and a ligand with increasing complexity and, as expected, the probability diminishes rapidly as the number of possible permutations increases. It is also the fact that as one increases the complexity of a molecule the probability of negative interactions increases, leading to suboptimal binding. This has a direct bearing on the size of a screening library, with studies suggesting that even a million compounds in an HTS collection barely scratches the surface of the number of possible molecules in drug-like chemical space[35]. Counter to the argument of reducing complexity to a bare minimum in order to maximise the probability of identifying hits, there is a lower limit imposed on the complexity due to the sensitivity of the screening protocol. Hence there is a balance between the probability of there being an exact match between receptor and ligand and the ability to detect that match in a screen. For a particular screening protocol, this leads to an optimum level of ligand complexity, which in turn dictates the size of the screening library necessary to maintain a sufficient hit rate.

Fragment-based drug discovery goes some of the way to compensating for our incomplete understanding of biological interactions and provides a complementary if not alternative route to finding chemical matter for discovery programmes:

- The smaller, less complex nature of fragments increases the probability of finding a match to a receptor; moreover, instances have shown that removing complexity by screening fragments can succeed where HTS has failed [6].
- Fragment libraries can be smaller than HTS libraries as hit rates are generally higher than in traditional HTS due to better sampling and the increased probability of identifying a match to the receptor. This has many advantages associated with the construction, storage and screening of fragment libraries versus HTS libraries.
- Fragment-based screening can also identify more optimal matches (higher ligand efficiency) to the receptor without the need first to deconstruct a hit compound.
- Fragment hits then provide greater scope for development when following a standard medicinal chemistry development strategy (Figure 1.2).
- Screening fragments requires more sensitive detection methods, but at the same time these methods provide invaluable information in the development from hits to drug candidates.

1.4 Practical Implications of Using Fragments in Drug Discovery

The primary application of fragments is in the identification of chemical matter to take forward into drug development. As the intention is to identify molecules that are typically in the micromolar to millimolar potency range, a method of detection is required that is more sensitive than a biochemical screen at 10–30 μM. Assays can be adapted to screen molecules at higher concentrations but, as discussed above and in following chapters on fragment library design, more stringent requirements are placed on the molecules. It is also paramount to obtain confirmation of the mechanism of action as screening at higher concentrations leads to higher false positive rates. Most if not all fragment screening strategies will then incorporate a biophysical technique either as the primary screen or to validate hits obtained from an alternative source. NMR and X-ray crystallography are undoubtedly the most popular approaches, as highlighted in ensuing chapters, requiring substantial investment in terms of skill base and technology. Drug discovery programmes, however, benefit immensely from access to structural information at this critical stage, when the chemical nature of the lead compounds is being decided, and thus structural biology can have a significant positive impact on the speed and success of the programme.[36] Effective dissemination of the information gained from a screen is also decisive both in understanding the activity of the hit fragments and in their development. Here computational chemistry and informatics tools can play a key role in integrating data across the different disciplines in addition to directing the design of follow-up compounds. Development can follow one of many strategies, from simple elaboration of single compounds, linking fragments from adjacent regions in a binding site, focused libraries around one or more hits or to more complex amalgamations of compounds observed to bind in overlapping regions of the binding site. Having access to multiple structures of fragment–protein complexes is an invaluable tool in this process; it can also provide a direct understanding of the protein binding site and guide optimisation throughout the lifetime of the project. The final chapters will review the use of computational methods in the overall process and then lead on to practical examples of fragment-based drug discovery.

References

[1] Leach, A. R., Hann, M. M., Burrows, J. N., and Griffen, E.J. (2006). Fragment screening: an introduction. *Mol. BioSyst.* **2**, 429–446.

[2] Rees, D. C., Congreve, M., Murray, C. W., and Carr, R. (2004). Fragment-based lead discovery. *Nat. Rev. Drug Discov.* **3**, 680–672.

[3] Zartler, E. R., and Shapiro, M. J. (2005). Fragonomics: fragment-based drug discovery. *Curr. Opin. Chem. Biol.* **9**, 366–370.

[4] Erlanson, D. A., McDowell, R. S., and O'Brien, T. (2004). Fragment-based drug discovery. *J. Med. Chem.* **47(14)**, 3463–3482.

[5] Mitchell, T., and Cherry, M. (2005). Fragment-based drug design. *Innov. Pharm. Technol.* **16**, 34–36.

[6] Boehm, H. J., Boehringer, M., Bur, D., Gmuender, H., Huber, W., Klaus, W., Kostrewa, D., Kuehne, H., Luebbers, T., Meunier-Keller N., and Mueller, F. (2000). Novel inhibitors of DNA gyrase: 3D structure based biased needle screening, hit validation by biophysical methods and 3D guided optimization. a promising alternative to random screening. *J. Med. Chem.* **43**, 2664–2674.

[7] Lipinski, C.A., Lombardo, F., Dominy, D. W., and Feeney P. J. (1997). Experimental and computational approaches to estimate solubility and permeability in drug discovery and development settings. *Adv. Drug Deliv. Rev*. **23**, 3–25.

[8] Landers, P. (2004). Drug industry's big push into technology falls short: testing machines were built to streamline research – but may be stifling it. *Wall Street J.* February 24.

[9] Gribbon, P., and Sewing, A. (2005). High-throughput drug discovery: what can we expect from HTS?. *Drug Discov. Today* **10**, 17–22.

[10] Bohacek, R. S., McMartin, C., and Guida, W. C. (1996). The art and practice of structure-based drug design: a molecular modeling perspective. *Med. Res. Rev*. **16**, 3–50.

[11] Erlanson, D. A., and Jahnke, W. (2006). In *Fragment-Based Approaches in Drug Discovery*, ed Jahnke, W., and Erlanson, D. A., Wiley-VCH Verlag GmbH, Weinheim, pp. 3–9.
[12] Jennings, A., and Tennant, M. (2005). Discovery strategies in a biopharmaceutical startup: maximising your chances of success using computational filters. *Curr. Pharm. Des.* **11**, 335–344.
[13] Teague, S. J., Davis A. M., Leeson, P. D., and Oprea, T. (1999). The design of leadlike combinatroial libraries. *Angew. Chem.* **38**, 3743–3748.
[14] Congreve, M., Carr, R., Murray, C., and Jhoti, H. (2003). A 'rule of three' for fragment-based lead discovery? *Drug Discov. Today* **8**, 876–877.
[15] Lipinski, C. A., Lombardo, F., Dominy, B. W., and Feeney, P. J. (2001). Experimental and computational approaches to estimate solubility and permeability in drug discovery and development settings. *Adv. Drug Deliv. Rev.* **46**, 3–26.
[16] Baurin, N., Baker, R., Richardson, C., Chen, I., Foloppe, N., Potter, A., Jordan, A., Roughley, S., Parratt, M., Greaney, P., Morley, D., and Hubbard, R. E. (2004). Drug-like annotation and duplicate analysis of a 23-supplier chemical database totalling 2.7 million compounds. *J. Chem. Inf. Comput. Sci.* **44**, 643–651.
[17] Blaney, J., Nienaber, V., and Burley, S. K. (2006). In *Fragment-Based Approaches in Drug Discovery*, ed. Jahnke, W. and Erlanson, D.A., Wiley-VCH Verlag GmbH, Weinheim, pp. 215–248.
[18] Fejzo, J., Lepre, C. A., Peng, J. W., Bemis, G. W., Ajay, Murcko, M. A., and Moore, J. M. (1999). The SHAPES strategy: an NMR-based approach for lead generation in drug discovery. *Chem. Biol.* **6**, 755–769.
[19] Bemis, G. W., and Murcko, M. A. (1996). The properties of known drugs. 1. Molecular frameworks. *J. Med. Chem.* **39**, 2887–2893.
[20] Bemis, G. W., and Murcko, M. A. (1999). Properties of known drugs. 2. Side chains. *J. Med. Chem.* **42**, 5095–5099.
[21] Lewell, X. Q., Judd, D. B., Watson, S. P., and Hann, M.M. (1998). RECAP – retrosynthetic combinatorial analysis procedure: a powerful new technique for identifying privileged molecular fragments with useful applications in combinatorial chemistry. *J. Chem. Inf. Comput. Sci.* **38**, 511–522.
[22] Hann, M. M., Leach A. R., and Harper G. (2001). Molecular complexity and its impact on the probability of finding leads for drug discovery. *J. Chem. Inf. Comput. Sci.* **41**, 856–864.
[23] Schuffenhauer, A., Ruedisser, S., Marzinzik, A. L., Jahnke, W., Blommers, M., Selzer, P., and Jacoby, E. (2005). Library design for fragment based screening. *Curr. Top. Med. Chem.* **5**, 751–762.
[24] Baurin, N., Aboul-Ela, F., Barril, X., Davis, B., Drysdale, M., Dymock, B., Finch, H., Fromont, C., Richardson, C., Simmonite, H., and Hubbard, R. E. (2004). Design and characterization of libraries of molecular fragments for use in NMR screening against protein targets. *J. Chem. Inf. Comput. Sci.* **44**, 2157–2166.
[25] Card, G. L., Blasdel, L., England, B. P., Zhang, C., Suzuki, Y., Gillette, S., Fong, D., Ibrahim, P. N., Artis, D. R., Bollag, G., Milburn, M. V., Kim, S.-H., Schlessinger, J., and Zhang, K. Y. J., (2005). A family of phosphodiesterase inhibitors discovered by cocrystallography and scaffold-based drug design. *Nat. Biotechnol.* **23**, 201–207.
[26] Hadjuk, P. J. (2006). Fragment-based drug design: how big is too big?. *J. Med. Chem.* **49**, 6972–6976.
[27] Makara, G. M. (2007). On sampling of fragment space. *J. Med. Chem.* **50**, 3214–3221.
[28] Shuker, S. B., Hajduk, P. J., Meadows, R. P., and Fesik, S. W. (1996). Discovering high-affinity ligands for proteins: SAR by NMR. *Science* **274**, 1531–1534.
[29] Murray, C., and Verdonk, M. L. (2006). In *Fragment-based Approaches in Drug Discovery*, ed. Jahnke, W. and Erlanson, D.A., Wiley-VCH Verlag GmbHWeinheim, pp. 55–66.

[30] Abad-Zapatero, C., and Metz, T. (2005). Ligand efficiency indices as guideposts for drug discovery. *Drug Discov. Today* **10**, 464–469.
[31] Andrews, P. R., Craik, D. J., and Martin, J. L. (1984). Functional group contributions to drug-receptor interactions. *J. Med. Chem.* **27**, 1648–1657.
[32] Kuntz, I. D., Chen, K., Sharp, K. A., and Kollman, P. A. (1999). The maximal affinity of ligands. *Proc. Natl. Acad. Sci. USA* **96**, 9997–10002.
[33] Reynolds, C. H., Bembenek, S. D., and Tounge, B. A. (2007). The role of molecular size in ligand efficiency. *Bioorg. Med. Chem. Lett.* **17**, 4258–4261.
[34] Oprea, T. I., Davis, A. M., Teague, S. J., and Leeson, P. D. (2001). Is there a difference between leads and drugs? A historical perspective. *J. Chem. Inf. Comput. Sci.* **41**, 1308–1315.
[35] Hann, M. M., and Oprea, T. I. (2004). Pursuing the leadlikeness concept in pharmaceutical research. *Curr. Opin. Chem. Biol.* **8**, 255–263.
[36] Stevens, R.C. (2004). Long live structural biology. *Nat. Struct. Mol. Biol.* **11**, 293–295.

2

Designing a Fragment Process to Fit Your Needs

Edward R. Zartler and Michael J. Shapiro

2.1 Fragment Definition

While the definition of a fragment is in the eye of the beholder, it is always considered a molecule of lower molecular weight than its corresponding drug.[1] Typically, fragments have molecular weights of 100–250 Da (7–20 heavy atoms), have less functionality than the daughter molecule and 'low' binding affinity. Typical binding affinities for fragments range from mM to low μM; although nanomolar fragment sized molecules do exist. A 'rule of three' for fragments has been prescribed for 'good fragments',[2] comparable to the 'rule of five' for drug-like compounds.[3] Fragments occupy a small region of total chemical space due to their low complexity, but reasonably sized libraries (several thousands) can theoretically explore a great majority of fragment space, whereas 'lead-like' molecules cannot possibly explore all of the 10^{60} possible molecules that exist in chemical space[4] or even a significant fraction of them. The diversity of a fragment library is encompassed in a smaller number of compounds than it would be for a diverse high-throughput screening library of lead-like compounds. The simple nature of fragments may even prove advantageous in a screening setting as fewer detrimental effects of 'nonfunctional' molecular appendages 'bump' into the protein surface.[5]

2.1.1 Overview of the Design Process

The design of a robust fragment-based drug discovery (FBDD) process can lead to large increases in productivity in lead generation and lead optimization.[6] It should be noted that

Fragment-Based Drug Discovery: A Practical Approach Edited by Edward R. Zartler and Michael J. Shapiro

there is not a one size fits all process; each target has different needs and presents different challenges. This chapter will discuss how to go about creating a workable framework for initiating FBDD efforts and discuss the options available at each step. It will be up to practitioners to develop processes specific to their individual needs.

For the purpose of this chapter, we have divided the FBDD process into three phases: Phase I is the assessment phase, Phase II is the screen/re-screen[1] phase and Phase III is the post-screen phase (Figure 2.1). Phase I involves three assessments: target, assay and compound. Phase II is initiated with hypothesis generation, followed closely by screening and iterative confirmation, rescreening and hypothesis evaluation. Phase III starts the moment the first compound comes out of Phase II and continues in parallel with Phase II efforts. Phase III efforts are the same as lead-like post-screen efforts, even though for fragment-based drug discovery they proceed with different rules and paradigms for evaluating success (discussed in this chapter and elsewhere in this book). The criteria for exit from Phase III are exactly the same for a compound that is initially found by more typical library-based discovery: a high-potency compound that shows *in vivo* activity. The key to FBDD is its highly iterative nature that occurs rapidly due to the low inherent complexity of the molecules. All of the individual parts must be seamlessly integrated; 'siloed' components do not work well together. The chief reasons for utilizing FBDD are greater diversity[2] with fewer compounds, higher hit rates leading to more possible avenues to explore and completely rational and deliberate medicinal chemistry efforts, ideally suited for 'undruggable' and novel targets.[7–12]

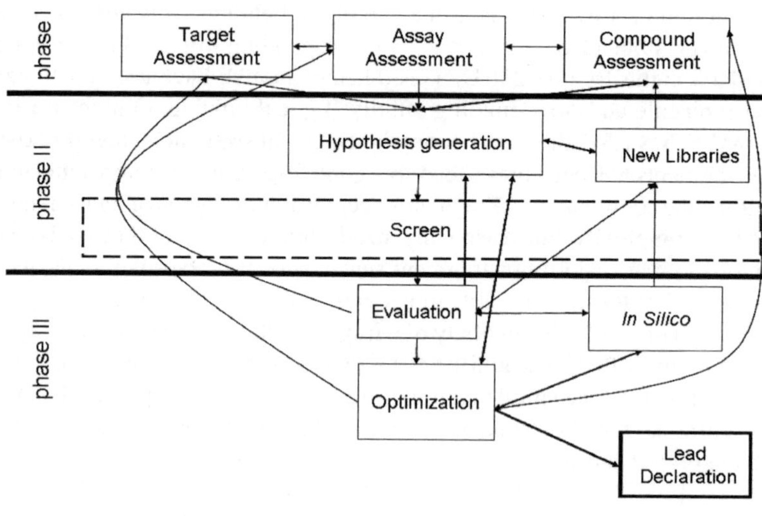

Figure 2.1 *Schematic representation of the FBDD process. Individual steps are discussed in the chapter.*

[1] For the purpose of this Chapter, screening will generally refer to biochemical and biophysical assays with the generic term assay or screen. When the differences are significant, the two will be differentiated.

[2] Diversity in this context refers to the coverage of available chemical space.

2.2 Phase I Activities

2.2.1 Target Assessment

It has been estimated that ~10 % of the entire human genome is involved in disease onset or progression,[13] resulting in several thousand potential targets suitable for therapeutic intervention. Most drug discovery targets are proteins, but this is not always the case.[14–17] We will focus on protein targets only for the rest of this chapter, as the concepts for nonprotein targets are the same. The first stage in the drug discovery process is target identification and validation. If a target is not validated with the disease, resources can be wasted in a fruitless search for a drug. Sometimes the result of FBDD (or LLDD) is de-validation of a target, which can be just as important as finding a lead compound against a target, just not nearly as glorified. Some people would argue for 'druggability' as a relevant part of target assessment but, as discussed below, we do not. An estimated 60 % of small molecule drug discovery projects fail because the biological target is found to be not 'druggable'.[18] The easy 'druggable' targets have already been the focus of intense drug discovery efforts; the future of drug discovery lies in drugging 'undruggable' or novel targets. The current lead-like drug discovery paradigm consists of the creation of libraries around previous work for a target;[1, 19–21] therefore, 'undruggable' and novel targets will be a difficult task. This also creates inherent issues with intellectual property; issues that are not inherent in fragments.

Once a validated target has been chosen, it is important to develop a detailed dossier on the protein. This information will affect both assay assessment (Figure 2.1) and compound assessment (Figure 2.1).[22] This obviously starts with the classification of the target into a given class: nuclear hormone receptor, membrane protein, kinase, protease, carrier protein, chaperone, metalloenzyme, etc. Some target classes are easier to find hits against than others, most notably enzyme targets.[23, 24] Soluble, single-domain proteins (or those that have isolatable enzymatic domains) are much easier to work with, in both fragment-based and lead-like drug discovery, e.g. kinases and proteases. Membrane proteins, which are the most difficult to work with, especially from a biophysical standpoint, make up 50 % of the pharmaceutically relevant targets.[25]

It is at this point that the target validation status of the target is determined. It is beyond the scope of this chapter to discuss the criteria used to determine this, but there are many excellent papers that describe some relevant aspects.[26, 27] One proposed method is to prioritize targets based upon their druggability. Figure 2.2 shows the calculated druggability of targets of high pharmaceutical interest.[28] Of particular note is the very high calculated druggability (10 μM) of the 'undruggable' target HIV integrase (a typical 'undruggable' case scenario). In 2007, Merck launched Isentress, which inhibits this target, showing that 'undruggable' in this context does not mean what one would think it means. We would argue that target druggability is irrelevant as long as the target is suitably well validated. Where there is a validated target and a will, there is a (drug discovery) way.

We feel that target validation is the key component of the target assessment step and many interesting approaches have been detailed.[29–31] An interesting approach has recently emerged for target assessment involving chemical genomics. Both forward and reverse chemical genomics can play a role in target validation.[32] These paradigms are depicted in Figure 2.3. Forward chemical genomics explores phenotypes by screening libraries to obtain

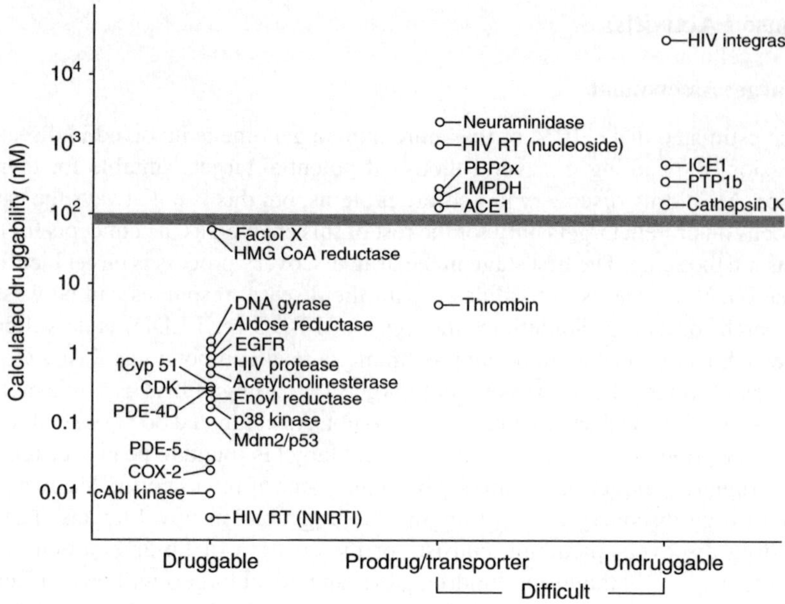

Figure 2.2 *Calculated druggability for a set of 27 target binding sites. Known druggable protein targets are shown on the left vertical, whereas known difficult targets (prodrug and 'undruggable') are shown in the right verticals. Difficult and druggable target binding sites are effectively separated by the gray bar. The predicted druggability is the MAPpod score calculated from the protein–ligand binding site structure. HMG-CoA, 3-hydroxy-3-methylglutaryl-CoA; EGFR, epidermal growth factor receptor kinase; CDK, cyclin-dependent kinase 2; PDE, phosphodiesterase; COX, cyclooxygenase; HIV RT, HIV reverse transcriptase; PBP2x, penicillin binding protein 2x; IMPDH, inosine ICE1, interleukin-1 converting enzyme 1; PTP1b, phosphotyrosine phosphatase 1b.*[28] *Reprinted by permission from Macmillan Publishers Ltd.*

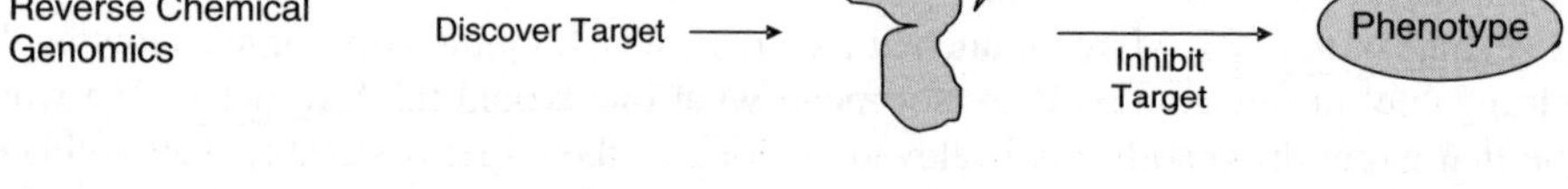

Figure 2.3 *Diagram representing forward and reverse chemical genomic paradigms.*[32]

an observable change in phenotype so that a target may be discovered. Small molecules interact with and alter the target and changes in phenotype are explored to connect it to the pathway of interest. Forward chemical genomics can be described as screening a ligand for a target. In reverse chemical genomics, targets without a known biochemical activity (such as genomics targets) are interrogated by a binding assay. The molecules that bind are then used in cellular assays to investigate the possible role of the target. Reverse chemical genomics can be defined as screening a target for a ligand, akin to the typical drug discovery paradigm. Both of these approaches have obvious advantages for novel pathways/targets and are imminently tractable to FBDD.[32–38]

The information that should be put into a complete target dossier on the target is as follows. Can the protein be produced in sufficient quantities and purity for biochemical and/or biophysical assay (microgram versus milligram quantities)? Is the protein soluble, monodisperse and stable under typical working conditions? Has it been a target of drug discovery before? If so, how was it done and what was the outcome? Are there structural data? If not, are there suitable surrogates? Questions of this type (not to be considered an exhaustive list) are important for designing an appropriate process and will be discussed in more detail under assay assessment. The more that is known about the target, the better the decision can be made on assembling the process. For example, if the protein cannot be produced in the amount and purity needed for biophysical screening,[39, 40] this affects decisions in assay and compound assessment. If the target has stability issues, it may be more amenable to one screening method over an other. If the target was previously the focus of drug discovery, it is important to ascertain whether those efforts were successful, what methods were applied, etc.

Another important component of the protein dossier is structural information. Do high-resolution structures (either NMR or X-ray) exist that can be used for library generation and hypothesis testing/generation[12, 41–44]? Are these structures with bound ligands? In many medicinal chemists' eyes, this is the most important of the criteria; in fact, to many it is so important that hit follow-up will not be pursued until suitable structures are in hand. Several companies' entire business model is based upon X-ray-based screening of fragment libraries.

If structures of the target do not exist, structures of isoforms or surrogates may exist. There are many reasons for using surrogates: the target cannot be produced in sufficient quantities for screening, no assay is developed for the target, there is no structure for the target, etc.[45] Surrogates can also be used as the first filter in the screening process, thereby reducing the need for protein for later follow-up screening. This can be especially useful if the target is in limited supply. One important caveat is that it is possible that something may be missed. However, as mentioned later in this chapter and in other chapters in this book, the hit rate for fragments is high relative to other screening methods, so although a real possibility, it is typically worth the risk. In many cases, particularly membrane proteins, there is a very small likelihood of ever having structural information, so this is a moot point. FBDD, like lead-like discovery, can progress just as efficiently without structural information as with it.

2.2.2 Assay Assessment

The decision on what type of screen to use in FBDD is affected by many different factors: availability of protein for screening, compound selection, throughput, turnaround and rate of false positives and negatives. The resolution of these questions from target assessment directly impact the possibilities in this section. If there is not sufficient protein for a biophysical screen, a biochemical screen is the only choice. Protein that is not stable for extended periods are not especially amenable to many biophysical screens. If the target was the focus of previous efforts that utilized primarily biochemical screens and resulted in no hits suitable for lead optimization, then the next iteration on this target may warrant a biophysical screen, similar to reverse chemical genomics. If there is structural data of the target or suitable surrogate, then any of the multitude of structure-based drug discovery (SBDD) or *in silico* design paradigms may be utilized.[46, 47] The breadth of these is far beyond the scope of this chapter and this topic has recently been the focus of many recent reviews.[48–50]

The primary choice in this assessment is whether to utilize a biochemical or biophysical screen as the primary filter. Although it seems like an either/or choice, this is a false dichotomy. The most successful fragment screens obtain orthogonal data, i.e. both biochemical and biophysical data in parallel or in quick succession. With orthogonal data, the probability of false positives (or negatives) is reduced. Most commonly biochemical and biophysical data are obtained. However, all the different biochemical and biophysical screens can be considered orthogonal. We would recommend that if two biophysical methods are to be used at least one should be a direct method (discussed below). As is noted many times in this book, rapid iterations among the various data sources are the key to a successful process.

Biochemical versus biophysical screens. As shown in Figure 2.1, the first three steps of FBDD are interdependent: the choice made in one assessment impacts the choices that are/can be made in the others. For example, a fluorescent biochemical screen requires compounds which do not quench the assay or give false positives. However, fluorescence quenching compounds can be easily run in a biophysical screen. On the other hand, a fluorescence biochemical screening looking for changes in fluorescence polarization anisotropy[51, 52] requires the exact opposite set of characteristics in a molecule. A biophysical screen using mass spectrometric (MS) detection requires moderately soluble compounds which will ionize,[53] whereas an NMR-based screen requires compounds with at least one-nonexchangeable proton and high solubility.[54–57] Most of the time the choice of biochemical versus biophysical screen is made out of comfort and available expertise. Cost is, of course, a concern at this point as many biophysical screens require equipment that have expensive upfront costs. However, it would be unusual to consider a screen unless the equipment was already available or a suitable collaboration partner could be found. The nature of FBDD requires such close commitment of primary and secondary screens. For FBDD these terms are really inadequate, but for now will have to suffice. The assumption that new biochemical assays need to be developed for fragments is generally unwarranted.. Fragments are typically screened at higher concentrations than lead-like compounds ($\geq$1 mM versus 10–25 μM) and therefore the highest DMSO concentration (fragments are typically solubilized at $\geq$100 mM in DMSO) in the biochemical screen is 1 %. As long as the assay's DMSO tolerance is known, it can be used for FBDD. A brief summary of each of the main types of screening approaches used in FBDD is given in Table 2.1 and discussed in Section 2.3.1.

2.2.3 Compound Assessment

Compound assessment and library design are part and parcel. However, one trap that can be tempting, but in the long run inefficient, is the belief that a universal fragment library

Table 2.1 *Different screening approaches in FBDD.*

Method	Throughput[a]	Type of information	Structural data[b]	Multiplexing of ligands
NMR Ligand-based	100s	Direct detection of binding events based on chemical identity of ligand	Yes: epitope map of ligand	Yes[115]
NMR Target-based	<10	Direct detection of binding-induced changes in target conformation	Yes: changes in target, 3D structure in some cases	Yes, but difficult
X-ray crystallography	10s	Interpretation of electron density	Yes: 3D structure	Yes, but no isosteres in mixture
MS	1000s	Direct detection of binding events based upon molecular weight changes	No	Yes, but MW must differ
SPR	1000s	Direct detection of binding events based upon MW changes	No	No
ITC	<10	Indirect changes in thermodynamic parameters of system	Indirect inference based on changes in thermodynamic parameters	No
Fluorescence[c]	10–1000s	Indirect changes in fluorescence properties of ligands or probes	No	No
Biochemical	> 10 000	Direct changes in functional activity of target.	No	No

[a] Number of compounds that can be routinely screened on a daily basis.
[b] Structural data are defined as information on how the ligand binds to the target at an atomic level.
[c] Includes all types of assays, including ThermoFluor,[116–118], fluorescence polarization, fluorescence quenching, etc.

will suit all screening needs. Although it is possible to create one diverse fragment library to be used as a primary library, it is more advisable to have screening libraries with many different fragment libraries available depending on the results of the assay assessment. A compromise between these two extremes is to use 'cascaded' libraries (discussed below). Screening libraries should be different depending upon a choice of biochemical versus biophysical screens, e.g. a biophysical library that contains all the fluorescence quenching fragments in your collection. The restrictions that can be placed upon generic, 'diverse' libraries based on the result of assay assessment can be onerous and lead to the out-of-hand removal of fragments that can be of great interest depending on the target of interest.

Fragment library design. The challenge for all approaches to drug discovery is twofold: one wants sufficient diversity to make sure one achieves adequate sampling of available space and one wants the library to be small enough to screen quickly. Fragments have a proven advantage over lead-like libraries in both of these areas. This is expertly presented in Chapters 3, 8 and 9. Here we will briefly give an overview of the major components and factors that need to be assessed in assembling this part of FBDD. The first component of the library to address is the level of diversity that it will encompass. There is always going to be a tradeoff between diversity and inherent SAR: one cannot reasonably have every compound possible in a library and one wants to have as many as possible! It is also recommended to have compounds which are closely related to each other so that SAR can be established. The tension between these competing needs is what drives library composition.[10] A diverse fragment library is sufficiently diverse if it provides good chemical starting points for hit optimization. Part of providing 'good' starting points is that the results lead to testable hypotheses from the outset; this means that there *must* be SAR that results from the initial screen. We believe that a sparse matrix approach with very limited SAR around each populated node is the optimal approach. The actual composition of the library is dictated by the results of target, assay and compound assessment. Another component of assembling the library is that the fragment should be suitable for medicinal chemistry elaboration. Many different approaches have been used in constructing fragment libraries to have the follow-up chemistry built into the fragments, including capped chemistry handles, built-in SAR and ready-to-use chemistry handles.[19, 58] If fragments do not have handles available for immediate elaboration, it should be possible to elaborate off the scaffold should it hit. This elaboration can use additional fragment libraries or newly synthesized (or purchased) compounds that address the budding SAR made evident by the screen. It is also advisable to have 'nested' libraries ready to follow upon the initial hit. This pyramidal approach has been shown to be quite expedient.[59] Another approach similar to 'nested' libraries is 'cascaded' libraries. In this approach, the initial library is set up to sample maximal fragment space with the fewest possible compounds, which is sometimes called a 'sparse matrix' approach. Follow-up libraries are prepared to sample a much tighter area of fragment space and are screened if a hit is within a certain distance to a trigger point. The initial hit(s) may trigger multiple follow-up libraries to be screened, but the results from the follow-up screening will triangulate on a given region of fragment space and lead to obvious further testing. At this point, the medicinal chemistry resources can be engaged having identified a very robust SAR hypothesis.

2.3 Phase II Activities

2.3.1 The Screen

The variety of biophysical and biochemical techniques now available that can be used as screens for fragment based studies is impressive.[9]

Biochemical techniques. Conceptually the most straightforward approach to identifying fragments is through a functional (biochemical) screen; if the target activity is affected by the compound, it is a 'hit'. However, since fragments usually have relatively low binding affinities (sometimes in the millimolar range), they are difficult to detect in typical HTS assays where compound libraries are usually screened at low micromolar concentrations. Additionally, with increasing targeting of nonactive site compounds, the possibility of easily read activity is low. In addition, functional screening (especially at high concentrations) is subject to a number of pitfalls.[60] However, biochemical data are inherently orthogonal to biophysical data and hence is an extremely useful type of data to have and, whenever possible, should be collected for all fragment-based screens. The greatest downside is that one must know the function of the target to develop the screen.

Biophysical techniques. In order to use fragments effectively in drug design, experimental methods are required that can rapidly and reliably screen up to thousands of low-MW test compounds for weak binding to the target protein. The most robust screens for fragments are binding assays (Table 2.1). Detection of binding to the target by fragments is the key to successful FBDD. The initial stages of FBDD can efficiently proceed without functional activity, but to exit from the LG phase functional activity must be correlated with binding activity. X-ray crystallography,[61, 62] NMR, surface plasmon resonance (SPR),[63] fluorescence spectroscopy,[52] MS[64] and isothermal titration calorimetry[65] have also been used alone or as part of fully integrated FBDD effort to deliver initial fragment hits. These methods can be divided into two groups: indirect methods which include low-resolution X-ray crystallography,[3] various fluorescence methods and isothermal titration calorimetry and direct methods which include SPR, MS, high-resolution X-ray crystallography and NMR spectroscopy. Direct methods (i.e. those that directly measure ligand–target interactions) are obviously superior. NMR is the most utilized direct method in FBDD because the protein does not need to be modified for analysis (like SPR), the samples can mimic the native environment as closely as needed (unlike MS), either ligand-based or target-based NMR experiments can be used to deliver data[39, 40] and the throughput can be relatively high (hundreds of compounds or more a day). NMR, however, is not inexpensive, spectrometers cost portions or multiples of millions of dollars, upkeep can be costly and it requires expert users for both analysis and maintenance. Also, in many instances, there is little institutional knowledge of how to interpret and utilize NMR data to progress medicinal chemistry efforts. This institutional inertia can be very difficult to overcome and quantitative measurement of NMR's contribution is impossible. Once all of these hurdles are overcome, NMR proves to be a most powerful drug discovery tool.[66, 67]

The other two main direct methods (SPR and MS) have been shown to be vital tools in FBDD.[53, 68, 69] These are less commonly utilized as each method has its own caveats. MS needs to have the sample in a volatile buffer (commonly ammonium acetate). The use of adjuncts and co-solvents is thus severely curtailed for MS; proteins that require them for stability will not be amenable to MS screens. SPR has its own needs,[68] most notably the need to have the target immobilized. This is both a bane and a boon. It is a bane because, depending on the immobilization technique chosen, the target could be adversely affected.

[3] X-ray crystallography is both an indirect and direct method because the quality of the model used to fit the electron density depends on the resolution of the data. At lower resolution, the fit between the model and the density may be good (in R-free terms), but lacking details in interactions between target and ligand due to poor or lacking electron density. Hence any interactions are inferred and not directly measured. At high resolution, the electron density is of such high resolution that side-chains can be unambiguously placed in the structure, in addition to the ligand. In this case, X-ray crystallography is a direct method.

It is a boon because the chip technology of SPR lends itself to miniaturization and thus great increases in throughput and decreases in material consumption.

X-ray crystallography is the most commonly used indirect biophysical technique.[61, 62, 70–73] It has the potential to yield very high-resolution atomic data of the ligand–target complex, but that is reliant upon obtaining high-quality crystals; this is not a trivial task. Certain classes of targets, such as membrane proteins, are simply not amenable to routine analysis by X-ray crystallography. If a ligand–target complex can be obtained, the resultant data are a crucial feed for the SBDD process,[74–77] which also includes the multitude of *in silico* methods. As noted earlier, sometimes the high-resolution structure of a surrogate target can be used as an entrée into the SBDD process. Isothermal titration calorimetry (ITC) is the other popular indirect biophysical method.[78, 79] It is especially powerful for FBDD because it measures the thermodynamic parameters of the system. This information must then be used to infer ligand–target binding information. However, this can be vital in hit confirmation and SAR evaluation.[65, 80] In hit confirmation, ITC can discriminate 'bad actors', especially those that aggregate. During SAR evaluation, ITC can monitor enthalpic and entropic changes in the hit series and determine if the line of SAR is proceeding in the right direction. The most commonly used indirect screening technique is the biochemical screen.[81] Discussion of this topic is well beyond the scope of this chapter; we simply point the reader to a recent review.[82]

An interesting, new method to screen for ligand–target interactions involves protein stability measurements.[83] The ability of ligand binding to enhance protein stability is a well-recognized phenomenon. The degree of stabilization can be systematically probed by observing the increased resistance of a protein to chemically or physically denaturing conditions, such as urea or temperature, in the presence of compounds. Many methods are able to monitor the extent of denaturation, e.g. specific enzyme activity, NMR, circular dichroism (CD) and, most recently, extrinsic fluorescence using a probe that binds selectively to unfolded protein.[83, 84] Care still must be exercised here as there are cases in which ligand biding actually causes protein stability to decrease.[85, 86]

Alex and Flocco, in a superb paper, have reviewed the pipeline of fragment-derived molecules.[87] In their analysis, they divide the FBDD approaches into three broad categories: linking, expansion and assembly. In terms of the three types of fragment approaches we define (see below), linking and assembly are the same and expansion corresponds to our 'anabolic' approach. There were data for 51 projects that we further analyzed for increase/decrease in ligand efficiency, the type of target (kinase, protease, other) and the primary method used (NMR, X-ray, other) (Tables 2.2 and 2.3). Of these 51 projects, 23 used linking/assembly as their hit-to-lead strategy and 28 used the anabolic (expansion) approach. Of the 23 linking/assembly projects, most (88 %) resulted in a compound that upon optimization was less ligand efficient. For the 28 anabolic projects, slightly more than half (54 %) resulted in less ligand-efficient compounds. From this, it can be concluded that the anabolic approach is better at maintaining ligand efficiency through the LG process compared with linking/assembly. Breaking down the hit-to-lead (H2L) strategy by screening method reveals the two main biophysical methods (NMR and X-ray) and a catch-all category (other). In both H2L strategies, NMR has the highest rate of producing more efficient lead compounds. In Table 2.2, we survey the rate of projects that maintain or increase ligand efficiency in going from hit to lead. NMR again has the highest rate of maintaining or increasing ligand efficiency. One caveat to this is that we have not explored the magnitude of the changes seen in these studies. A small decrease in a very ligand-efficient compound still results in a very ligand-efficient compound.

Table 2.2 *Breakdown of hit-to-lead strategies in published FBDD.*[87]

	Total	Increase in LE (%)	Decrease in LE (%)	Unknown LE change (%)[a]
Linking/assembly	23	22	61	17
NMR	9	33	67	0
X-ray	4	25	50	25
Other	10	10	60	30
Anabolic	28	46	54	0
NMR	5	60	40	0
X-ray	8	25	75	0
Other	15	53	47	0

[a] Due to a lack of data, a comparison of ligand efficiencies is not possible for projects in this category.

Table 2.3 *Analysis of screening methods used in published FBDD.*[87]

Screen	Total	Increase in LE (%)	Decrease in LE (%)	Unknown change in LE (%)[a]
NMR	14	43	57	0
X-ray	12	25	75	0
Other	25	36	52	12

[a] Due to a lack of data, a comparison of ligand efficiencies is not possible for projects in this category.

A pitfall of screening fragments by all biophysical methods, such as those already described, is that they do not reveal whether a fragment that is found to bind to a target protein has an effect on the normal function of the protein (see below). This is why biochemical data should be collected in parallel with biophysical data whenever possible.

2.3.2 Screen Confirmation

Many drug designers feel that functional activity must be coupled to binding of the fragment. This paradigm is slowly being changed, especially as non-traditional interactions (allosteric and/or non-active site binding sites are becoming of increasing interest and 'druggability'.[88, 89] In fact, SAR driven solely by binding information is much easier to interpret than biochemical data because it is a direct measurement. Biochemical assays can lead to false avenues of interest, especially with novel modes of binding/inhibition where interpretation of the biochemical effect may be difficult. A screen should be viewed as confirmed, not upon duplication of the original result, but when a second, orthogonal type of data is obtained. Often it is easiest to obtain biochemical data as the orthogonal data to biophysical data. We reiterate that biochemical data and biophysical data must be correlated for a hit to advance out of the LG phase. After the hit has been confirmed, the SAR hypothesis must be evaluated.

2.4 Phase III Activities

2.4.1 Evaluation of Fragments

The key to any successful drug discovery effort, whether lead like or fragment based, is the evaluation of the screen hits and the development of a path forward. There are two main components of this evaluation: (1) compositional SAR and (2) ligand efficiency. Compositional SAR is the traditional SAR that most chemists are familiar with: how do changes in the hit molecules correlate with changes in their potency? FBDD has several advantages over LLDD in developing compositional SAR. First, the number of compounds typically screened is in the range of a few hundred to a few thousand, which is smaller than most typical lead-like libraries.[58, 90–92] This make evaluation simpler than for lead-like libraries. One of the important aspects that is often overlooked in SAR evaluation is the SAR of what was not a hit (inverse SAR). With relatively small libraries, both compositional SAR and inverse SAR can be evaluated fairly quickly. The first point for evaluating SAR is to ensure that compounds (both hits and non-hits) pass quality standards. There are many anecdotal accounts of SAR being developed when the original hit(s) had poor quality (wrong compound, impurity, etc.). It is also at this time that the hypotheses generated during compound assessment are evaluated. This is important because of the highly iterative nature of FBDD. If the output of the screen does not meet preconceived notions it does not mean that it is untrue. Serendipity is the best friend of drug discovery.

Fragment libraries are typically smaller because lower complexity compounds (fragments) have a higher probability of matching a target protein-binding site, resulting in higher hit rates.[5] Even with much smaller libraries, the higher hit rate means that data for a fewer number of hits have to be scrutinized. Lower complexity compounds have their own inherent problems in developing SAR, i.e. how does one quantify the complexity of non-complex molecules? As discussed in Chapters 8 and 9, there are abundant and robust computational tools available to help in this process.

So, what can be expected from general screening of fragment-like molecules? It has recently been shown that the observed hit rates for fragment screens were 10–1000 times higher than conventional biochemical screens.[58] This bodes well for 'blind' screening of orphan, undruggable or unvalidated targets. Thus, even from a small library, there are usually sufficient data to generate compositional SAR or hypothesize SAR, that can be rapidly evaluated by further virtual and/or real-world screening. It is at this point that orthogonal screens are typically initiated. 'Nested' libraries have their greatest impact during iteration and in generating further SAR. Compositional SAR can thus be rapidly generated and evaluated, but again, this can only be carried so far without intense medicinal chemistry efforts. Culling of fragments for elaboration becomes an important step in the process.

Ligand efficiency. Ligand efficiency is the other main component for the evaluation of FBDD efforts. When multiple lines of SAR exist to be followed up, it makes the most sense to prioritize highest the most ligand-efficient compounds.[93] In order to evaluate or rank fragments or any compound, a metric called 'ligand efficiency' has been proposed.[93, 94] There exists a plethora of ligand efficiency evaluators,[94] but they all work on the same basic principle, the logarithm of the activity divided by molecular size. Molecular size can be molecular weight or heavy atom (i.e. non-hydrogen) count (HAC). We prefer HAC because halogens or non-second row elements (such as sulfur) do not bias the results, with false exclusions. Although not a new idea,[95] this method has proven valuable to assess fragment hits.[93] We prefer the simple term –log(activity)/HAC because it lends itself most easily to 'back of an envelope' evaluations. A 100 μM compound with 10 heavy atoms is more ligand efficient than a 1 μM compound with 20 heavy atoms. Starting from a compound with only 10 heavy atoms is advantageous because it gives more available chemical space to optimize the activity. Based upon what is known about orally available compounds, there is a ceiling for 30–35 heavy atoms.

With the availability of structural information, rapid improvements in potency can be rapidly obtained.[52, 65, 85, 86] SBDD is not a panacea. It is simply another method to develop and evaluate compositional SAR. In the final analysis, compounds must be made based on SBDD guidance that shows significant improvement from the initial hit for SBDD to have value. Approximately 50 % of all FBDD uses NMR and/or X-ray methods as the key component[87] (Table 2.2). For that to happen, the ligand efficiency index of the fragment hits does not have to be extremely high: molecules of 200–300 Da (14–20 HA) with potency ranges of 0.5–20 μM have a LE range of 0.23–0.45 [using then term –log(activity)/HAC]. The median ligand efficiency for all methods discussed in Tables 2.2 and 2.3 is 0.40 (ΔG/HAC). This seems to indicate that good fragments (ones that advance) should start within this range of ligand efficiency.

2.4.2 Screen Follow-up

The fragment hit follow-up process is no different to lead-like hit follow-up: the results must be confirmed with orthogonal data, the SAR hypothesis of the screen must be evaluated and modified and iterative screening must commence to support further or disprove the SAR hypothesis. To this end, additional libraries are assembled, screened and SBDD and *in silico* efforts are engaged at a higher level.[7, 48, 96–100] As long as it is understood that a different paradigm for evaluation of the fragments quality needs to be employed (ligand efficiency), there is no difference between lead-like and fragment-based hit optimization.

2.4.3 Computational Efforts/Modeling

In silico efforts reside in Phase III activities, but can be used in every facet of the FBDD process and comprise many computational tools.[101] SBDD approaches that have shown promise in recent years involve the use of computational methods to find putative binding proteins for a given compound from either genomic or protein databases and subsequently to use experimental procedures to validate the computational result. As developed over the past decade and summarized in a recent paper,[26] NMR-based screening of a variety of protein targets with a large compound library demonstrates that the hot-spot regions bind a large variety of small molecules. It was found that small compounds bind almost exclusively to well-defined, localized regions of proteins, independent of their affinity. Once these hot-spots have been identified, binding interactions with adjacent regions of the protein surface can be subsequently explored to increase selectivity and improve affinity. Druggable binding pockets generally contain small regions that are crucial to the binding of functional groups, making them the prime targets in drug design.[102]

Computation tools are now fairly abundant for the design of libraries, some of which will be described briefly below. RECAP (retrosynthetic combinatorial analysis procedure) electronically fragments molecules based on chemical knowledge.[103] When applied to databases of biologically active molecules, this allows the identification of building block fragments rich in biologically recognized elements and privileged motifs and structures. This allows the design of building blocks and the synthesis of libraries rich in biological motifs. Flux (fragment-based ligand builder reaxions) analysis demonstrated that the fragmentation of drug-like molecules by applying simplistic pseudo-retrosynthesis results in a stock of chemically meaningful building blocks for *de novo* molecule generation.[104] Resultant flux-designed molecules were chemically reasonable and contained essential substructure motifs.[105] SHAPES is another method of fragment analysis.[17, 106] It turns out that the acronym SHAPES does not stand for anything but compounds were classified by shape descriptors that included their ring structure and linker with no regard for atom type or bond order. At the basic level, SHAPES encompasses 32 frameworks which include structures from half of all known drugs. The program DAIM (decomposition and identification of molecules) has been developed virtually to deconstruct compounds in small-molecule libraries for fragment-based docking and also database analysis. An advantage of using DAIM-generated fragments, instead of filtering existing libraries for compounds with low MW, is that it keeps track of the covalent bonds 'cleaved' in the deconstruction. This information is useful for estimating the ease of synthesis of *de novo*-designed molecules.[107] DAIM has been successfully used in *in silico* screening for inhibitors of β-secretase and EphB4 kinase by fragment-based high-throughput docking. Analysis of the outcome of high-throughput screenings in addition to approved drug molecules showed recurring moieties in the active compounds. These moieties were separated from their parent molecules and termed 'privileged fragments'.[108]

Optimization of fragments. Different terminologies have been used to identify approaches for the elaboration and optimization processes in FBDD: SAR by NMR,[109] fragment expansion, fragment linking, assembly and analogue-by-catalogue (database mining), to name just a few.[52] These approaches can be sorted into three general categories, 'anabolic' (bond making), 'catabolic' (molecule breaking) and 'linking' strategies (combining fragments), and it is possible to have different hybrids of these three approaches. It should be noted that the goal of all three approaches is a lead molecule with *in vitro* activity that maintains a desirable LE (>0.4).

In the anabolic approach, fragment hits with the highest ligand efficiency are optimized through deliberate medicinal chemistry. Since the medicinal chemistry is deliberate, fragments optimized by this approach should not be encumbered by adverse properties. As shown above, this process has the best history of maintaining the LE for a fragment. A rule of thumb is that for every three heavy atoms added, the activity should increase by an order of magnitude (LE ≈ 0.4).

In a catabolic or deconstructive, process, a higher-affinity inhibitor is decomposed into fragments by retrosynthetic means. The catabolic approach is no different from typical LLDD and does not need to be evaluated here (although, as noted below, a different paradigm, ligand efficiency, is needed to drive the process forward). This is no different to what is routinely done to optimize 'lead-like' molecules. Often it is assumed that the catabolized fragments replicate, or remain very close to, the binding geometry of the original molecule. However, this is a simplistic view and should be used with great caution, especially in the absence of structural data. There is a report that shows that co-crystal structures of fragments catabolized from a known β-lactamase inhibitor do not bind where expected.[97] This result suggests that there will be gaps in the molecules created through the catabolic approach, possibly missing good lead molecules. One way to minimize this is to use progressive catabolism. Progress catabolism is the logical way to catabolize a molecule so that binding orientations can be tested and the portion(s) of the molecule responsible for activity (including binding orientation) can be determined. Figure 2.4 shows an example of progressive catabolism.

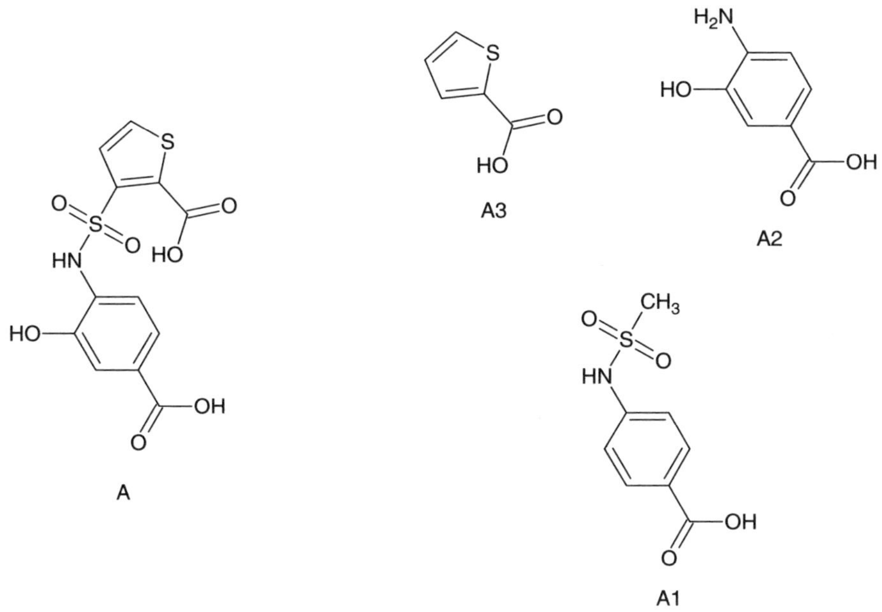

Figure 2.4 *Catabolism of a known inhibitor of thymidylate kinases (A) into component parts A1, A2 and A3.*

In a linking strategy, affinity is enhanced by joining two independent fragments together, thereby realizing the gains of synergy.[44, 110, 111] This approach can use such tactics as dynamic combinatorial chemistry and click chemistry or be the result of multiple site fragment screening.[112] The linked hits are assumed to adopt the same geometry as the original fragment hit(s). As with the catabolic approach, this may not always be the case. How the fragments are linked plays a very significant role, as the wrong linker can nullify any potential synergy gains. As shown by Alex and Flocco[87] (Table 2.1), linking is the H2L approach most likely to result in a decrease in LE. SBDD can have a huge impact in this approach, as it can confirm or deny the proper geometry of linked fragments.[111] This represents the 'home run' of FBDD. It can be extremely fast and efficient at producing very active molecules. However, this is the least likely of the approaches to work without significant expenditure of resources (typically in modifying the target). In cases where this is most routinely shown to work, the researchers start with one fragment guaranteed to bind (a warhead specific to the target or a covalently modified target–ligand complex) and the screen is initiated to find a second fragment that binds in proximity to the first fragment. This is reviewed in Chapter 10.

Still, there is no strong theoretical reason to expect that anabolic, catabolic or even linking strategies will generally apply and it is easy to imagine how non-additive effects could combine in a molecule whose binding and affinity emerge only once a critical number of functional groups are present. With all of these caveats to consider, FBDDis an essential method by which to start a new drug design project.

2.4.4 Keys to FBDD Success

The concepts that underpin the chemical fragments approach can be traced back to the pioneering work of Jencks[113] and Farmer and Ariens,[114] who showed that drug-like molecules can be regarded as the combination of two or more individual binding epitopes (fragments). The success of FBDD ultimately resides in the hands of the medicinal chemist. This is not to underestimate the importance of designing a robust process as outlined here. In the end, no matter what the quantity and quality of data around a given target are, if chemists do not make molecules the project dies. Therefore, for all the steps of this process, it is best to have the end-users (chemists) involved in the design. The interaction between chemist, biologist and structure biologist should allow for rapid, iterative collaboration. Project teams should include all of these aspects from the beginning such that alignment can be obtained constructively. FBDD is not a stand-alone endeavor.

2.5 Abbreviations

ADME/TOX	adsorption, disposition, metabolism and toxicity
Da	dalton
DMSO	dimethyl sulfoxide
EffCo	efficiency coefficient
FBDD	fragment-based drug discovery
FPA	fluorescence polarization anisotropy
H2L	hit-to-lead
HTS	high-throughput screening
ITC	isothermal titration calorimetry
LE	ligand efficiency
LLDD	lead-like drug discovery
MS	mass spectrometry
MW	molecular weight
NMR	nuclear magnetic resonance
SAR	structure–activity relationship
SBDD	structure-based drug discovery
SPR	surface plasmon resonance

References

[1] Siegel, M. G., and Vieth, M. (2007). Drugs in other drugs: a new look at drugs as fragments. *Drug Discovery Today* **12**, 71–79.

[2] Congreve, M., Carr, R., Murray, C. W., and Jhoti, H. (2003). A 'rule of three' for fragment-based lead discovery? *Drug Discovery Today* **8**, 876–877.

[3] Lipinski, C. A. (2001). Drug-like properties and the causes of poor solubility and poor permeability. *Journal of Pharmacological and Toxicological Methods* **44**, 235–249.

[4] Bohacek, R. S., McMartin, C., and Guida, W. C. (1996). The art and practice of structure-based drug design: a molecular modeling perspective. *Medicinal Research Reviews* **16**, 3–50.

[5] Hann, M. M., Leach, A. R., and Harper, G. (2001). Molecular complexity and its impact on the probability of finding leads for drug discovery. *Journal of Chemical Informatics and Computational Science* **41**, 856–864.

[6] Hajduk, P. J., and Greer, J. (2007). A decade of fragment-based drug design: strategic advances and lessons learned. *Nature Reviews Drug Discovery* **6**, 211–219.

[7] Bartoli, S., Fincham, C. I., and Fattori, D. (2007). Fragment-based drug design: combining philosophy with technology. *Current Opinion in Drug Discovery and Development* **10**, 422–429.

[8] Leach, A. R., Hann, M. M., Burrows, J. N., and Griffen, E. J. (2006). Fragment screening: an introduction. *Molecular Biosystems* **2**, 430–446.

[9] Bleicher, K., Bohm, H.-J., Muller, K., and Alanine, A. (2003). Hit and lead generation: beyond high-throughput screening. *Nature Reviews Drug Discovery* **2**, 369–378.

[10] Kubinyi, H. (2003). Drug research: myths, hype and reality. *Nature Reviews Drug Discovery* **2**, 665–668.

[11] Gill, A. (2004). New lead generation strategies for protein-kinase inhibitors – fragment based screening approaches. *Mini-Reviews in Medicinal Chemistry* **4**, 301–11.

[12] Rees, D. C., Congreve, M., Murray, C. W., and Carr, R. (2004). Fragment-based lead discovery. *Nature Reviews Drug Discovery* **3**, 660–672.

[13] Brown, D., and Superti-Furga, G. (2003). Rediscovering the sweet spot in drug discovery. *Drug Discovery Today* **8**, 1067–1077.

[14] Mayer, M., and James, T. L. (2005). Discovery of ligands by a combination of computational and NMR-based screening: RNA as an example target. *In Nuclear Magnetic Resonance of Biological Macromolecules, Part C, Methods in Enzymology*, **394**, 571–587 Ed. Thomas L. James.

[15] Kreutz, C., Kahlig, H., Konrat, R., and Micura, R. (2006). A general approach for the identification of site-specific RNA binders by F-19 NMR spectroscopy: proof of concept. *Angewandte Chemie International Edition* **45**, 3450–3453.

[16] Chung, F., Tisne, C., Lecourt, T., Dardel, F., and Micouin, L. (2007). NMR-Guided fragment-based approach for the design of tRNA(Lys3) ligands. *Angewandte Chemie International Edition* **46**, 4489–4491.

[17] Johnson, E. C., Feher, V. A., Peng, J. W., Moore, J. M., and Williamson, J. R. (2003). Application of NMR SHAPES screening to an RNA target. *Journal of the American Chemical Society* **125**, 15724–15725.

[18] Vazquez, J., Tautz, L., Ryan, J. J., Vuori, K., Mustelin, T., and Pellecchia, M. (2007). Development of molecular probes for second-site screening and design of protein tyrosine phosphatase inhibitors. *Journal of Medicinal Chemistry* **50**, 2137–2143.

[19] Golebiowski, A., Klopfenstein, S. R., and Portlock, D. E. (2003). Lead compounds discovered from libraries: Part 2. *Current Opinion in Chemical Biology* **7**, 308–325.

[20] Hann, M. M., and Oprea, T. I. (2004). Pursuing the leadlikeness concept in pharmaceutical research. *Current Opinion in Chemical Biology* **8**, 255–263.

[21] Wenlock, M. C., Austin, R. P., Barton, P., Davis, A. M., and Leeson, P. D. (2003). A comparison of physiochemical property profiles of development and marketed oral drugs. *Journal of Medicinal Chemistry* **46**, 1250–1256.
[22] Shelat, A. A., and Guy, R. K. (2007). The interdependence between screening methods and screening libraries. *Current Opinion in Chemical Biology* **11**, 244–251.
[23] Robertson, J. G. (2005). Mechanistic basis of enzyme-targeted drugs. *Biochemistry* **44**, 5561–5571.
[24] Robertson, J. G. (2007). Enzymes as a special class of therapeutic target: clinical drugs and modes of action. *Current Opinion in Structural Biology* **17**, 674–679.
[25] Drews, J. (2000). Drug discovery: a historical perspective. *Science* **287**, 1960–1964.
[26] Hajduk, P. J., Huth, J. R., and Fesik, S. W. (2005). Druggability indices for protein targets derived from NMR-based screening data. *Journal of Medicinal Chemistry* **48**, 2518–2525.
[27] Hajduk, P. J., Huth, J. R., and Tse, C. (2005). Predicting protein druggability. *Drug Discovery Today: Targets* **10**, 1675–1682.
[28] Cheng, A. C., Coleman, R. G., Smyth, K. T., Cao, Q., Soulard, P., Caffrey, D. R., Salzberg, A. C., and Huang, E. S. (2007). Structure-based maximal affinity model predicts small-molecule druggability. *Nature Biotechnology* **25**, 71–75.
[29] Oslob, J. D., and Erlanson, D. A. (2004). Tethering in early target assessment. *Drug Discovery Today: Targets* **3**, 143–150.
[30] Wunberg, T., Hendrix, M., Hillisch, A., Lobell, M., Meier, H., Schmeck, C., Wild, H., and Hinzen, B. (2006). Improving the hit-to-lead process: data-driven assessment of drug-like and lead-like screening Hits. *Drug Discovery Today* **11**, 175–180.
[31] Egner, U., Kratzschmar, J., Kreft, B., Pohlenz, H. D., and Schneider, M. (2005). The target discovery process. *ChemBioChem* **6**, 468–479.
[32] Becattini, B., and Pellechia, M. (2006). SAR by ILOEs: an NMR-based approach to reverse chemical genetics. *Chemistry: a European Journal* **12**, 2658–2662.
[33] Spring, D. R. (2005). Chemical genetics to chemical genomics: small molecules offer big insights. *Chemical Society Reviews* **34**, 472–482.
[34] Allen, J. J., and Shokat, K. M. (2006). Chemical genomics: dialed in transcriptional network control with non-steroidal glucocorticoid receptor modulators. *ACS Chemical Biology* **1**, 139–140.
[35] Kwon, H. J. (2003). Chemical genomics-based target identification and validation of anti-angiogenic agents. *Current Medicinal Chemistry* **10**, 717–736.
[36] Kwon, H. J. (2006). Discovery of new small molecules and targets towards angiogenesis via chemical genomics approach. *Current Drug Targets* **7**, 397–405.
[37] Willson, T. (2003). Chemical genomics of orphan nuclear receptors. *Ernst Schering Research Foundation Workshop*, 29–42.
[38] Caron, P. R. (2005). Introduction to chemical genomics. *Methods in Molecular Biology* **310**, 3–10.
[39] Zartler, E. R., and Shapiro, M. J. (2006). Protein NMR-based screening in drug discovery. *Current Pharmaceutical Design* **12**, 3963–3972.
[40] Zartler, E. R., Yan, J., Mo, H., Kline, A. D., and Shapiro, M. J. (2003). 1D NMR methods in ligand–receptor interactions. *Current Topics in Medicinal Chemistry* **3**, 25–37.
[41] Card, G. L., Blasdel, L., England, B. P., Zhang, C., Suzuki, Y., Gillette, S., Fong, D., Ibrahim, P. N., Artis, D. R., Bollag, G., Milburn, M. V., Kim, S.-H., Schlessinger, J., and Zhang, K. Y. J. (2005). A family of phosphodiesterase inhibitors discovered by cocrystallography and scaffold-based drug design. *Nature Biotechnology* **23**, 201–207.
[42] Jhoti, H. (2005). A new school for screening. *Nature Biotechnology* **23**, 184–6.
[43] Sanders, W. J., Nienaber, V., Lerner, C. G., McCall, J. O., Merrick, S. M., Swanson, S. J., Harlan, J. E., Stoll, V. S., Stamper, G. F., Betz, S. F., Condroski, K. R., Meadows, R. P.,

Severin, J. M., Walter, K., Magdalinos, P., Jakob, C. G., Wagner, R., and Beutel, B. A. (2004). Discovery of potent inhibitors of dihydroneopterin aldolase using CrystaLEAD high-throughput X-ray crystallographic screening and structure-directed lead optimization. *Journal of Medicinal Chemistry* **47**, 1709–1718.

[44] Howard, N., Abell, C., Blakemore, W., Chessari, G., Congreve, M., Howard, S., Jhoti, H., Murray, C. W., Seavers, L. C. A., and van Montfort, R. L. M. (2006). Application of fragment screening and fragment linking to the discovery of novel thrombin inhibitors. *Journal of Medicinal Chemistry* **49**, 1346–1355.

[45] Bright, H., Watts, P., Carroll, T., and Fenton, R. (2003). The validation of GBV-B as a surrogate model for HCV in the drug discovery process. *Antiviral Research* **57**, A85.

[46] Orry, A. J. W., Abagyan, R. A., and Cavasotto, C. N. (2006). Structure-based Development of target-specific compound libraries. *Drug Discovery Today* **11**, 261–6.

[47] Combs, A. P. (2007). Structure-based drug design of new leads for phosphatase research. *Idrugs* **10**, 112–115.

[48] Hubbard, R. E., Chen, I., and Davis, B. (2007). Informatics and modeling challenges in fragment-based drug discovery. *Current Opinion in Drug Discovery, and Development* **10**, 289–297.

[49] Villar, H. O., and Hansen, M. R. (2007). Computational techniques in fragment based drug discovery. *Current Topics in Medicinal Chemistry* **7**, 1509–1513.

[50] Reddy, A. S., Pati, S. P., Kumar, P. P., Pradeep, H. N., and Sastry, G. N. (2007). Virtual screening in drug discovery – a computational perspective. *Current Protein and Peptide Science* **8**, 329–351.

[51] Hesterkamp, T., Barker, J., Davenport, A., and Whittaker, M. (2007). Fragment based drug discovery using fluorescence correlation spectroscopy techniques: challenges and solutions. *Current Topics in Medicinal Chemistry* **7**, 1582–1591.

[52] Barker, J., Courtney, S., Hesterkamp, T., Ullman, D., and Whittaker, M. (2006). Fragment screening by biochemical assay. *Expert Opinion in Drug Discovery* **1**, 225–236.

[53] Annis, D. A., Nickbarg, E., Yang, X., Ziebell, M. R., and Whitehurst, C. E. (2007). Affinity selection–mass spectrometry screening techniques for small molecule drug discovery. *Current Opinion in Chemical Biology* **11**, 518–526.

[54] Jahnke, W. (2007). Perspectives of biomolecular NMR in drug discovery: the blessing and curse of versatility. *Journal of Biomolecular NMR* **39**, 87–90.

[55] Klages, J., Coles, M., and Kessler, H. (2007). NMR-based screening: a powerful tool in fragment-based drug discovery. *Analyst* **132**, 693–705.

[56] Papeo, G., Giordano, P., Brasca, M. G., Buzzo, F., Caronni, D., Ciprandi, F., Mongelli, N., Veronesi, M., Vulpetti, A., and Dalvit, C. (2007). Polyfluorinated amino acids for sensitive F-19 NMR-based screening and kinetic measurements. *Journal of the American Chemical Society* **129**, 5665–5672.

[57] Taylor, J. D., Gilbert, P. J., Williams, M. A., Pitt, W. R., and Ladbury, J. E. (2007). Identification of novel fragment compounds targeted against the pY pocket of v-Src SH2 by computational and NMR screening and thermodynamic evaluation. *Proteins: Structure Function and Bioinformatics* **67**, 981–990.

[58] Schuffenhauer, A., Ruedisser, S., Marzinzik, A. L., Jahnke, W., Blommers, M. J. J., Selzer, P., and Jacoby, E. (2005). Library design for fragment based screening. *Current Topics in Medicinal Chemistry* **5**, 751–62.

[59] Baurin, N., Aboul-Ela, F., Barril, X., Davis, B., Drysdale, M., Dymock, B., Finch, H., Fromont, C., Richardson, C., Simmonite, H., and Hubbard, R. E. (2004). Design and characterization of libraries of molecular fragments for use in NMR screening against protein targets. *Journal of Chemical Information and Computer Sciences* **44**, 2157–2166.

[60] McGovern, S. L., Caselli, E., Grigorieff, N., and Shoichet, B. K. (2002). A Common mechanism underlying promiscuous inhibitors from virtual and high-throughput screening. *Journal of Medicinal Chemistry* **45**, 1712–1722.

[61] Blundell, T. L., and Patel, S. (2004). High-throughput X-ray crystallography for drug discovery. *Current Opinion in Pharmacology* **4**, 490–496.

[62] Hartshorn, M. J., Murray, C. W., Cleasby, A., Frederickson, M., Tickle, I. J., and Jhoti, H. (2005). Fragment-based lead discovery using X-ray crystallography. *Journal of Medicinal Chemistry* **48**, 403–413.

[63] Borch, J., and Roepstorff, P. (2004). Screening for enzyme inhibitors by surface plasmon resonance combined with mass spectrometry. *Analytical Chemistry* **76**, 5243–5248.

[64] Moy, F. J., Haraki, K., Mobilio, D., Walker, G., Powers, R., Tabei, K., Tong, H., and Siegel, M. M. (2001). MS/NMR: a structure -based approach for discovering protein ligands and for drug design by coupling size exclusion chromatography, mass spectrometry and nuclear magnetic resonance spectroscopy. *Analytical Chemistry* **73**, 571–581.

[65] Ciulli, A., Williams, G., Smith, A. G., Blundell, T. L., and Abell, C. (2006). Probing hot spots at protein-ligand binding sites: a fragment-based approach using biophysical methods. *Journal of Medicinal Chemistry* **49**, 4992–4500.

[66] Zartler, E. R., Yan, J., Mo, H., Kline, A. D., and Shapiro, M. J. (2003). 1D NMR Methods in ligand–receptor interactions. *Curr Top Med Chem* **3**, 25–37.

[67] Zartler, E. R., and Shapiro, M. J. (2006). Protein NMR-based screening in drug discovery. *Current Pharmaceutical Design* **12**, 3963–3972.

[68] Neumann, T., Junker, H. D., Schmidt, K., and Sekul, R. (2007). SPR-based fragment screening: advantages and applications. *Current Topics in Medicinal Chemistry* **7**, 1630–1642.

[69] Giannetti, A. M., Koch, B. D., and Browner, M. F. (2008). Surface plasmon resonance based assay for the detection and characterization of promiscuous inhibitors. *Journal of Medicinal Chemistry* **51**, 574–580.

[70] Lesuisse, D., Lange, G., Deprez, P., Benard, D., Schoot, b., Delettre, G., Marquette, J.-P., Broto, P., Jean-Baptiste, V., Bichet, P., Sarubbi, E., and Mandine, E. (2002). SAR and X-ray. A new approach combining fragment-based screening and rational drug design: application to the discovery of nanomolar inhibitors of Src SH2. *Journal of Medicinal Chemistry* **45**, 2379–2387.

[71] Congreve, M., Aharony, D., Albert, J., Callaghan, O., Campbell, J., Carr, R. A. E., Chessari, G., Cowan, S., Edwards, P. D., Frederickson, M., McMenamin, R., Murray, C. W., Patel, S., and Wallis, N. (2007). Application of fragment screening by X-ray crystallography to the discovery of aminopyridines as inhibitors of beta-secretase. *Journal of Medicinal Chemistry* **50**, 1124–1132.

[72] Jhoti, H., Cleasby, A., Verdonk, M., and Williams, G. (2007). Fragment-based screening using X-ray crystallography and NMR spectroscopy. *Current Opinion in Chemical Biology* **11**, 485–493.

[73] Murray, C. W., Callaghan, O., Chessari, G., Cleasby, A., Congreve, M., Frederickson, M., Hartshorn, M. J., McMenamin, R., Patel, S., and Wallis, N. (2007). Application of fragment screening by X-ray crystallography to beta-secretase. *Journal of Medicinal Chemistry* **50**, 1116–1123.

[74] Edwards, P. D., Albert, J. S., Sylvester, M., Aharony, D., Andisik, D., Callaghan, O., Campbell, J. B., Carr, R. A., Chessari, G., Congreve, M., Frederickson, M., Folmer, R. H. A., Geschwindner, S., Koether, G., Kolmodin, K., Krumrine, J., Mauger, R. C., Murray, C. W., Olsson, L. L., Patel, S., Spear, N., and Tian, G. (2007). Application of fragment-based lead generation to the discovery of novel, cyclic amidine beta-secretase inhibitors with nanomolar potency, cellular activity and high ligand efficiency. *Journal of Medicinal Chemistry* **50**, 5912–5925.

[75] Geschwindner, S., Olsson, L. L., Albert, J. S., Deinum, J., Edwards, P. D., de Beer, T., and Folmer, R. H. A. (2007). Discovery of a novel warhead against beta-secretase through fragment-based lead generation. *Journal of Medicinal Chemistry* **50**, 5903–5911.

[76] Williams, B. (2007). Fragment based drug discovery - from crystal to clinic. *Journal of Pharmacy and Pharmacology* **59**, A77.

[77] Kuglstatter, A., Stahl, M., Peters, J.-U., Huber, W., Stihle, M., Schlatter, D., Benz, J., Ruf, A., Roth, D., Enderle, T., and Hennig, M. (2008). Tyramine fragment binding to BACE-1. *Bioorganic and Medicinal Chemistry Letters* **18**, 1304–1307.

[78] Talhout, R., Villa, A., Mark, A. E., and Engberts, J. B. F. N. (2003). Understanding binding affinity: a combined isothermal titration calorimetry/molecular dynamics study of the binding of a series of hydrophobically modified benzamidinium chloride inhibitors to trypsin. *Journal of the American Chemical Society* **125**, 10570–10579.

[79] Turnbull, W. B., and Daranas, A. H. (2003). On the value of *c*: can low affinity systems by studied by isothermal titration calorimetry. *Journal of the American Chemical Society* **125**, 14859–14866.

[80] Trosset, J. Y., Dalvit, C., Knapp, S., Fasolini, M., Veronesi, M., Mantegani, S., Gianellini, L. M., Catana, C., Sundstrom, M., Stouten, P. F. W., and Moll, J. K. (2006). Inhibition of protein–protein interactions: the discovery of druglike beta-catenin inhibitors by combining virtual and biophysical screening. *Proteins: Structure, Function and Bioinformatics* **64**, 60–67.

[81] Inglese, J., Johnson, R. L., Simeonov, A., Xia, M. H., Zheng, W., Austin, C. P., and Auld, D. S. (2007). High-throughput screening assays for the identification of chemical probes. *Nature Chemical Biology* **3**, 466–479.

[82] Pereira, D. A., and Williams, J. A. (2007). Origin and evolution of high throughput screening. *British Journal of Pharmacology* **152**, 53–61.

[83] Cummings, M. D., Farnum, M. A., and Nelen, M. I. (2006). Universal screening methods and applications of ThermoFluor(R). *Journal of Biomolecular Screening* **11**, 854–863.

[84] Koblish, H. K., Zhao, S., Franks, C. F., Donatelli, R. R., Tominovich, R. M., LaFrance, L. V., Leonard, K. A., Gushue, J. M., Parks, D. J., Calvo, R. R., Milkiewicz, K. L., Marugan, J. J., Raboisson, P., Cummings, M. D., Grasberger, B. L., Johnson, D. L., Lu, T., Molloy, C. J., and Maroney, A. C. (2006). Benzodiazepinedione inhibitors of the Hdm2:p53 complex suppress human tumor cell proliferation *in vitro* and sensitize tumors to doxorubicin *in vivo*. *Molecular Cancer Therapeutics* **5**, 160–169.

[85] Homans, S. W. (2005). Probing the binding entropy of ligand–protein interactions by NMR. *ChemBioChem* **6**, 1585–1591.

[86] Perozzo, R., Folkers, G., and Scapozza, L. (2004). Thermodynamics of protein-ligand interactions: history, presence and future aspects. *Journal of Receptor and Signal Transduction Research* **24**, 1–52.

[87] Alex, A. A., and Flocco, M. M. (2007). Fragment-based drug discovery: what has it achieved so far? *Current Topics in Medicinal Chemistry* **7**, 1544–1567.

[88] Milligan, G., and Smith, N. J. (2007). Allosteric modulation of heterodimeric G-protein-coupled receptors. *Trends in Pharmacological Sciences* **28**, 615–620.

[89] Beher, D. (2008). Gamma-secretase modulation and its promise for Alzheimer's disease: a rationale for drug discovery. *Current Topics in Medicinal Chemistry* **8**, 34–37.

[90] Wong, C.-H., Hendrix, M., Manning, D. D., Rosenbohm, C., and Greenberg, W. A. (1998). A library approach to the discovery of small molecules that recognize RNA: use of a 1,3-hydroxyamine motif as core. *Journal of the American Chemical Society* **120**, 8319–8327.

[91] Lepre, C. A. (2001). Library design for NMR-based screening. *Drug Discovery Today* **6**, 133–140.

[92] Siegal, G., Ab, E., and Schultz, J. (2007). Integration of fragment screening and library design. *Drug Discovery Today* **12**, 1032–1039.

[93] Hopkins, A. L., Groom, C. R., and Alex, A. (2004). Ligand efficiency: a useful metric for lead selection. *Drug Discovery Today* **9**, 430–431.

[94] Abad-Zapatero, C., and Metz, J. T. (2005). Ligand efficiency indices as guideposts for drug discovery. *Drug Discovery Today* **10**, 464–469.

[95] Kuntz, I. D., Chen, K., Sharp, K. A., and Kollman, P. A. (1999). The maximal affinity of ligands. *Proceedings of the National Academy of Sciences of the nUnited States of America* **96**, 9997–10002.

[96] Vangrevelinghe, E., and Rudisser, S. (2007). Computational approaches for fragment optimization. *Current Computer-Aided Drug Design* **3**, 69–83.

[97] Babaoglu, K., and Shoichet, B. K. (2006). Deconstructing fragment-based inhibitor discovery. *Nature Chemical Biology* **2**, 720–723.

[98] Davis, A. M., Keeling, D. J., Steele, J., Tomkinson, N. P., and Tinker, A. C. (2005). Components of successful lead generation. *Current Topics in Medicinal Chemistry* **5**, 421–439.

[99] Keseru, G. M., and Makara, G. M. (2006). Hit discovery and hit-to-lead approaches. *Drug Discovery Today* **11**, 741–748.

[100] Tsao, D. H. H., Sutherland, A. G., Jennings, L. D., Li, Y. H., Rush, T. S., Alvarez, J. C., Ding, W. D., Dushin, E. G., Dushin, R. G., Haney, S. A., Kenny, C. H., Malakian, A. K., Nilakantan, R., and Mosyak, L. (2006). Discovery of novel inhibitors of the ZipA/FtsZ complex by NMR fragment screening coupled with structure-based design. *Bioorganic and Medicinal Chemistry* **14**, 7953–7961.

[101] Poppe, L., Harvey, T. S., Mohr, C., Zondlo, J., Tegley, C. M., Nuanmanee, O., and Cheetham, J. (2007). Discovery of ligands for Nurr1 by combined use of NMR screening with different isotopic and spin-labeling strategies. *Journal of Biomolecular Screening* **12**, 301–311.

[102] DeLano, W. L. (2002). Unraveling hot spots in binding interfaces: progress and challenges. *Current Opinions in Structural Biology* **12**, 14–20.

[103] Lewell, X. Q., Judd, D. B., Watson, S. P., and Hann, M. M. (1998). RECAP – retrosynthetic combinatorial analysis procedure: a powerful new technique for identifying privileged molecular fragments with useful applications in combinatorial chemistry. *Journal of Chemical Informatics and Computer Science* **38**, 511–522.

[104] Fechner, U., and Schneider, G. (2006). Flux (1): a virtual synthesis scheme for fragment-based *de novo* design. *Journal of Chemical Information and Modeling* **46**, 699–707.

[105] Fechner, U., and Schneider, G. (2007). Flux (2): comparison of molecular mutation and crossover operators for ligand-based de novo design. *Journal of Chemical Information and Modeling* **47**, 656–667.

[106] Fejzo, J., Lepre, C. A., Peng, J. W., Bemis, G. W., Ajay, Murcko, M. A., and Moore, J. M. (1999). The SHAPES strategy: an NMR-based approach for lead generation in drug discovery. *Chemistry and Biology* **6**, 755–769.

[107] Kolb, P., and Caflisch, A. (2006). Automatic and efficient decomposition of two-dimensional structures of small molecules for fragment-based high-throughput docking. *Journal of Medicinal Chemistry* **49**, 7384–7392.

[108] Vieth, M., Siegel, M. G., Higgs, R. E., Watson, I. A., Robertson, D. H., Savin, K. A., Durst, G. L., and Hipskind, P. A. (2004). Characteristic physical properties and structural fragments of marketed oral drugs. *Journal of Medicinal Chemistry* **47**, 224–232.

[109] Shuker, S. B., Hajduk, P. J., Meadows, R. P., and Fesik, S. W. (1996). Discovering high-affinity ligands for proteins: SAR by NMR. *Science* **274**, 1531–1534.

[110] Rohrig, C. H., Loch, C., Guan, J. Y., Siegal, G., and Overhand, M. (2007). Fragment-based synthesis and SAR of modified FKBP ligands: influence of different linking on binding affinity. *ChemMedChem* **2**, 1054–1070.

[111] Olejniczak, E. T., Hajduk, P. J., Marcotte, P. A., Nettesheim, D. G., Meadows, R. P., Edalji, R., Holzman, T. F., and Fesik, S. W. (1997). Stromelysin Inhibitors designed from weakly bound fragments: effects of linking and cooperativity. *Journal of the American Chemical Society* **119**, 5828–5832.

[112] Huth, J. R., Park, C., Petros, A. M., Kunzer, A. R., Wendt, M. D., Wang, X. L., Lynch, C. L., Mack, J. C., Swift, K. M., Judge, R. A., Chen, J., Richardson, P. L., Jin, S., Tahir, S. K., Matayoshi, E. D., Dorwin, S. A., Ladror, U. S., Severin, J. M., Walter, K. A., Bartley, D. M., Fesik, S. W., Elmore, S. W., and Hajduk, P. J. (2007). Discovery and design of novel HSP90 inhibitors using multiple fragment-based design strategies. *Chemical Biology and Drug Design* **70**, 1–12.

[113] Jencks, W. P. (1981). On the attribution and additivity of binding energies. *Proceedings of the National Academy of Sciences of the United States of America* **78**, 4046–4050.

[114] Farmer, P. S., and Ariens, E. J. (1982). Speculations on the design of nonpeptidic peptidomimetics. *Trends in Pharmacological Science* **3**, 362–365.

[115] Zartler, E. R., Hanson, J., Jones, B. E., Kline, A. D., Martin, G., Mo, H., Shapiro, M. J., Wang, R., Wu, H., and Yan, J. (2003). RAMPED-UP NMR: multiplexed NMR-based screening for drug discovery. *Journal of the American Chemical Society* **125**, 10941–10946.

[116] Cummings, M. D., Farnum, M. A., and Nelen, M. I. (2006). Universal screening methods and applications of ThermoFluor (R). *Journal of Biomolecular Screening* **11**, 854–863.

[117] Houston, J. G., Banks, M. N., Binnie, M., Brenner, S., O'Connell, J., and Petrillo, E. W. (2008). Case study: impact of technology investment on lead discovery at Bristol-Myers Squibb, 1998–2006. *Drug Discovery Today* **13**, 44–51.

[118] Niesen, F. H., Berglund, H., and Vedadi, M. (2007). The use of differential scanning fluorimetry to detect ligand interactions that promote protein stability. *Nature Protocols* **2**, 2212–2221.

3

Assembling a Fragment Library

Mark Brewer, Osamu Ichihara, Christian Kirchhoff, Markus Schade and Mark Whittaker

3.1 Introduction

The current popularity of fragment-based drug discovery (FBDD) represents a shift in philosophy from the random screening of molecules with higher molecular weights and physical properties more akin to those of drug-like compounds to the screening of smaller, less complex molecules. This is because it has been recognised that fragment hit molecules can be efficiently optimised into leads particularly if the binding mode to the target protein has been first determined by 3D structural elucidation.

Several studies have shown that medicinal chemistry optimisation results in a final compound with increased molecular weight compared with the starting structure. The evolution of a low molecular weight fragment hit represents an attractive approach to optimisation and may be more efficient than pruning back a higher molecular weight hit compound discovered by conventional high-throughput screening of drug-like compound libraries. Fragment hits represent a simpler molecular entry point to the drug discovery process compared with, say, conventional high-throughput screening (HTS) hit molecules but owing to their inherent simplicity often exhibit lower potencies than larger drug-like hit molecules. Furthermore, there are cases where the screening of fragment libraries has yielded hits where standard HTS has proven challenging; a recent example is the Alzheimer's disease target β-secretase.[1, 2] It has been argued, by consideration of 'ligand efficiency',[3, 4] that fragments offer a more practical starting point for hit-to-lead and lead optimisation programmes.[4] Ligand efficiency has been developed from the concept of Kuntz *et al.*[5] on

Fragment-Based Drug Discovery: A Practical Approach Edited by Edward R. Zartler and Michael J. Shapiro

the maximum affinity of ligands and represents the binding energy per heavy atom in a molecule. This is equal to the free energy of binding of a ligand to a target protein divided by the non-hydrogen atom count (NHC) of the ligand[3] (this may be approximated to by $[-RT\ln(IC_{50})]$/NHC. Therefore, rather than focusing on the potency of a hit molecule, fragment-based drug discovery gives access to low molecular weight compounds where the optimisation process can concentrate on improvements in potency and other desirable attributes without an immediate concern of increasing molecular weight. Not all fragment hits will display high ligand efficiency in their interaction with a particular target protein, so the concept of ligand efficiency is particularly useful in selecting which molecules to take forward into optimisation.

A key question is how fragments differ from drugs. There is considerable variation in the literature over the definition of fragments. Commonly fragment molecules are defined in terms of their chemoinformatic and calculated physical properties and in a similar vein to Lipinski's rule of five for oral availability of molecules[6] [molecular weight $\leq$500 Da, Clog*P* $\leq$5, hydrogen bond donors (HBD) $\leq$5; hydrogen bond acceptors (HBA) $\leq$10] a rule of three (molecular weight $\leq$300 Da, Clog*P* $\leq$3, HBD $\leq$3 with optional additional criteria of rotatable bonds $\leq$3 and polar surface area $\leq$60 Å^2) has been put forward for molecules that are used in high-throughput crystallography fragment-based screening (Table 3.1).[7] That a variety of different approaches have been adopted for the assembly of fragment libraries is evidenced by Table 3.2, which provides an overview of the general characteristics of fragment libraries from a variety of research groups. The preferred fragment property profiles of many of these libraries are related to the rule of three. During the generation of a fragment library, such rules for property profiles can be easily applied to short list fragment-compliant compounds to be purchased from commercial collections or, alternatively, to be synthesised from virtual libraries. When considering fragment molecules in the context of screening libraries in general, we have found it informative to consider the molecules in terms of a molecular weight spectrum ranging from small solvent molecules at the low end to large drug-like molecules at the high end. Lead-like molecules occupy a space between the fragment and drug-like molecules and some groups have also defined 'scaffolds' as occupying this intermediate region of the spectrum.

Table 3.1 *Comparison of rule of three (RO3)*[7] *with criteria for compounds of reduced complexity and for lead-like compounds.*[27]

	Rule of three	Lead-like	Reduced complexity
MW	<300	≤460	<350
Log*P* (o/w)	≤3	≥–4 and ≤4.2	≤2.2
Log*S* (water)	NA	≥–5	NA
Rotatable bonds	≤3	≤10	≤6
Rings	NA	≤4	NA
H-bond donors	≤3	≤5	≤3
H-bond acceptors	≤3	≤9	≤8
Heavy atoms	NA	NA	≤22
TPSA (Å^2)	≤60	NA	NA

NA, not applicable.

Table 3.2 *Comparison of selected fragment libraries reported in the literature and on company WEB sites.*

Company	MW (mean)	ClogP (mean)	H-bond donors	H-bond acceptors	Rotatable bonds	Polar surface area (mean) (Å²)	No. of compounds	Terminology
Abbott	(220)	(1.5)					~10 000	
ActiveSite	90–220 (142)						384	Small, Soluble, Stiff, Synthetically AcceSSible
Astex	≤300	≤3	≤3				~1000	Pyramid
AstraZeneca	(220)	(1.6)					2000	
Evotec	150–310 (256)	0.5–3.5 (2.2)	≤4	≤7	≤5		20 063	EVOlution NMR Library
Evotec	150–350 (247)	0–3 (1.6)	1–3	≤4	≤5	70	21 869	EVOlution Biochemical Library
GSK	<22 heavy atoms	≤2.2	≤3	≤8	≤6		?	Reduced complexity set[a]
Graffinity	≤300	≤3	≤3				~20 000	RAISE
Plexxikon	120–350 (250)						20 360	Scaffolds
Roche	≤300	≤3	≤3		≤3	≤60	~2000	Needles
SGX	≤16 heavy atoms (174)[b]	≤2[c]	≤3[d]	≤3[d]	≤3[d]		~1000	FAST
Sunesis	<200						~7000	Tethering
Vernalis	100–250 (190 ± 41)	(0.3 ± 1.3)	(1.9 ± 1.3)	(3.2 ± 1.2)	(2.3 ± 1.6)		~1400	SeeDs
Vertex	68–341 (194)	−2.2 to 5.5					132	SHAPES

[a] Diversity selection made based on 3D gridding and partitioning pharmacophores and all compounds include a synthetic handle.
[b] Bromine treated as methyl for calculation of mean MW. All compounds have ≥1 ring and ≥2 synthetic handles; 50 % of compounds contain aromatic bromine.
[c] CLog*D* applies to 90 % of molecules.
[d] Applies to 90–94 % of molecules.

A key question is what is the optimum number of fragments that are required to make up an effective fragment screening library. Chemical space for fragment molecules is expected to be considerably smaller than the chemical space of molecules with drug-like physical properties, so a fragment library should sample fragment chemical space more efficiently than a drug-like library of the same size can sample drug-like chemical space. It is hard to obtain a good fix on the true size of druggable chemical space as there is significant variation in the estimates reported in the literature.[8] Recently, Reymond and co-workers have built a database of all possible organic molecules with up to 11 atoms from a restricted set of C, N, O and F.[9, 10] This contains 26.4 million compounds (110.9 million stereoisomers) with an average molecular weight (MW) of 153 Da. It is extrapolated that there should be approximately 10^{27} different molecules at 25 atoms, which is the size of average drug molecules (MW 340), since the number of molecules in the Reymond database increases exponentially with the square of the number of atoms.[10] It has been further estimated by Guida and co-workers that the universe of organic compounds containing up to 30 C, N, O and S atoms is in excess of 10^{60} different molecules.[11] Based on analyses such as these, it has been argued that the universe of potential molecules is very large and it is by no means certain that any screening library will adequately sample this chemical space. The universe of possible molecules decreases significantly in size with decreasing molecular weight and so even a relatively small fragment library will do a better job of sampling fragment chemical space than a very large screening library samples drug-like chemical space. Indeed, Hann *et al.* have argued that lower molecular weight molecules exhibit reduced complexity compared with the molecules in drug-like collections and have developed a model to rationalise ligand–receptor interactions in the molecular recognition process.[12] According to this analysis, it was shown that the probability of observing a useful interaction falls dramatically with increasing molecular complexity of the ligand. It can be argued that drug-like collections such as typical HTS screening decks contain molecules that are too complex to yield useful binding events. Several analyses have shown that as leads are optimised into drugs, certain properties, such as MW and log*P*, almost always increase during this process.[13, 14] It would therefore be more efficient to start by screening molecules that are more lead-like rather than a set of molecules that have property profiles more akin to mature, optimised drug molecules.

A key consideration in fragment-based drug discovery[15] is the construction of a fragment library that is fit for purpose in relation to the types of targets to be screened and the screening method itself. As with the construction of HTS libraries,[16–18] there are different approaches that can be taken and philosophies to be followed.[19] However, it is usually a matter of conjecture as to whether one approach for assembling a screening library is superior to another.[20] The ideal outcome from an HTS campaign is multiple series of hit compounds with robust, reproducible activity, target specificity (i.e. not promiscuous inhibitors), novel and diverse with respect to any known chemical matter for the target of interest, providing some initial structure–activity relationship (SAR) and suitable for rapid optimisation. The ideal outcome from fragment screening is broadly similar with the additional requirement that the active fragments should provide sufficient ligand efficiency and give rise to X-ray and/or NMR structures in complex with the target protein to merit taking forward into fragment-to-lead (F2L) optimisation. However, the number of fragment molecules that are currently available from commercial sources is considerably smaller than the number of molecules with more drug-like property profiles, and this constrains the pool of compounds from which a fragment library can be built.

In addition to the selection of the compounds themselves, there needs to be a detailed consideration of compound logistics. Although most fragment compound collections are relatively small in size compared with large high-throughput screening compound collections, the management of samples can still become complex and time consuming if not effectively planned from the outset.[21]

3.2 Selection of Fragments

In this section we review the considerations that are necessary in the selection of molecules for a fragment screening library. There will be instances when a fragment library is assembled from scratch but frequently, and in common with screening collections in general, the construction is performed on an iterative basis. New chemical matter might be introduced to increase the overall chemical diversity of the library or to add sets of compounds that are focused towards a particular class of target or a specific target.

Some of the initial considerations when selecting molecules for a fragment library may be of a practical nature and as such could apply equally well to the construction of any screening collection. For example, it will be necessary to determine the number of compounds that are required. The throughput of the fragment screening assay will help determine this number and also, of course, budgetary considerations. In addition, it will also be necessary to determine the amount of each compound that will be required; this can be estimated from the amount of material consumed each time an assay is carried out and the number of times assays are likely to be performed. With practical considerations addressed, attention can be turned to more detailed aspects of the building process, in particular the properties of the molecules that will constitute the fragment library. A decision will need to be made whether to acquire a sufficient quantity of each compound to allow immediate follow-up of fragment hits (e.g. by verification in an orthogonal assay and by X-ray structural studies) or to acquire 'fresh' samples by repurchase or synthesis. This will have a significant impact on the time taken to follow up fragment hits.

There are two key underlying factors in determining the properties of molecules in a fragment library: first, the assay(s) in which the library is to be employed and any concomitant restriction on the properties of molecules that can be successfully screened, and second, the nature of the target(s) against which the fragment library will be screened.

There is now a broad variety of different assay techniques that can be used to screen fragments and each has its own set of requirements for the properties of molecules that can be successfully screened.[22, 23] For example, since fragment molecules are often screened at high concentration, compound solubility can be an important consideration,[24] the tethering technique requires the presence of a thiol functionality in the screening molecules[25] and fragment molecules for high-throughput crystallography can benefit from the presence of bromine atoms (both due to its unique dispersion signal and the possibility of hit follow-up with Suzuki cross-coupling protocols (see Section 3.2.5)).[26]

With regard to the second of the two key factors, the purpose of the library will require careful consideration. At the outset of library construction it is often the case that all the targets against which it will be tested will not have been determined. Indeed, in our experience, screening collections are challenged by a diverse range of targets during their useful lifetime and on this basis it can be beneficial to have a set of chemically diverse molecules at the heart of the library that can be augmented with focused or hypothesis-driven sets of molecules on a target-by-target basis. We advocate that a screening collection should be regarded as a dynamic rather than a static entity.

3.2.1 Sources of Molecules

Before selecting molecules for a fragment library, a pool of candidate molecules will be required for which there are several possible sources. As outlined in the introduction, numerical estimates for the possible number of molecules within the drug-like portion of chemical space are vast.[8–11] To help with a further classification of chemical space, Hann and Oprea introduced a concept of the virtual, the tangible and the real worlds of molecules.[27] The virtual world consists of molecules that are possible in theory, the tangible world includes molecules that are chemically feasible (and the number of molecules within its realm will increase as new chemistries become available) and, finally, the real world contains molecules that are physically available. In principle, the selection of fragment molecules can be made from the tangible and real worlds of molecules. Once the molecular structures are available for tangible compounds, these can be processed in much the same way as real world molecules, subject to the caveat that any molecules that are selected can in fact be synthesised. In practical terms, the tangible and real worlds of molecules will vary to some extent between different organisations, with some having access to novel chemistry, synthetic routes and building blocks, etc., in addition to each having access to their own internal compound collections.

Commercially available molecules occupy both the tangible and real worlds of molecules. Many organisations maintain copies of these vendor catalogues in internal databases which can be searched and filtered as required. Evotec maintains a database of commercial screening compounds which is compiled from the currently available collections of over 35 different compound vendors and is updated on a regular basis. After removal of duplicate structures, the database contains over 3.8 million chemical structures and is therefore representative of the real world of commercially available screening compounds. As such, this database provides a useful resource for the selection of fragment molecules in library enhancement initiatives and the generation of focused sets, and also for the selection of molecules in fragment hit follow-up activities.

3.2.2 Physical Property and Substructure Filters

Physical property and substructure filters are straightforward to apply to large databases and are often used as one of the initial steps in a fragment selection process. Fragment molecules are sometimes defined in terms of their physical properties, which in turn can be used to help shortlist molecules in fragment selection processes. In a similar manner to Lipinski's rule of five for the identification of molecules with poor absorption or permeability,[6] Congreve *et al.* proposed a rule of three for fragment-based lead discovery.[7] The criteria for the rule of three are shown in Table 3.1 and were deduced from the analysis of a diverse set of fragment hits that were obtained against a range of targets. Hann and Oprea have designed a set of criteria for reduced complexity screening and these are also shown in Table 3.1 along with an example set of criteria for lead-like molecules for comparison.[27] When applying molecular weight-based filters, there are some subtleties that can be applied to account for what chemists often term 'real molecular weight' to avoid penalising compounds that feature substituents such as sulfonamide groups, trifluoromethyl groups and bromine atoms.[26, 28] An alternative approach would be to filter on the basis of the NHC rather than apply a molecular weight-based filter.

As mentioned previously, compound solubility can be an important consideration since fragment molecules have low potencies and high concentrations will be used during screening. For example, insoluble compounds can lead to artefacts in fluorescence and ligand-based NMR fragment screening methods. Several groups have used computational models for aqueous solubility to help favour the selection of soluble fragment molecules during the construction of fragment libraries.[24, 28]

In addition to physical properties, substructure-based filters can be applied to reduce further the number of molecules, for instance molecules with undesirable functionality; for example, reactive or toxic groups can be removed and molecules with particular features (or atoms) can be actively selected. There may be particular functionality that it is desirable to avoid due to assay format, such as fluorophores in fluorescence-based approaches. Structural features for these inclusion and exclusion criteria can be readily formulated using SMILES-based procedures and this type of substructure-based compound selection technique can also be employed in the generation of focused sets of fragment molecules.

For reference we have profiled the drug-like portion of the Evotec supplier database to provide the reader with guideline numbers of compounds that satisfy the rule of three, reduced complexity and lead-like definitions as provided in Table 3.1. The drug-like portion of this database has been derived from the 3.8 million compounds by applying 85 proprietary substructure filters in conjunction with physical property filters and contains just over 2.2 million compounds. Approximately 65 000 molecules (2.8 %) in the database of commercially available screening compounds are rule of three compliant (including the rotatable bond and polar surface area criteria), just under 178 000 molecules (7.7 %) fulfil the criteria for reduced complexity screening molecules and 1.2 million molecules (50 %) fall within the lead-like definition. In practice, there will be even fewer reduced complexity molecules than this because this set has not been further filtered using the GaP approach or profiled for the presence of synthetic handles.[27] Whatever the case, however, it is clear from these numbers that there are at present far fewer molecules available from commercial sources which satisfy fragment-like properties than molecules which have more lead-like or drug-like properties. This trend may well change in the future, given the current interest in FBDD methods.

3.2.3 Diverse Selections

Once a database of candidate molecules has been prepared, it may be desirable to select a diverse set of molecules. Diversity algorithms are designed to select sets of molecules in such a way that the chemical space from which they have been extracted is sampled democratically.[29] Molecules are represented in this space using molecular descriptors and dissimilarity between them is quantified using metrics derived from the value of the descriptors. In terms of descriptors that have been used for fragment molecules,

Schuffenhauer *et al.* have used Similog keys in molecular complexity studies,[30] and other groups have used pharmacophore-based descriptors.[28]

3.2.4 Focused Sets

Focused sets of fragment molecules may be derived for a particular class of targets, for a specific target or from the structures of molecules with desirable biological activity. These selections are often performed using *in silico* virtual screening techniques. As with any virtual screening process, the tools that can be applied will depend critically on the information that is available; information about the structure of the protein target and the chemical structures of active molecules is of particular relevance.

Models of protein binding sites can be used in molecular docking studies to identify fragment molecules with favourable binding characteristics. For example, in the identification of novel inhibitors of DNA Gyrase, Boehm *et al.* used molecular docking in an *in silico* screening approach to identify a focused set of 3000 molecules from an initial set of 350 000 compounds,[31] and researchers at Astex used a modified form of GOLD docking software to select focused sets of fragment molecules from a large database of commercially available compounds.[32]

The two-dimensional structures of known active molecules can be analysed to identify constituent fragments, which may form the basis of substructure- and similarity-based cheminformatics searches. During this deconstruction process, we have found it useful not to be limited to a purely retrosynthetic approach but also to consider other approaches to deconstructing the molecule. After all, molecular recognition does not occur on the basis of how the ligand can be synthesised. If active conformations are known, e.g. from X-ray crystallography, these or components of them can form three-dimensional queries which could be used in similarity searches to identify fragment molecules with low-energy conformations that can adopt similar shapes.[33] In addition, three-dimensional conformations can be used in more conventional pharmacophore searches. For instance, workers at Vernalis used a CDK2-hymenialdisine complex crystal structure to derive a kinase pharmacophore to select a set of focused fragment for kinase targets.[28]

Focused sets of fragment molecules have also been derived from the two-dimensional structures of known drugs. Frequently occurring molecular features can be identified in known drugs using computational approaches[34] and fragment molecules which exhibit these features can then be selected from vendor collections. It has been suggested that the chemotypes derived from known drugs may possess good properties for development into new drugs. For example, proven chemistries should be applicable and physicochemical properties should be more favourable. Although these arguments may be true to some extent, one has to bear in mind that it is the physiochemical properties of the whole drug molecule that are most important, not its constituents. In addition, chemotypes that could lead to novel drugs might be overlooked if they are not represented in the screening collection[35] and, finally, there could be implications for the establishment of intellectual property if this style of approach is followed exclusively.

Analysis of the hit molecules that are obtained in different fragment screening campaigns can also yield useful information regarding active structural motifs. For example, an analysis of NMR-based screening data indicated that compounds which contain a privileged biphenyl substructure show an enhanced probability of target binding.[36]

3.2.5 Synthetic Aspects

The low potency of the hits obtained from the fragment or reduced complexity screening (RCS), compared with those of HTS hits, makes significant optimisation of the ligand potency necessary. Several different synthetic strategies can be applied to achieve this, typically by fragment evolution or fragment linking. Target-guided fragment self-assembly (dynamic combinatorial chemistry) has also been reported.[37] Whatever strategy is chosen, it is highly desirable for the fragment hits to have at least one, or preferably a few, synthetic handles (functional groups) for rapid structural expansion/modification.

The most simple and effective but labour-intensive way of including this synthetic expandability criterion in the fragment library design process is visual inspection of the fragments by medicinal chemists. Scientists at Vernalis visually inspected their fragment set to identify potentially insoluble fragments and to ensure that most fragments contained suitable functional groups to allow rapid chemical evolution of each fragment scaffold.[28] Astex also employed a similar approach for ensuring the presence of selected functional groups to make their fragments chemically tractable and suitable for rapid optimization.[32]

In order to ensure that every fragment represents a good synthetic starting point, it would be ideal to use building blocks such as those used for combinatorial and parallel synthesis. Many chemical suppliers offer a wide range of compounds for this purpose. However, it would be advantageous if such selections of building blocks could be enriched with substructures of biologically active molecules or privileged motifs found in known drugs. Using the Retrosynthetic Combinatorial Analysis Procedure (RECAP), Lewell *et al.* performed fragmentation of the compounds in the Derwent World Drug Index (WDI) at 11 predefined bond types in a retrosynthetic fashion. These bond types were carefully chosen so that all the resulting fragments were amenable to combinatorial chemistry using standard synthetic procedures.[34] An alternative approach, similar to RECAP, referred to as SHAPES, has been described by the group at Vertex. The original SHAPES library has been further expanded to the SHAPES linking library. Synthetic accessibility aspects have been taken into account in the design of the new library so that the linkers and side-chains contained in the compounds in the library are compatible with the reactions implemented by the Vertex combinatorial chemistry group.[38]

The drawback of selecting building blocks as screening fragments is their reactivity. Very reactive compounds such as acid chlorides, isocyanates and alkyl halides are not suitable for fragment screening. Even functional groups of low reactivity that are often used as synthetic handles in building blocks, such as carboxylic acids and amines, can cause problems. Analysis of the literature carried out by scientists at Novartis revealed that hits containing these groups, especially carboxylic acids, often lose binding when these groups are used as a synthetic starting point during fragment evolution.[39] Functional groups such as the amino group or the carboxylic group, not only offer potential for chemical modification, but also can be a key motif for fragment–protein interaction due to their ability to form strong hydrogen bonds and ionic interactions.

Functional group masking is a strategy to circumvent this problem, in which fragments containing chemically inert products of reactive functional groups (masked functional groups) are used for screening. Once hit compounds have been identified, the unmasked chemical handle is used for further chemical elaboration. Ellman and co-workers used *O*-methyloximes as the masked equivalent of aldehydes in their development of c-Src inhibitors.[40] A set of 305 aldehydes was selected and were capped with *O*-methylhydroxylamine to make a library of *O*-methyloximes. The library was screened against c-Src and hit fragments were identified. The parent aldehydes, corresponding to the *O*-methyloximes found to be active, were then used for the synthesis of a combinatorial library of linked binding fragments using flexible linking groups. Screening of this library successfully identified a highly potent inhibitor (IC_{50} 64 nM). It is noteworthy that the IC_{50}s of the original fragment hits were only in the region of 40 μM.

The unmasked versions of each screening fragment can also be assembled as a separate complementary set to the fragment library. This fragment pairing concept has been championed by the group at Novartis.[39] A matching pair of synthesis and screening fragments are selected using a dictionary of reactions suitable for functional group interconversion (Figure 3.1).

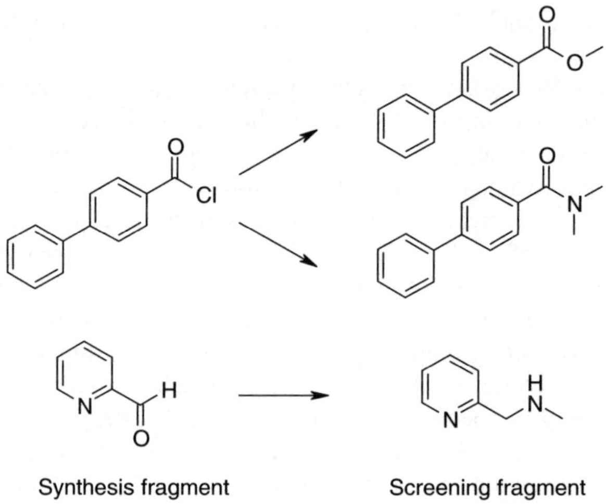

Figure 3.1 *Examples for pairs of screening and synthesis fragments.*

The masked functionality does not necessarily have to have an unmasked counterpart. It can also be a simple synthetic handle which is inert under screening conditions but can be activated in a controlled manner, for example, by catalysis. Examples of such catalytic reactions include Suzuki coupling, which is a palladium-catalysed cross-coupling reaction between an organoborane and an aryl halide and is used widely as an extremely powerful tool for C–C bond formation. The team at SGX Pharmaceuticals designed their fragment library for X-ray crystallographic screening so that approximately half of the compounds in the fragment library contain one or more bromine atoms (aryl bromides) which can be used as synthetic handles. Further, the presence of bromine not only facilitates synthetic elaboration of the screening hits, but also permits unambiguous bromine atom identification in experimental electron density maps. The X-ray wavelength can be adjusted during the diffraction data collection to allow detection of anomalous dispersion signals unique to bromine.[26] Another example of such a catalytic reaction that satisfies the requirements described above is the 1,3-dipolar cycloaddition of organic azides and alkynes to form 1,2,3-triazoles. Sharpless and co-workers demonstrated the use of this completely bio-orthogonal reaction (termed 'click chemistry') as a useful tool for lead generation by a fragment linking approach.[41]

In the database of 3.8 million commercially available compounds that Evotec maintains, the number of compounds that qualify as a 'fragment' by rule of three criteria is only 65 000 (see above). Further incorporation of the criterion of synthetic expandability (to have multiple synthetic handles) in fragment library design inevitably reduces the maximum size of the fragment library considerably. This may be acceptable if the screening method chosen is NMR or X-ray crystallography due to the relatively low throughput of the methodology. However, if the library is to be screened by high-concentration biochemical assays, potentially all the available 65 000 fragments could be screened within a reasonable cost and time irrespective of synthetic tractability. Although there is clearly merit in the synthetic handle approach, ultimately it is how a fragment specifically interacts with its biological target that will determine whether the potential for synthetic elaboration can be exploited for a specific fragment. A key issue with this approach is that for relatively low molecular weight species it is likely that any masked chemical functionality may be intimately involved in fragment binding to the target protein and so may not provide a suitable vector for fragment decoration. Furthermore, scientists at Novartis also pointed out that the concept of a potential chemical linker not involved in the protein binding of the fragment, but available for chemical modification, conflicts with the aim of a maximum ligand efficiency requiring that all parts of the fragment contribute to protein binding.[39]

3.3 Evotec's NMR Fragment Library

From the previous section, it is apparent that there are many considerations to be taken into account when assembling a fragment library and a variety of different approaches and philosophies. Evotec has two separate fragment libraries which were constructed independently and on the basis of different criteria. One has been built up for fragment screening by high-concentration biochemical assays whereas the other was built up for fragment screening by NMR. Each library comprises approximately 20 000 compounds but there are fewer than 1000 compounds in common between the two libraries. In this section we describe the process that was followed for the construction of the library for NMR-based fragment screening (Figure 3.2) and in the following section we describe the assembly of a fragment library for use in high-concentration biochemical screening (Figure 3.3).

Due to its high sensitivity for detecting weak binders and its unique binding site resolution, protein-observed NMR screening is ideal for progressing to lead molecules from fragment hits, as demonstrated by the pioneering work of Fesik's group at Abbott Laboratories over a decade ago.[42] Being significantly lower in molecular weight than leads, fragment hits occupy sub-pockets within the active or allosteric site of interest rather than the entire pocket. As nicely exemplified by a 168 Da fragment bound to PDE4d,[43] the most binding efficient fragment hits will preferentially sit in the most druggable sub-pocket(s) of the target, mediate attractive target interactions *via* the majority of its atoms and possess a low entropic barrier to binding. Following this concept of fragment–target interaction,

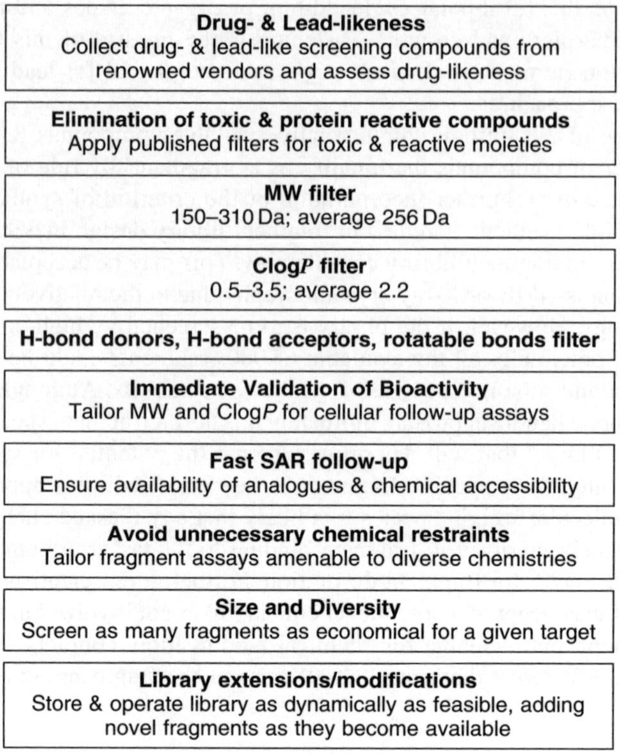

Figure 3.2 *Process for the assembly and use of Evotec's NMR fragment library.*

the NMR fragment library has to contain optimal chemical matter for the most druggable sub-pocket(s) that allow(s) for fragment elaboration at one or a few positions not involved in target interactions. Fragment moieties that mediate highly efficient target interactions are ideally maintained throughout F2L optimization because 'scaffold hopping' of such moieties is challenging. In our opinion, optimal chemical matter for NMR fragment screening should meet the following criteria, which we have applied by using simplified, but very accessible parameters.

3.3.1 Drug- and Lead-likeness

Fragments should be composed of drug-like chemical matter because substantial parts of a fragment hit will end up in the final lead candidate for *in vivo* assays. Therefore, we selected our NMR screening library from commercially available compound repositories that have a proven track record in lead discovery (e.g. by many years of use in HTS) and also as they include structures and substructures that occur in known marketed drugs, as well as in clinical and preclinical candidates. This ensures that the fragment library is built on several decades of experience in drug discovery.

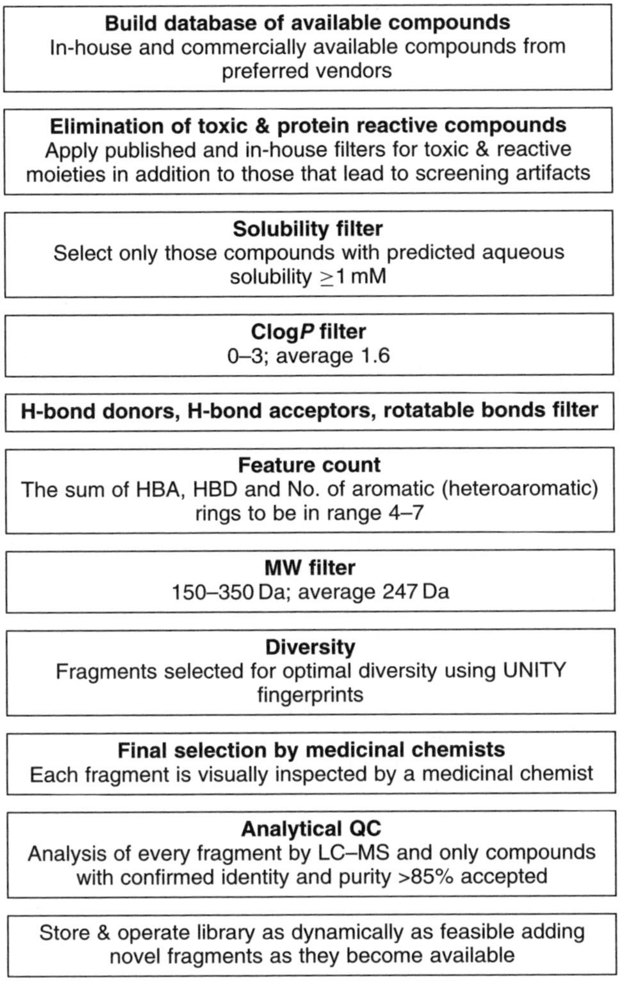

Figure 3.3 *Process for the assembly and use of Evotec's fragment library for high-concentration biochemical screening.*

Second, we analysed such compound repositories with respect to the frequency of functional groups and scaffolds most abundant in known drugs, as published by Murcko's group at Vertex.[44, 55] For example, the two most abundant pairs of functional groups, carbonyl plus carbonyl and carbonyl plus methyl, occur in 13 % and 10 % of the 5090 known drugs with assigned therapeutic classes in the Comprehensive Medicinal Chemistry (CMC) database, respectively. By using 2D substructure filters, we prioritised compound repositories towards similar percentages of functional groups and scaffolds by filtering out over-represented groups and scaffolds. This drug-likeness prioritisation was repeated at the end of the filtering process.

3.3.2 Elimination of Toxic and Reactive Chemistry

Although present in several approved drugs, protein reactive chemical groups are known to produce screening artefacts and are undesirable starting points for F2L optimization.[46] We eliminated potentially protein reactive chemical groups, such as aldehydes, phosphonate esters and activated ketones, in order to avoid identification of covalent, irreversible fragment binders. Furthermore, we also removed potential toxicophores.

3.3.3 Molecular Weight (MW)

MW is the most obvious, but also the most stringent, filter for fragment-like compounds within commercial screening compound collections, reducing the number of available compounds immediately by a factor of 10–20. We set the upper MW limit of our NMR fragment library to 310 Da and targeted an average MW of ~250 Da. This also allows for the merging or linking of two fragments binding at adjacent sites, as demonstrated in the SAR-by-NMR study of Fesik and co-workers,[42] to provide linked/merged compounds that conform to drug-like criteria.

In addition, we set a lower MW limit of 115 Da and accepted less than 0.4 % of all compounds smaller than 150 Da because fragments of significantly lower MW become similar to solvents, such as DMSO (94 Da). Such small fragments are less likely to overcome the energetic hurdles of desolvation and of inducing small conformational changes in amino acid side-chains, that have been demonstrated for many protein–small molecule complexes.[47] Moreover, small fragment binders can change their binding geometry during F2L optimization, as was shown experimentally in a recent crystallographic study. The two 128 and 153 Da constituents of a 343 Da lead compound adopt completely different binding geometries when co-crystallised one at a time as compared with their pose in the co-crystal of the lead.[48] Consequently, utilizing fragments of too low MW puts binding mode conservation and hence rational fragment elaboration at risk.

3.3.4 Solubility

Solubility is another critical and stringent filter for fragment libraries. Compared with HTS libraries, fragments require additional hydrophilicity to be soluble to a minimum NMR assay concentration of 200–1000 μM in aqueous buffers containing 2–10 % DMSO. On the other hand fragments should contain sufficient hydrophobicity to make effective van der Waal interactions with the target protein and to provide the basis for optimisation into lead compounds with satisfactory permeability in cellular assays and *in vivo* efficacy models. Hydrophobicity is essential for preferentially targeting protein pockets with strong hydrophobic contributions to binding because protein pockets with primarily hydrophilic interactions have a lower probability of leading to orally available drugs.

Balancing these opposing requirements, we estimated fragment solubility by utilising the parameters calculated log$P$ (Clog$P$) and calculated logS_w (solubility in water, ClogS_w) and set the upper and lower limit to Clog$P$s of 3.5 and 0.5, respectively. In comparison with most other fragment libraries for biochemical, X-ray and other biophysical assays, this Clog$P$ range uniquely shifts the NMR library towards higher hydrophobicity.

Since solubility prediction from Clog$P$ or ClogS_w is well known to be unreliable for many compound classes, we tested our library compounds experimentally for solubility at 500 μM concentration under aqueous assay conditions. It was found that that the majority of compounds were soluble and that many of the precipitate-forming compounds still showed significant partial solubility. Protein-observed NMR assays can tolerate such partially soluble compounds, which can still give rise to strong signals. In consequence, the NMR technology permits the screening of fragment libraries which reflect the Clog$P$ profile of known drugs.

3.3.5 Physicochemical Parameters

The above-mentioned constraints on MW and Clog$P$ already set the framework for H-bond donor (HBD) and H-bond acceptor (HBA) filters. We found that the majority of compounds passing our MW and Clog$P$ filters also met our HBD and HBA filters, which we defined as ≤4 HBDs, ≤7 HBAs, resulting in average values of 1.2 HBDs and 3.6 HBAs. Very few compounds (<0.4 %) were allowed to exceed these limits.

The number of rotatable bonds (rotB) reflects the rigidity of compounds, which is thought to correlate with a low entropic penalty of binding and improved target selectivity. By contrast, a few rotatable bonds allow fragments to flexibly adapt to the protein pocket, thereby sampling a wider range of 3D chemical space. As a compromise, we set an upper limit of five rotatable bonds and targeted an average of three rotatable bonds.

3.3.6 Immediate Validation of Bioactivity

Ultimately aiming at *in vivo* active target modulators, we designed our fragment library so that the most efficient fragment binders can be immediately tested for bioactivity in biochemical or cellular assays at low throughput. Filtering fragment binders immediately for bioactivity avoids misallocating valuable F2L chemistry and structure determination resources on bioinactive binders. We have validated this fast-into-cells approach in several anti-infective discovery projects and found that bioactivity is indeed a very stringent filter for fragment hits, showing drop-out rates of 80–90 % for antibacterial targets. Furthermore, by testing the entire NMR fragment library in a robust biochemical HTS assay, the feasibility of finding several biochemically active fragments with 2–100 μM IC_{50} values was experimentally confirmed.[49]

3.3.7 Fast SAR Follow-up

Fast and cost-efficient hit follow-up with respect to SAR is an important advantage of working with fragments. We made sure that we had sufficient chemical matter in stock for confirmative secondary assays. Further, we laid emphasis on timely commercial availability of ≥10 mg follow-up material for co-crystallisation and cellular assays by choosing reliable vendors. Likewise, quick commercial access to analogues enables fragment hit validation by comparatively low-cost, preliminary SAR studies. As the next step, we required fragment scaffolds to be accessible by one or very few synthetic steps and to contain sufficient functionality for fragment elaboration. For these reasons, we have avoided chiral compounds and complex natural products.

3.3.8 Avoidance of Unnecessary Restraints

In order to map thoroughly the available chemical space, we avoided the introduction of assay-related chemical restraints, which are frequently used for other fragment assays, such as bromine libraries for X-ray assays, fluorine libraries for ^{19}F-detected NMR assays, fluoro/chromophore elimination for biochemical assays, lengthy linkers for chip-based assays, chemical tags for transient immobilisation, methyloxime-based linking or *in situ* click chemistry. It is our view that starting points for small molecule drugs should be as compact and as binding efficient as the target allows. This goal is ideally supported by our NMR assays, which impose solely the following, nonstringent restraints on the chemical matter:

1. $\geq$1 non-water-exchangeable H atom for ligand-detected NMR assays;
2. preference for $\geq$1 aromatic ring which ensures stronger chemical shift signals in protein-detected NMR assays and better spectral resolution in ligand-detected NMR assays.

3.3.9 Size and Diversity

Finally, the fragment collection filtered for the above-mentioned criteria needs to be prioritised for the desirable size and diversity. For several biophysical fragment screening techniques, particularly crystallography, the library size is limited by throughput and the cost of protein consumption. Nevertheless, library size and diversity need to be balanced within an economical range, which we approximate to be between 500 and 25 000 fragments for NMR techniques. We intentionally decided to assemble a library of 20 000 compounds, which is the largest NMR screening library published so far, because we wish to sample thoroughly the available fragment space in order to identify the most efficient binders. This screen-big strategy reduces the impact of false negatives in fragment assays and minimises the number of singleton hits. A small and highly diverse fragment library may miss out attractive hits, which show up as hits of analogues in a larger, but less diverse, fragment collection.

In practice, the size of fragment library screened will vary between protein targets due to more than 10-fold differences in protein expression yield and substantial cost differences between bacterial and mammalian expression systems. Therefore, we sorted our entire collection of 20 000 fragments into a maximum diverse subset of 5000 fragments for costly proteins and a second subset of the remaining 15 000 fragments by using a chemical dissimilarity algorithm.[50]

Protein targets, which are even too costly for the 5000 subset, can be prescreened by virtual screening (VS) or a biochemical assay and the NMR assay restricted to an economical library size of 100–1000 hit compounds. Taken together, we believe that it pays off to screen experimentally as many fragments as economical for a target. This notion appears to have been adopted by other proponents in NMR screening, such as Abbott with an NMR library of approximately 10 000 compounds (Table 3.2).

3.3.10 Library Extensions, Modifications

Since the most favourable chemical fragment space for any given class of targets varies over time, we recommend planning ahead for dynamic library extensions, which may comprise target-focused fragments, new chemistries, new privileged chemical classes or simply enhanced sets identified by virtual screening. At the opposite end, some fragments may need to be removed from the original collection because they become unavailable, IP protected or otherwise undesirable. Hence storage and compound handling automation should allow flexible management of individual compounds wherever possible, rather than being restricted to presorted compound mixtures or multiple compound containers.

3.4 Evotec's Library for HTFS

For our fragment screening activities by biochemical assays, we have built a library of over 20 000 fragment compounds that have been carefully selected in three iterations by applying a set of computational filters followed by review by a panel of medicinal chemists. Evotec has coined the term High-Throughput Fragment Screening (HTFS) to describe the screening of fragment molecules by high-concentration biochemical assays performed on the EVOscreen uHTS platform. These techniques have been described elsewhere[24] and, in comparison with other fragment screening methods, library size is not limited by throughput, enabling large fragment collections to be screened. Based on promising early results with this technology, a decision was made to assemble a fragment library of 20 000 molecules. The team assigned to this effort was fortunate to be able to rely on long-term experience and a well-established infrastructure for library assembly and maintenance for the management of the in-house corporate HTS collection (currently over 250 000 unique screening compounds).

The assembly of the fragment library was conducted in three iterations: (i) the Stage 1 library consisted of ~1200 fragments; (ii) the Stage 2 library was expanded to ~ 5000 fragments; and (iii) the Stage 3 library consists of over 20 000 fragments. The process that was adopted for the selection (Figure 3.3) is related to that described above for the NMR library. The similarities and differences in the approaches are apparent from the discussion below.

3.4.1 Vendor Selection

The process for the selection of the fragment library for biochemical screening is based on a database of Evotec's proprietary building blocks and compounds from selected commercial vendors that is updated on a regular basis. Different subsets of vendors were selected for the three phases, mainly according to their reliability, quality of their compounds and our experience regarding delivery timelines. Although containing 35 different vendors, the list for the final phase was restricted to the best 13 suppliers. More than 1 million records (~715 000 structures when duplicates were taken into account) with MW $\leq$350 Da were fed into the selection process. The vendors that were prioritised for the selection of fragments for the fragment library for biochemical screening were different to those selected for the NMR library.

3.4.2 Substructure Filters

An extensive range of in-house filters was applied to remove toxic, reactive and undesirable functionality. This set of filters has been built up based on years of experience with HTS screening and medicinal chemistry optimisation of hits to leads. A special focus was made on the fact that the compounds are to be used in the context of fluorescence readouts so a subset of the filters avoid readout artefacts due to the screening compounds themselves. The filters were applied as substructures encoded in SMILES notation to automatically identify undesirable compounds in the database.

3.4.3 Solubility

A solubility filter using an in-house *in silico* model to select compounds predicted to have an aqueous solubility in excess of 1 mM was applied. Aqueous solubility is an important factor for the success of HTS screening in general and fragment screening in particular. Due to the low affinity binding of fragments at protein interaction sites, the screening concentration range is typically at least 20-fold higher than that used for general HTS screening. Depending on the target, the primary fragment screens at Evotec are usually run at concentration ranges between 250 μM and 1 mM final concentration. Dose–response curves can reach final concentrations up to 2 mM. The stock solutions of the fragments are stored in 100 % DMSO. In a proprietary process (NanoStore), the solvent is removed on the assay plates prior to screening to generate a DMSO-free assay environment.[51] We validated the application of NanoStore to fragment screening by high-concentration bioassay early on in a pilot study of a subset of the Stage 1 fragment library. This fragment set was screened against PTP1B, for which we achieved superior results using NanoStore in comparison with presentation of the fragments in DMSO.[52] The NanoStore method of presentation of compounds for screening avoids many assay artefacts but still requires fragments to have reasonable aqueous solubility. Setting the cut-off of the predicted solubility at 1 mM gives a good compromise between reliability of the *in silico* predictions and constraining the coverage of chemical space.

3.4.4 Physicochemical Parameters, Features and Molecular Weight

Calculated properties were restricted to: Clog*P* $\leq$3, the number of rotatable bonds $\leq$5, the number of hydrogen bond acceptors (HBA) $\leq$4 and the number of hydrogen bond donors (HBD) between 1 and 3. The topological polar surface area (TPSA) was set to $\leq$70 Å^2. In addition, a special feature count excludes structures that are too functionalised (feature rich) or dull (absence of pharmacophoric interaction points). This feature count is the sum of HBA, HBD and the number of five-and six-membered aromatic rings and selection was restricted to fragments with a feature count in the range 4–7.

In order to meet our requirements for a relatively large fragment screening library to maximise the potential of the HTFS concept, it was necessary to relax the upper limit on MW in order to access sufficiently diverse compounds. A final MW filter was applied to allow only compounds with MW in the range 150–350 Da. This MW range is slightly broader than used by others (Table 3.2).

Application of all of the above filters resulted in a list of approximately 32 000 unique structures that are available from commercial vendors. During the assembly of the Stage 1 and 2 iterations of the fragment library, an overlap check to the current screening collection was applied trying to avoid overlap in those cases where a target would be addressed by both full-deck HTS screening and the fragment-based approach. This concern was dropped later for the Stage 3 iteration, since the exclusion of the chemical space from the full HTS deck would significantly reduce the coverage of the fragment chemical space and because the fragment molecules in the full HTS deck could not, for logistical reasons, be screened at high concentration.

3.4.5 Diversity and Expert Inspection

The set of 32 000 theoretically available structures was reduced to a specific number to be manually assessed by experienced medicinal chemists. During the assembly of the Stage 1 fragment library special focus was set on the selection of proprietary building blocks from Evotec, a random diverse set from the commercial vendors list and a subset of carboxylic acids and their isosteres targeted towards phosphatases. During the assembly of the Stage 2 library the main focus was set on maximum diversity (less than 75 % Tanimoto similarity based on UNITY fingerprints). For the target size of 4000 fragments to be added to the Stage 1 fragment library, a total set of close to 10 000 was assessed by a panel of three medicinal chemists. The structures were categorised into three groups: (a) must have, (b) nice to have and (c) rejected fragments. This was based on their subjective opinion on each fragments to be a suitable starting point for structure-based medicinal chemistry optimisation in terms of lead-like features and possibilities of synthetic elaboration. The 'must have fragments' were clustered in diversity space using UNITY fingerprints and a representative set was selected. The holes in diversity space were then filled by a selection of 'nice to have' fragments. The 'rejected fragments' set was analysed and additional rules were derived and added to the automated filtering process.

For the compounds added during Stage 3, the log$P$ constraint (see above) was revised to Clog$P = 0$–3 as the medicinal chemistry inspection at Stages 1 and 2 had tended to remove a significant number of highly hydrophilic compounds. For the Stage 3 fragment library, the manual exclusion rate was estimated to be about 10 %. The diversity selection cut-off was raised from 75 to 93 % maximum Tanimoto similarity generating 18 000 structures to be reviewed manually (this time by a group of 18 medicinal chemists). Raising the cut-off to numbers above 90 % indicates two points: first, the almost complete coverage of the chemistry space available, and second, initial SAR information can be deduced from small series of similar fragments.

3.4.6 Compound Purchase and Quality Control Checks

The final order lists were optimised towards a minimum number of different vendors. In a vendor analysis of our database, we could observe that a significant proportion of screening compounds are offered by more than one vendor, sometimes even with identical or similar order codes. Such compounds were sourced from one major supplier to benefit from single transaction costs and possible discounts. Both after obtaining the quotes and receiving the final deliveries, we had to account for a certain percentage of dropouts due to compound unavailability. It is our experience that this varies from supplier to supplier, but typically only 90 % of desired compounds are available. Therefore, it is advisable to order in excess of the final library target size.

At Evotec, all compounds delivered are quality control (QC) checked by LC–MS. This is a major effort, but it is of the utmost importance to collect the initial purity data and to eliminate those fragments with unconfirmed structures (typically this is 5 % of the compounds ordered). In general, the purities are consistently satisfactory and the number of dropouts due to unconfirmed structure was spread evenly over the prioritised list of vendors that were used. However, this may be due to the fact that vendors for which we previously have had poor experience were not considered to be suitable suppliers in the first place or it might be due to the industry-wide efforts to improve compound quality in recent years.

3.5 Conclusion

The techniques for fragment screening by NMR and high-concentration biochemical assay are complementary.[53] Evotec's NMR fragment library was built up independently from the fragment library for high-concentration biochemical assays when the NMR group was part of another organisation. Evotec has assembled one of the largest fragment collections to be used on a routine basis for NMR fragment screening and high-concentration biochemical assay screening. The philosophy for library construction for NMR fragment screening is somewhat different to that adopted for the library for biochemical screening, but this has led to two highly complementary libraries with minimal overlap (less than 1000 compounds in common). Both libraries have been tailored to allow rapid fragment hit follow-up and both collections are used for internal and external projects on a number of targets, so far with highly promising results. We have found that the use of these two fragment libraries has favourably shifted lead discovery activities at Evotec towards lower MW starting points, where chemical optimisation is faster and less resource consuming, especially when advantage is taken of structural insights gained though detailed NMR analysis and/or X-ray crystallography of fragments in complexes with target proteins.

3.6 Abbreviations

CDK2	cyclin-dependent kinase-2
CMC	comprehensive medicinal chemistry
Da	dalton
DMSO	dimethyl sulfoxide
DNA	deoxyribonucleic acid
F2L	fragment-to-lead
FBDD	fragment-based drug discovery
GaP	gridding and partitioning
HBA	hydrogen bond acceptor
HBD	hydrogen bond donor
HTFS	high-throughput fragment screening
HTS	high-throughput screening
IP	intellectual property
LC–MS	liquid chromatography–mass spectrometry
MW	molecular weight
NHC	non-hydrogen atom count
NMR	nuclear magnetic resonance
PTP1B	protein tyrosine phosphatase 1B

QC	quality control
RCS	reduced complexity screening
RECAP	retrosynthetic combinatorial analysis procedure
rotB	rotatable bonds
SAR	structure–activity relationship
SMILES	simplified molecular input line entry specification
TPSA	topological polar surface area
uHTS	ultra-high-throughput screening
VS	virtual screening

3.7 Acknowledgements

The authors wish to thank John Barker, Oliver Barker and Thomas Hesterkamp for their contributions to the fragment-based drug discovery initiative at Evotec and for useful discussions.

References

[1] C.W. Murray, O. Callaghan, G. Chessari, A. Cleasby, M. Congreve, M. Frederickson, M.J. Hartshorn, R. McMenamin, S. Patel and N. Wallis, Application of fragment screening by X-ray crystallography to β-secretase, *J. Med. Chem.*, **50**, 1116–1123 (2007).

[2] M. Congreve, D. Aharony, J. Albert, O. Callaghan, J. Campbell, R.A.E. Carr, G. Chessari, S. Cowan, P.D. Edwards, M. Frederickson, R. McMenamin, C.W. Murray, S. Patel and N. Wallis, Application of fragment screening by X-ray crystallography to the discovery of aminopyridines as inhibitors of β-secretase, *J. Med. Chem.*, **50**, 1124–1132 (2007).

[3] A.L. Hopkins, C.R. Groom, and A. Alex, Ligand efficiency: a useful metric for lead selection, *Drug Discov. Today*, **9**, 430–431 (2004).

[4] C. Abad-Zapatero and J.T. Metz, Ligand efficiency indices as guideposts for drug discovery, *Drug Discov. Today*, **10**, 430–431 (2005).

[5] I.D. Kuntz, K. Chen, K.A. Sharp and P.A. Kollman, The maximal affinity of ligands, *Proc. Natl. Acad. Sci. USA*, **96**, 9997–10002 (1999).

[6] C.A. Lipinski, F. Lombardo, B.W. Dominy and P.J. Feeney, Experimental and computational approaches to estimate solubility and permeability in drug discovery and development settings, *Adv. Drug Deliv. Rev.*, **23**, 3–25 (1997).

[7] M. Congreve, R. Carr, C. Murray and H. Jhoti: A 'Rule of Three' for fragment-based lead discovery, *Drug Discov. Today*, **8**, 876–877 (2003).

[8] A.-D. Gorse, Diversity in medicinal chemistry space, *Curr. Top. Med. Chem.*, **6**, 3–18 (2006).

[9] T. Fink, H. Bruggesser and J.-L. Reymond, Virtual exploration of the small-molecule chemical universe below 160 daltons, *Angew. Chem. Int. Ed.*, **44**, 1504–1508 (2005).

[10] T. Fink and J.-L. Reymond, Virtual exploration of the chemical universe up to 11 atoms of C, N, O, F: assembly of 26.4 million structures (110.9 million stereoisomers) and analysis for new ring systems, stereochemistry, physicochemical properties, compound classes, and drug discovery, *J. Chem. Inf. Model.*, **47**, 342–353 (2007).

[11] R.S. Bohacek, C. Martin and W.C. Guida, The art and practice of structure-based drug design: a molecular modelling approach, *Med. Res. Rev.*, **16**, 3–50 (1996).

[12] M.M. Hann, A.R. Leach and G. Harper, Molecular complexity and its impact on the probability of finding leads for drug discovery, *J. Chem. Inf. Comput. Sci.*, **41**, 856–864 (2001).

[13] T.I. Opera, A.M. Davies, S.J. Teague and P.D. Leeson, Is there a difference between leads and drugs? A historical perspective. *J. Chem. Inf. Comput. Sci.*, **41**, 1308–1315 (2001).

[14] J.R. Proudfoot, Drugs, leads, and drug-likeness: an analysis of some recently launched drugs, *Bioorg. Med. Chem. Lett.*, **12**, 1647–1650 (2002).

[15] P.J. Hajduk and J. Greer, A decade of fragment-based drug design: strategic advances and lessons learned, *Nat. Rev. Drug Discov.*, **6**, 211–219 (2007).

[16] G. Harper, S.D. Pickett and D.V.S. Green, Design of a compound screening collection for use in high throughput screening, *Comb. Chem. High Throughput Screen.*, **7**, 63–70 (2004).

[17] M.M. Olah, C.G. Bologa and T.I. Oprea, Strategies for compound selection, *Curr. Drug Discov. Technol.*, **1**, 211–220 (2004).

[18] J.J. Irwin, How good is your screening library?, *Curr. Opin. Chem. Biol.*, **10**, 352–356 (2006).

[19] R.E. Hubbard, I. Chen and B. Davis, Informatics and modelling challenges in fragment-based drug discovery, *Curr. Opin. Drug Discov. Dev.*, **10**, 289–297 (2007).

[20] A.A. Shelat and R.K. Guy, Scaffold composition and biological relevance of screening libraries, *Nat. Chem. Biol.*, **3**, 442–446 (2007).

[21] U. Schopfer, C. Engeloch, J. Stanek, M. Girod, A. Schuffenhauer, E. Jacoby and P. Acklin, The Novartis compound archive – from concept to reality, *Comb. Chem. High Throughput Screen.*, **8**, 513–519 (2005).

[22] D.A. Erlanson, R.S. Mcdowell and T O'brien, Fragment-based drug discovery. *J. Med. Chem.* **47**, 3463–3482 (2004).

[23] S. Bartoli, C.I. Fincham and D. Fattori, Fragment-based drug design: combining philosophy with technology, *Curr. Opin. Drug Discov. Dev.*, **10**, 422–429 (2007).

[24] J. Barker, S. Courtenay, T. Hesterkamp, D. Ullmann and M. Whittaker, Fragment screening by biochemical assay, *Expert Opin. Drug Discov.*, **1**, 225–236 (2006).

[25] D. A. Erlanson, M.D. Ballinger and J.A. Wells, Tethering, in *Fragment-based Approaches in Drug Discovery* (ed. W. Jahnke and D.A. Erlanson), *Methods and Principles in Medicinal Chemistry*, Vol. 34, Wiley-VCH Verlag GmbH, Weinheim, pp. 285–310 (2006).

[26] J. Blaney, V. Nienaber and S.K. Burley, Fragment-based lead discovery and opitization using X-ray crystallography, computational chemistry, and high-throughput organic synthesis, in *Fragment-based Approaches in Drug Discovery* (ed. W. Jahnke and D.A. Erlanson), *Methods and Principles in Medicinal Chemistry*, Vol. 34, Wiley-VCH Verlag GmbH, Weinheim, pp. 215–248 (2006).

[27] M.M. Hann and T.I. Oprea, Pursuing the leadlikeness concept in pharmaceutical research, *Curr. Opin. Chem. Biol.*, **8**, 255–263 (2004).

[28] N. Baurin, F. Aboul-Ela, X. Barril, B. Davis, M. Drysdale, B. Dymock, H. Finch, C. Fromont, C. Richardson, H. Simonite and R.E. Hubbard, Design and characterization of libraries of molecular fragments for use in NMR screening against protein targets, *J. Chem. Inf. Comput. Sci.*, **44**, 2157–2166 (2004).

[29] M. Snarey, N.K. Terrett, P. Willett and D.J. Wilton, Comparison of algorithms for dissimilarity-based compound selection, *J. Mol. Graph. Model.*, **15**, 372–385 (1997).

[30] A. Schuffenhauer, N. Brown, P. Selzer, P. Ertl and E. Jacoby, Relationships between molecular complexity, biological activity, and structural diversity, *J. Chem. Inf. Model.*, **46**, 525–535 (2006).

[31] H.J. Boehm, M. Boehringer, D. Bur, H. Gmuender, W. Huber, W. Klaus, D. Kostrewa, H. Kuehne, T. Luebbers, N. Meunier-Keller and F. Mueller, Novel inhibitors of DNA gyrase: 3D structure based biased needle screening, hit validation by biophysical methods, and 3D guided optimization. A promising alternative to random screening, *J. Med. Chem.*, **43**, 2664–2674 (2000).

[32] M.J. Hartshorn, C.W. Murray, A. Cleasby, M. Frederickson, I.J. Tickle and H. Jhoti, Fragment-based lead discovery using X-ray crystallography, *J. Med. Chem.*, **48**, 403–413 (2005).

[33] P.C.D. Hawkins, A.G. Skillman and A. Nicholls, Comparison of shape-matching and docking as virtual screening tools, *J. Med. Chem.*, 50, 74–82 (2007).

[34] X.Q. Lewell, D.B. Judd, S.P. Watson and M.M. Hann, RECAP – retrosynthetic combinatorial analysis procedure: a powerful new technique for identifying privileged molecular fragments with useful applications in combinatorial chemistry, *J. Chem. Inf. Comput. Sci.*, **38**, 511–522 (1998).

[35] A.A. Shelat and R.K. Guy, Scaffold composition and biological relevance of screening libraries, *Nat. Chem. Biol.*, **3**, 442–446 (2007).

[36] P.J. Hajduk, M. Bures, J. Praestgaard and S.W. Fesik, Privileged molecules for protein binding identified from NMR-based screening, *J. Med. Chem.*, **34**, 3443–3447 (2000).

[37] M.S. Congreve MS, D.J. Davis, L. Devine, C. Granata, M. O'Reilly, P.G. Wyatt and H. Jhoti, Detection of ligands from a dynamic combinatorial library by X-ray crystallography. *Angew Chem Int Ed.*, **42**, 4479–4482 (2003).

[38] C.A. Lepre, Library design for NMR-based screening, *Drug Discov. Today*, **6**, 133–140 (2001).

[39] A. Schuffenhauer, S. Ruedisser, A.L. Marzinzik, W. Jahnke, M. Blommers, P. Selzer and E. Jacoby, Library design for fragment based screening, *Curr. Top. Med. Chem.*, **5**, 751–761 (2005).

[40] D.J. Maly, I.C. Choong and J.A. Ellman, Combinatorial target-guided ligand assembly: Identification of potent subtype-selective c-Src inhibitors, *Proc. Natl. Acad. Sci. USA*, **7**, 2419–2424 (2000).

[41] M. Whiting, J. Muldoon, Y.C. Lin, S.M. Silverman, W. Lindstrom, A.J. Olson, H.C. Kolb, M.G. Finn, K.B. Sharpless, J.H. Elder and V.V. Fokin, Inhibitors of HIV-1 protease by using *in situ* click chemistry, *Angew. Chem. Int. Ed.*, **45**, 1435–1439 (2006).

[42] S.B. Shuker, P.J. Hajduk, R.P. Meadows and S.W. Fesik, Discovering high-affinity ligands for proteins: SAR by NMR, *Science*, **274**,1531–1534 (1996).

[43] G.L. Card, L. Blasdel, B.P. England, C. Zhang, Y. Suzuki, S. Gillette, D. Fong, P.N. Ibrahim, D.R. Artis, G. Bollag, M.V. Milburn, S.H. Kim, J. Schlessinger and K.Y. Zhang, A family of phosphodiesterase inhibitors discovered by cocrystallography and scaffold-based drug design, *Nat. Biotechnol.*, **23**, 201–207 (2005).

[44] G.W. Bemis and M.A. Murcko, The properties of known drugs. 1. Molecular frameworks, *J. Med. Chem.*, **39**, 2887–2893 (1996).

[45] G.W. Bemis and M.A. Murcko, The properties of known drugs. 2. Side chains, *J. Med. Chem.*, **42**, 5095–5099 (1999).

[46] G.M. Rishton, Reactive compounds and *in vitro* false positives in HTS, *Drug. Discov. Today*, **2**, 382–384 (1997).

[47] M. Joshi, C. Vargas, P. Boisguerin, A. Diehl, G. Krause,P.Schmieder, K. Moelling, V. Hagen, M. Schade and H. Oschkinat, Discovery of low-molecular-weight ligands for the AF6 PDZ domain, *Angew. Chem. Int. Ed.*, **45**, 3790–95 (2006).

[48] K. Babaoglu and B.K. Shoichet, Deconstructing fragment-based inhibitor discovery, *Nat. Chem. Biol.*, **2**, 720–723 (2006).

[49] M. Manger, M. Scheck, H. Prinz, J.P. von Kries, T. Langer, K. Saxena, H. Schwalbe, A. Furstner, J. Rademann and H. Waldmann, Discovery of *Mycobacterium tuberculosis* protein tyrosine phosphatase A (MptpA) inhibitors based on natural products and a fragment-based approach, *ChemBioChem*, **6**, 1749–1753 (2005).

[50] S.V. Trepalin, V.A. Gerasimenko, A.V. Kozyukov, N.P. Savchuk and A.A. Ivaschenko, New diversity calculation algorithms used for compound selection., *J. Chem. Inf. Comput. Sci.*, **42**, 249–258 (2002).

[51] N. Benson, H.F. Boyd, J.R. Everett, J. Fries, P. Gribbon, N. Haque, K. Henco, T. Jessen, W.H. Martin, T.J. Mathewson, R.E. Sharp, R.W. Spencer, F. Stuhmeier, M.S. Wallace and

D. Winkler, NanoStore: a concept for logistical improvements of compound handling in high-throughput screening, *J. Biomol. Screen.*, **10**, 573–580 (2005).

[52] T. Hesterkamp, J. Barker, A. Davenport and M. Whittaker, Fragment based drug discovery using fluorescence correlation spectroscopy techniques: challenges and solutions, *Curr. Top. Med. Chem.*, **7**, 1582–1591 (2007).

[53] J. Barker, T. Hesterkamp, M. Schade and M. Whittaker, Fragment screening: Biochemical assays versus NMR, *Innov. Pharm. Tech.*, **23**, 19–22 (2007).

[54] P.J. Hajduk, J.R. Huth and S.W. Fesik, Druggability indices for protein targets derived from NMR-based screening data, *J. Med. Chem.*, **48**, 2518–2525 (2005).

[55] http://www.active-sight.com/

[56] R.E. Armer and P.M. Cowley, Trends in medicinal chemistry 2004, *Drug News Perspect.*, **18**, 142–148 (2005).

[57] http://www.graffinity.com/

[58] M.R. Arkin, M. Randal, W.L. DeLano, J. Hyde, T.N. Luong, J.D. Oslob, D.R. Raphael, L. Taylor, J. Wang, R.S. McDowell, J.A. Wells and A.C. Braisted, Binding of small molecules to an adaptive protein–protein interface, *Proc. Natl. Acad. Sci. USA*, **100**, 1603–1608 (2003).

[59] R.E. Hubbard, presented at the Fragment-Based Drug Design Workshop, San Diego, CA, 13 May 2007.

[60] J. Fejzo, C.A. Lepre, J.W. Peng, G.W. Bemis, Ajay, M.A. Murcko and J.M. Moore, The SHAPES strategy: an NMR-based approach for lead generation in drug discovery, *Chem. Biol.*, **6**, 755–769 (1999).

4

Practical Aspects of Using NMR in Fragment-based Screening

Johan Schultz

4.1 Introduction

Fragment-based screening (FBS) has attracted much attention in the pharmaceutical industry since its emergence 10 years ago and is now widely used as an alternative or complementary approach to traditional screening.[1–6] The basic idea is to

- screen only a small number of small molecules ('fragments'), typically somewhere between 100 and a few thousand fragments;
- find weakly binding but efficient (high binding energy per atom) binders;
- develop the binding fragments into more potent compounds using fragment expansion, merging or linking while maintaining a high ligand efficiency.

Only a relatively small number of fragments need to be screened since the hit rate will be much higher than with traditional screening. This is a consequence of lower complexity fragments having a higher probability of matching a target protein binding site.[7] Further, since the number of theoretically possible molecules increases exponentially with the number of atoms in the molecules, screening a small fragment library, typically 10^2–10^4 compounds, samples substantially more chemical diversity space than a conventional HTS, where typically 10^5–10^6 larger compounds are screened.[8] The detected fragment binders will, however, bind to the target macromolecule with a relatively low affinity since they contain relatively few features that can interact with the target. Fortunately, there are several biophysical techniques that are well suited to detect weak binding events reliably.

Fragment-Based Drug Discovery: A Practical Approach Edited by Edward R. Zartler and Michael J. Shapiro

A good way to compare ligands with different affinities/potencies and molecular weights is to normalize the affinity/potency with the ligand size.[9, 10] This is called ligand efficiency (LE), with the following definition:[10]

$$\mathrm{LE} = -\Delta G_{\mathrm{bind}}/N_{\text{non-hydrogen atoms}} = -RT\ln K_{\mathrm{d}}/N_{\text{non-hydrogen atoms}} \tag{4.1}$$

where ΔG_{bind} is the free energy of binding, $N_{\text{non-hydrogen atoms}}$ is the number of heavy atoms in the ligand and K_{d} is the dissociation constant. In practice, K_{d} is often replaced with IC_{50} values. Since an orally available drug should adhere to the Lipinski 'rule-of-five',[11] including a maximum molecular weight (MW) of 500, the ligand efficiency should be at least 0.29 kcal mol^{-1} assuming a desired potency of 10 nM and that the mean molecular mass for a non-hydrogen atom in drug-like compounds is 13.3.[10] In terms of ligand efficiency, this compound is then comparable to a fragment with a 1 mM potency and 14 non-hydrogen atoms (MW $\approx$ 200) or a compound with a 1 μM potency and 28 heavy atoms (MW $\approx$ 400), both with a ligand efficiency of 0.29 kcal mol^{-1}. For comparison, a study of 160 ligand–protein complexes showed that the maximum binding affinity per heavy atom for organic compounds is approximately 1.5 kcal mol^{-1}.[12] Extending the ligand efficiency concept, the polar surface area has also been used as the normalizing factor instead of the number of non-hydrogen atoms.[13]

Compared with biochemical assays used in HTS, there are considerable advantages to using biophysical binding techniques such as NMR. The superior control of experimental parameters and the state of the sample components minimizes artifacts. Also, the sample in itself is less complex than in a typical HTS assay well. When using an NMR technique as a binding assay, a good practice is always to collect also a reference ^{1}H 1D spectrum in which it will be immediately obvious (i) if the compound (or an important buffer component) for some reason is not present in the solution, (ii) if the compound is pure and intact or not or (iii) if the target macromolecule has unfolded, aggregated or precipitated.[14] An additional attractive feature is that it is possible to start the fragment screening campaign as soon as 1–2 mg of the target has been produced since generic binding assays are used. No time- and resource-consuming assay development and formatting are necessary. A spin-off from an early fragment screen could also be identification of reference compounds to be used as tools in an HTS assay development.

When developing the weakly but efficiently binding fragment hits, it is important to ascertain that each feature added to the fragment contributes substantially to the binding energy. Structural information on fragment interactions with the target, preferably high-resolution crystal or NMR complex structures, will dramatically increase the chances of developing the fragment hits to potent compounds while keeping the ligand efficiency high.[4] Due to the high solubility of the fragments, the probability of obtaining crystal structures of the fragment–target complex is high. Experience has shown that fragments bind specifically and give well-resolved electron densities. Hits from HTS campaigns may, of course, be both highly potent and efficient binders, but very often the hit-to-lead process starts off with relatively large molecules with potencies in the high nM to low μM range and low to moderate ligand efficiencies. The development of such hits will in general require more synthetic chemistry resources than in the more focused fragment-based screening approach. It is difficult to identify the parts of the larger molecule responsible for the important interactions without synthesizing a large number of molecules and testing them. Fragments are also less likely to contain interfering moieties that block an otherwise attractive ligand–protein interaction. Fragment hits developed to potencies equivalent to HTS hits, usually below 10 μM, are in general smaller and with higher ligand efficiencies than the HTS hits and will provide better starting points for medicinal chemistry to develop 'lead-like' leads.[15] There is more freedom to add features, molecular weight and lipophilicity to obtain the necessary potency of a drug candidate without exceeding the limits of the physicochemical properties (e.g. molecular weight, polar surface area and Clog*P*) that are known to correlate with oral bioavailability.[11] Thus, the compounds originating from the fragment screening will have a head start compared with HTS hits. In addition, if structures of the ligand–target complex have been determined, key interactions in the binding pocket have been identified. The importance of keeping the molecular weight low and the ligand efficiency high is underlined by statistics from clinical studies.[16] Compounds in Phase 1 with an MW <400 have a 50 % better chance of reaching the market than compounds in Phase 1 with an MW >400. In this context, it is also noteworthy that marketed oral drugs have an average MW of not more than 340.

It should be pointed out that FBS and the more resource-intensive HTS often are complementary approaches[4] and will almost certainly produce hit and lead series distinct from each other. Highly prioritized targets should be subjected to both screening approaches if possible. Confirmed HTS hit series and fragment hits developed to low μM or better potencies could very well be fed into a common chemistry optimization program. This often leads to a fruitful cross-fertilization, where the HTS hit series may give hints on what features to add on to the compounds originating from the fragment screen and the knowledge gained from the fragment screen will be useful when parts of HTS hits are to be exchanged. A schematic overview of the development of FBS hits with respect to typical potencies is presented in Figure 4.1.

4.2 Fragment-based Screening Strategy

The target macromolecule is usually a protein and that will be assumed in the remainder of the chapter if nothing else is specifically pointed out. A typical workflow of a fragment-based screening project when a biophysical technique such as an NMR technique is used for primary screening is shown in Figure 4.2. Fragment hits are identified from primary screening of a library consisting of carefully selected fragments. The project is then entered into an iterative phase. Competition experiments using a previously known binder are employed in order to obtain information on the binding site of the fragment hits and to rank the hits with respect to affinity. As soon as possible, attempts to obtain structural information on fragment–target complexes by X-ray crystallography and/or NMR spectroscopy should be initiated. If high-resolution fragment–target complex structures are obtained, this will be very valuable to drive the chemistry and those fragments will be highly prioritized for further development. If no fragment–target complex structures are obtained, then there are still several possibilities for obtaining lower resolution structural information by NMR. This include epitope mapping of binding sites on the target protein by 2D $^{1}H–^{15}N$ (or $^{1}H–^{13}C$) correlation spectra and identification of binding surfaces on the ligand using ligand-detected methods, and also interligand NOEs between fragments binding simultaneously to adjacent sites. A biochemical activity assay should be employed at some stage to find out if the binding fragments or developed fragments not only bind but also modulate the

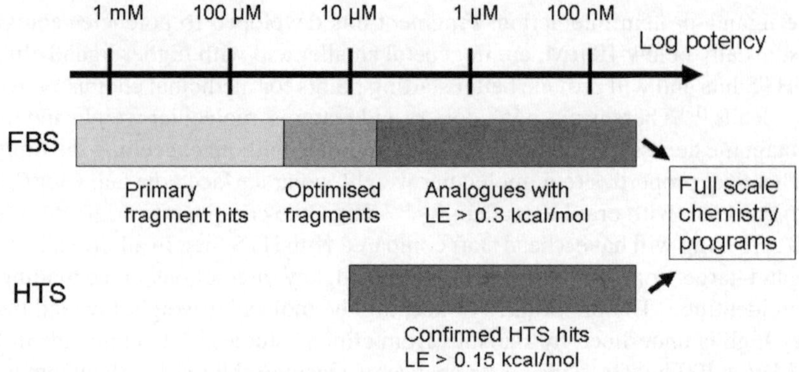

Figure 4.1 *A schematic overview of the development of FBS (and HTS) hits with respect to typical potencies achieved at different stages. The primary fragment hits typically have potencies in the range 0.1–1 mM. Selected fragment hits are optimized with respect to affinity without substantially increasing the molecular weight. The optimized fragments are then slightly expanded, usually by testing in-house or commercially available analogous compounds or by applying one-step chemistries if the fragment library is designed to facilitate this. Alternatively, fragments are linked or merged. After a couple of iterations, compounds with low micromolar or better potencies and with high ligand efficiencies have normally been identified. These compounds are good starting points for a full-scale chemistry program. Meanwhile, the HTS assay has been developed and formatted. Confirmed HTS hits refer to hits with high purity and stability and good dose–response curves. Further, they should be confirmed in an orthogonal assay, preferably a biophysical binding assay, to ascertain that they are active via a 'sound' mechanism.*[17, 18] *Finally, hit series should be identified and prioritized and commercially available analogues purchased. From this stage, further development of compounds originating from the fragment screen and the confirmed HTS hits does not differ significantly and could be fed into the same medicinal chemistry optimization program.*

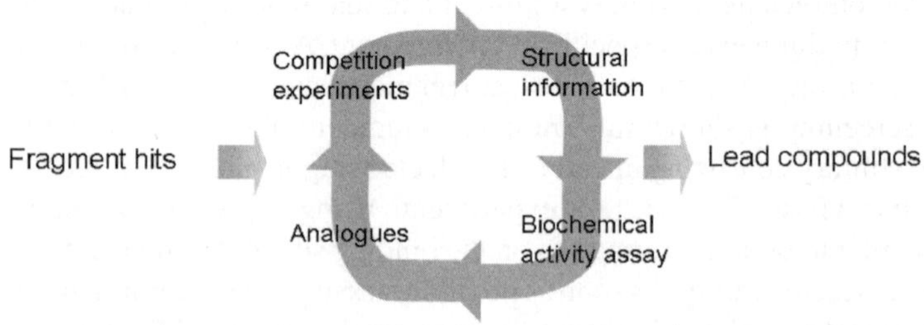

Figure 4.2 *The workflow in a fragment-based screening project.*

function of the target protein in the desired way. The biochemical assay does not have to be employed during the first iteration, but could wait until the affinity of the developed fragments has increased. This could very well be a low-throughput assay; in many instances it is possible to use NMR as an enzymatic activity assay.[19] As discussed in Figure 4.1, analogues to fragment hits are either synthesized by one-step chemistries or acquired from in-house compound collections or commercial sources. In the first iteration, the fragment is usually optimized by testing close analogues with no substantial increase in molecular weight. In the following iterations, the fragments become larger either by expansion of the fragment or by linking or merging different fragments.

The expansion strategy is the most generally applicable. Optimized fragments are expanded with features (ring systems, functional groups) that interact with the target protein in a favorable way without disrupting the binding of the core fragment. Selection of appropriate expansions is based on affinity (from a biophysical method) and/or potency (from a biochemical activity assay) and on structural information (from X-ray crystallography or NMR), and also binding hypotheses from modeling. The SAR obtained from each iteration is used for the selection of the next generation of molecules, which subsequently is tested for affinity and/or potency to the target protein in order to confirm or reject the direction of the design. In a successful iterative design cycle supported by detailed structural information, the process converges rapidly to more potent and complex molecules with retained high ligand efficiencies. There are many published examples of this approach.[4, 20–31]

An elegant approach is to identify fragments binding to separate but adjacent sites on the target protein and then link the fragments into one high-affinity compound. For the full potential of the linking to be fulfilled in terms of potency, the relative orientations of the bound fragments that are to be linked must be determined and the linker must be designed so that it preserves the relative orientations of the bound fragments. This is far from a trivial task but several successful cases have been published.[4, 29, 32–39] An interesting variant is so-called 'self-assembly' where reactive fragments link together to form an active inhibitor in the presence of the protein target.[40, 41] Instead of linking, fragments with overlapping but not identical binding sites can be merged into one hybrid molecule.[42–44]

A variation of the schematic workflow in Figure 4.2 is to combine virtual screening of compound databases and experimental verification by a biophysical method, usually NMR.[21, 45–47] An additional use of fragment screening is to identify fragments that can replace part of an existing lead compound responsible for a non-desired characteristic such as low solubility, poor bioavailability, high albumin binding, metabolic instability or lack of novelty.[48–50]

4.3 Techniques for the Primary Fragment Screen

The function of the primary screening is to identify good starting points, i.e. compounds that are chemically attractive (in terms of 'synthetic handles' and/or with many analogues available) and have a high likelihood of success in co-crystallization experiments (or structure determination of ligand–target complexes by NMR) with the target macromolecule. The requirements of the assay to be used in the primary fragment screen differ substantially from the requirements of an assay used in HTS. Most importantly, the assay should be able to reliably detect weakly binding fragments. Further, it should be possible to apply the assay (or a selection of assays) to a wide range of target proteins with a minimum of assay development. Due to the relatively small sizes of fragment libraries, questions of throughput and reagent consumption are much less important compared with HTS. Different types of assays for the primary fragment screen along with their respective strengths and weaknesses are listed below. The amount of protein necessary for the primary screening differs between the techniques. However, in many cases the success of fragment-based drug discovery projects is critically dependent on structures of ligand–target complexes to drive the chemistry. If a successful development of fragments to potent inhibitors is defined as reaching a potency <100 nM, statistics from Abbott Laboratories show that the success rate increases from 33 % to 93 % with the aid of structure-based design.[4] Thus, milligram amounts of target protein will in principle always be required, which renders the differences in protein consumption between the primary screening techniques insignificant in practice.

Different techniques require different degrees of ligand occupancy of the target protein for a binding event to be reliably detected. One-dimensional (1D) ligand-detected NMR techniques are very sensitive in this respect, where the binding fragment needs to bind to only a very small fraction of the target protein molecules for binding to be detected. For example, a fragment concentration of 20–40 μM in a transverse relaxation-filter experiment (see below) is sufficient to reliably detect a weak binding event with a K_d of several hundred μM, depending on the rotational correlation time of the target protein. This translates to a required ligand occupancy of <10 % with the ligand occupancy (p_B) defined as $p_B = [L]/([L] + K_D)$.[6] Two-dimensional (2D) protein-detected NMR methods and a biochemical assay require higher ligand occupancies for reliable detection and X-ray crystallography requires nearly 100 % ligand occupancy. Hence the fragment concentrations and consequently the solubility requirements of the fragments will be much higher when using a biochemical assay or X-ray crystallography compared with ligand-detected NMR techniques for the detection of fragments with a given affinity. This partly explains why it is usually not possible to obtain crystal structures of fragment–target complexes from all fragment hits detected by ligand-detected NMR techniques.

4.3.1 NMR

NMR techniques have dominated fragment screening from the beginning. The wide choice of different ligand-detected NMR techniques is especially suited to primary fragment screening. Very weak binders are reliably detected and these generic binding assays can be applied to a wide array of different therapeutic targets. Further, there are no serious limitations on buffer choice and information on compound solubility and integrity, and also on the state of the target protein, is obtained as a part of the experiment. The main disadvantage is that nonspecific binders will also be picked up, but they can be filtered out by competition experiments using known, specific binders. The greatest advantage with protein-detected NMR techniques is that it is possible to obtain direct information on the binding site. The drawbacks include that it is necessary to produce large amounts of isotopically labeled target protein and the difficulties associated with target proteins of molecular weights above 30 kDa. A survey of NMR techniques for fragment screening is given later in this chapter.

4.3.2 Nondenaturing Electrospray Ionization Mass Spectrometry (ESI-MS)

Reversible (noncovalent) binding events can be detected with nondenaturing ESI-MS.[51–54] A considerable advantage of the method is that the stoichiometry of the binding event is measured directly. As a consequence, it is possible to observe directly if two fragments bind simultaneously. There is also direct information on compound solubility and integrity, in addition to the state of the target protein. The method suffers, however, from a few major disadvantages. First, high homogeneity of the target protein is required and the method is limited to protein targets that are happy in a volatile buffer such as ammonium acetate. Second, the protein–ligand complex is detected in vacuum and electrostatic interactions will be strengthened whereas hydrophobic interactions will be weakened compared with an aqueous solution. Another approach utilizing mass spectrometry is 'tethering'.[23] This technique relies on the formation of a disulfide bond between the fragment and a cysteine (naturally occurring or introduced by mutagenesis) at the site of interest in the target protein. Therefore, only thiol-containing fragments can be used.

4.3.3 Surface Plasmon Resonance (SPR)

Protein–ligand interactions are detected by changes in the refractive index at the surface of a solid support on which the target protein or the fragments are immobilized.[55, 56] The analytes are then injected in a continuous flow and the real-time binding response (sensorgram) is recorded. The signal increases during injection of the analyte at a rate related to the on-rate, reaching a plateau at the equilibrium level determined by the K_d and the concentrations of fragment and protein. When the injection ends, the signal decreases back to the original level at a rate related to the off-rate. Hence a distinct advantage of the method is that it is possible to obtain both equilibrium (K_d) and kinetic parameters (on- and off-rates). Since the magnitude of the signal is directly related to the mass of the binding ligand, it is also possible to obtain stoichiometries from the signal at saturation level. The main disadvantage is the necessity to immobilize one of the two components. Consequently, this is not a true in-solution technique. Immobilization of the target protein is often protein specific and it could prove difficult to immobilize the protein so that it remains fully functional. Further, conditions in which the protein remains stable over a long period must be identified. This assay development phase may require a long time, but the throughput will be high once everything is set up. If the fragments are immobilized instead, every fragment must be synthesized with a long, flexible linker attached. The linker will then serve as an attachment point for the covalent immobilization on the solid support, usually via a thiol group. Another alternative is to immobilize a previously known high-affinity ligand which acts as a probe for a defined binding site and then inject a mixture of the target protein and the fragment(s).[57] If a fragment binds and competes for binding with the immobilized known ligand, the signal will be lower than when the fragment is absent.

4.3.4 Biochemical Activity or Binding Assay

A biochemical assay can be used directly in the primary screen at very high fragment concentrations.[58, 59] Modulation of the enzymatic function is detected by monitoring the rates of either the decreasing substrate or increasing product concentrations. Alternatively, a binding assay can be used where the displacement of (typically) a fluorophore-tagged ligand is monitored. For detection of weakly binding fragments, the fluorophore-tagged ligand should bind fairly weakly but still specifically. Advantages with using a biochemical assay in the primary fragment screen include direct information on potency, high throughput and very low protein consumption. The low protein consumption and the possibility of directly using membrane preparations (lysed and homogenized cells with overexpressed membrane protein) make it feasible to screen also a wide range of membrane proteins.

However, the use of a biochemical assay in the primary fragment screen is likely to lead to a comparatively high rate of false positives. The control of the assay system is much lower than with a method such as NMR and also the assay system in itself is more complex. There is no information on the state of the assay components, such as solubility and integrity of the compounds, or on the state of the protein. Further, the risk of compound-induced fluorescence artifacts increases with the very high fragment concentrations necessary. On the other hand, conjugated unsaturated systems are present to a lesser extent in fragments than larger compounds. Another drawback is that fragments binding in the vicinity of the active site without modulating the function or displacing the tagged ligand will not be detected. Such fragments are very interesting since they provide information on how to expand fragments binding to the active site.

4.3.5 X-ray Cystallography

Recent technological advances have led to automation of most steps in the protein crystallographic process. The development of crystallization robotics and the automation of data collection, processing and analysis have enabled X-ray crystallography to be used as a primary screening technique.[60] Cocktails of fragments are soaked into preformed crystals of the target protein. The fragments in a cocktail should preferably be distinguishable by shape in order to avoid deconvolution. The fragment concentration must be very high, not only because the fragments bind weakly, but also since the protein concentration in the crystal typically is on the order of 10 mM. The advantage of using X-ray crystallography as the primary screening technique is that detailed structural information is obtained directly. However, large amounts of target protein are necessary and the throughput is lower than with the other techniques. It is much more efficient to use another method, such as NMR, to 'prescreen' fragment cocktails. The soaking attempts will then be performed on binding fragments only.

4.3.6 Choice of Technique for Primary Fragment Screening

In our opinion, the most reliable and versatile primary screening techniques are NMR methods. X-ray crystallography will in general be necessary in a fragment screening campaign, but is much more efficient to employ after the initial fragment screening. Also, a biochemical activity assay to measure potency will be needed at some stage in the development of the fragment hits. When used for primary screening, the comparatively high number of false positives will make the subsequent crystallography work less efficient, unless a biophysical binding assay such as NMR is employed as a filter in between. It may be more suitable to apply a biochemical activity assay when the primary fragments have been developed to slightly larger compounds with a higher affinity.

4.4 Practical Aspects of NMR-based Fragment Screening

4.4.1 Fragment Library

In our experience, typical hit rates for fragment libraries when using ligand-detected NMR techniques are on the order of 2–10 %. Consequently, the fragment library used for the primary screen does not have to be very large in numbers: 500–1000 chemically diverse fragments are in general sufficient. Such relatively small libraries will not need the infrastructure necessary for large HTS compound collections that are two to three orders of magnitude larger in number. Fragment libraries will also be easier to keep at high quality standards with respect to purity, not only due to the smaller number of compounds but also since fragments contain fewer features that could cause stability problems. Rather than having a large fragment library, it is more important to have access to a large number of analogues to the fragments in the primary fragment screening library. Analogues are preferably already present in-house or it should be possible to synthesize them quickly from fragments with chemical handles for straightforward expansion using one-step chemistries. Analogues could also be ordered from commercial suppliers, but that will introduce a lag time waiting for the compounds to arrive. As the number of performed fragment screening campaigns increases for a given fragment library, the number of fragment analogues that are immediately available will accumulate.

The literature on fragment library design is limited following the publications on the SHAPES strategy from Vertex.[61–67] Fragment library design is essential since the physical properties of fragments remain largely the same when incorporated into larger drug-like molecules.[64] The size of the fragments in the library should be relatively small in order to have a high probability of binding to the target, but at the same time complex enough so that the probability that the fragment binds to the target in one way only is high.[7, 68] These requirements would be fulfilled by fragments containing approximately 10–20 heavy atoms. Further, it is very important that the fragments are highly soluble (preferably >0.5 mM in buffer solution) since one main purpose of the primary screening is to identify binding fragments that have a high likelihood of producing structures of the fragment–target complex. Another point to consider is that compounds with a low hydrogen bonding capability are less likely to be active.[66] In order to increase throughput and decrease protein consumption, the fragments are mixed into cocktails. For ligand-detected NMR experiments, the cocktails usually contain 6–10 fragments and with a maximum of chemical diversity within each cocktail. In order to avoid any chemical reactions taking place within the cocktail, separation of acids from bases and electrophiles from nucleophiles is important. For protein-detected NMR experiments, the number of compounds in a cocktail can be much higher and the diversity within a cocktail should be small, in order to minimize the number of cocktails necessary to deconvolute.

Typical restrictions applied to fragments at iNovacia include that they should contain 9–20 heavy atoms and have Clog*P* <2. All fragments are available as powder and most stock solutions of individual fragments are 0.5 M in deuterated DMSO. This permits high fragment concentrations in co-crystallization and soaking attempts without excessive DMSO concentration. The main part of the iNovacia fragment library is a diversity set where the fragments have many analogues among in-house scaffolds in addition to many larger compounds from the in-house compound collection where the fragments are present as substructures. A large fraction of the fragments have masked synthetic handles to allow quick expansion with one-step chemistries. For example, diverse fragments containing an amino group were purchased and converted into acetamides using parallel one-step chemistry. The acetamides are entered into the fragment library and if one of them is a hit, the acetamide group can easily be exchanged into a larger amide by the use of the corresponding amino fragment. Another example is the purchase of aryl bromides and boronic acids from which biaryls are produced by means of Suzuki couplings. If one of the biaryls is a hit, one of the aryls can then easily be exchanged with another aryl. The iNovacia fragment library also contains molecular scaffolds that occur frequently in marketed drug molecules.

4.4.2 NMR Instrumentation

For adequate resolution and sensitivity, the NMR spectrometer used should operate at a ^{1}H frequency of 500 MHz or higher and preferably be equipped with a cryogenically cooled probe that reduces the thermal noise in the receiver circuitry ('cryoprobe'). Automated sample changing is convenient and it should be possible to keep the samples cold while they are waiting to be measured. At iNovacia, we are using a 600 MHz spectrometer equipped with a cryoprobe with a flow cell insert. The samples are loaded into 96-well plates placed on a Peltier cooling device on a Gilson robot. It should be noted that the freezing point of pure D_2O is 3.82 °C.

4.4.3 Selection of Buffer

Most important is that the target protein is soluble, stable and monodisperse in the selected buffer. This will be checked in the limited protein characterization (see below). Phosphate buffer is very convenient since there are no ^{1}H signals, and is often the first choice for ligand-detected NMR techniques. However, there are several proteins that bind phosphate (notably phosphatases) for which phosphate buffer is not a good choice. Fortunately, there is now a large selection of commercially available buffers in deuterated form.

4.4.4 Protein Characterization

It is very important to perform at least a limited protein characterization study before starting the NMR screening. Here it is assumed that the identity and purity of the target protein produced has been checked by methods such as MALDI-MS and N-terminal sequencing.

Protein stability. It should be clear before starting the screen that the target protein stays folded in the selected screening buffer during the measurement time in the spectrometer and while in the queue waiting to be measured. Probably the simplest way to check the stability is to observe the protein methyl signals during e.g. 24 h, by repeatedly acquiring ^{1}H 1D spectra. By the appearance of the methyl signals it will be clear if the protein precipitates or unfolds during this time. If a protein-detected NMR technique is to be used in the primary screening and a stable isotope-labeled target protein is available, repeatedly acquired ^{1}H–^{15}N (or ^{1}H–^{13}C) correlation spectra should be collected instead.

Protein homogeneity. The target protein should be monodisperse in the selected buffer and at the protein concentration to be used in the screening. The presence of large, soluble aggregates may cause the hit rate to be very high due to nonspecific interactions with aggregates. The oligomeric state of the target protein in a given buffer can be checked by dynamic light scattering (DLS) or analytical ultracentrifugation. DLS is a very useful technique for buffer optimization.[69]

Fortuitous ligands. Ligands from the expression host (e.g. fatty acids from *Escherichia coli*) could be noncovalently bound, e.g. in the active site of the overexpressed target protein. This is a serious, but rare, problem and may require an additional refolding step in the protein purification protocol. The presence of any noncovalently bound ligands in the target protein can be checked by nondenaturing ESI-MS.

4.5 NMR Techniques for Fragment-based Screening

NMR spectroscopy is a versatile tool for assaying binding of a small molecule to a macromolecule such as a target protein. There are many excellent reviews describing NMR techniques used in fragment screening.[6, 70, 71] The available techniques may be divided into two different classes, ligand-detected and protein-detected methods. The protein-detected methods yield more information on the ligand–protein interaction, but the ligand-detected methods are more widely used in the pharmaceutical and biotech industries since they can be applied to a much wider range of target proteins and require less upfront resources. Therefore, this part will focus on the ligand-detected methods. Examples of NMR techniques for monitoring protein–ligand interactions are given in Table 4.1.

Table 4.1 *Examples of NMR techniques for monitoring protein–ligand interactions.*

NMR technique or observable	Nonbinding ligand	Protein	Bound ligand	Representative references
Saturation transfer difference	Unsaturated	Saturated	Saturated	72
WaterLOGSY	Positive NOE	Negative NOE	Negative NOE	73, 74
Linewidth	Narrow	Broad	Broadened	43
Transverse relaxation filter/selective longitudinal relaxation	Slow	Rapid	Increased	75, 76
Translational diffusion	Rapid	Slow	Slowed	75, 77
Transferred NOE	Weak positive NOE	Strong negative NOE	Weak negative NOE	78–80
Paramagnetic spin label	Slow relaxation	—	Rapid relaxation	81, 82
Protein chemical shift perturbation	—	Unshifted	Shifted	32

4.5.1 Ligand-detected Techniques

The ligand-detected techniques rely on detecting changes in the NMR observables of a ligand upon exposure to the target protein. The ligand will transiently adopt NMR parameters characteristic of the typically much larger target protein. Hence the sensitivity of these methods will in general increase with increasing size of the target protein. Two main classes of ligand detected NMR binding assays can be distinguished:

1. methods that rely on transfer of magnetization between target and ligand;
2. methods that rely on the detection of an altered hydrodynamic property (i.e. molecular tumbling rate or diffusion rate) upon binding to the target.

Methods that represent the first class of experiments are saturation transfer difference (STD)[72] and WaterLOGSY.[73, 74] These techniques are closely related in that they rely on dipole–dipole interaction between ligand and target protein spins. Diffusion and relaxation filter-type experiments belong to the second class of methods.[75, 77] When a small ligand binds to a macromolecule, its apparent rates of translational diffusion and reorientation are slowed. The transferred NOE technique[78, 83] also falls into this class. It relies on detecting intra-ligand NOEs that develop in the bound state, where the dipole–dipole interaction due to the decreased molecular tumbling rate is much more efficient than in the free state, by observing the free ligand. Experiments that do not fall in either category are the use of paramagnetic spin labels to increase the relaxation of nearby spins[81, 82, 84] and fluorine NMR to detect binding or displacement of fluorinated compounds to a target.[85–88] Selected ligand-detected NMR techniques are presented in more detail below.

In the ligand-detected NMR techniques, the ligands are usually present in molar excess compared with the target protein in the samples. As always when detection of binding depends on exchange between the bound and free states of the ligand, a minimum off-rate ($k_{\rm off}$) will be required for detection of binding. This should not pose a problem for fragments, however, since they are expected to have diffusion controlled on-rates ($k_{\rm on}$) and will not bind with very high affinities. Alternatively, a 'spy' molecule binding specifically at the binding site of interest can be employed.[45, 76, 86, 89, 90] The binding signals of the spy molecule are observed and ligands that are able to displace the spy molecule will be detected as binders. There are distinct advantages with using a spy molecule and indirect detection of binding via competition experiments for screening: there is no minimum off-rate necessary for detection of binding and no purely nonspecific binders will be picked up. On the other hand, the requirements for the spy molecule are high: it has to bind specifically to the desired site with a low enough affinity to be readily displaced by weakly binding fragments. The detection cut-off for the fragment affinity will be determined by the affinity of the spy molecule to the target protein. Further, it must display a clear binding signal in the chosen NMR technique and preferably cover the entire binding pocket of interest. A drawback is that no fragments binding to adjacent subsites to the binding site of the spy molecule will be detected. Such fragments could be of interest for linking or expansion of the fragment hit.

Saturation transfer difference. The origins of the STD experiment[72] can be traced to the spin-saturation transfer experiment or Forsén–Hoffman experiment from the 1960s.[91] In the STD experiment, a subset of the protein ^{1}H resonances are saturated by means of a train of frequency-selective radiofrequency pulses applied to a narrow spectral region devoid of ligand resonances. The saturation is transferred by spin diffusion (^{1}H–^{1}H cross-relaxation pathways) to the rest of the protein, a process that becomes more efficient with increasing rotational correlation time of the protein. As a consequence, the experiment will become more sensitive with increasing molecular weight of the target protein. The saturation of the protein will then be transferred, via intermolecular ^1H–^1H cross-relaxation, to weakly binding ligands at the ligand–protein interface. After the ligand has dissociated, the long spin–lattice relaxation times (T_1) of the free ligand protons ensure that it is possible to detect the transferred saturation as an attenuation of the ligand signals. During the time of the saturation, more unsaturated ligand molecules will bind and dissociate, leading to an increasing population of saturated ligands. The saturated spectrum is then subtracted from a spectrum obtained under nonsaturating conditions to obtain an STD spectrum showing only the signals from compounds binding reversibly to the target protein; see Figure 4.3 for an example. This simplicity of the resulting spectrum is a very attractive feature of the STD experiment. In practice, the pulse sequence is written in such a way that the subtraction is performed automatically in every other scan, i.e. the individual spectra are never observed.

The STD experiment allows for very high ligand:protein ratios to be used, often as high as 50 or 100. Thus, the protein concentration can be kept very low, in fragment screening it is possible to use less than 1 μM target protein in favorable cases. The protein signals are not visible in the STD spectrum due to the low protein concentration and/or the presence of a spinlock in the pulse sequence. The sensitivity of the method depends critically on how efficiently the protein resonances have been saturated.[92] The duration of the saturating selective pulse train is typically 1–4 s, where the longer saturation times in this interval can be used for smaller proteins. The upper limit of the saturation time for an efficient saturation transfer is determined by the T_1 of the ligand in the bound state. A spectral region suitable for efficient saturation is the protein methyl groups which usually has a maximum intensity close to 0.7 ppm. In order to avoid direct saturation of ligands, the saturation is usually applied further upfield where most target proteins have upfield shifted methyl resonances. The STD experiment is preferably performed in samples containing 100 % D_2O. This makes optimal water suppression less crucial and minimizes exchange-mediated saturation leakage via dipole–dipole interactions between saturated protein protons and hydration water molecules.[93, 94]

The intensity of the STD signals contains information on the ligand affinity and it should in principle be possible to rank fragment hits directly with respect to affinity. There are, however, several caveats to be aware of and caution should be exercised when interpreting data. First, the individual STD peak heights should be normalized to the corresponding peak heights in a reference spectrum. This could either be the spectrum obtained under nonsaturating conditions (collected as part of the STD experiment) or a T_2-filtered ^1H 1D spectrum of the same sample, using the same length of the spinlock and interscan delay as in the STD experiment. The parameters in the STD experiments, such as saturation frequency, saturation time and relaxation delay, should be identical for all samples. For the normalized STD response to be directly related to the relative affinity of different ligands to the target protein, there are a number of assumptions that should be fulfilled: the contribution to the STD signals should come from binding to a single site only, the exchange must be fast (should be fulfilled for weakly binding fragments) and the T_1 values for the proton signals to be compared must be similar. The last condition is important since the STD responses are highly dependent on the T_1 value for the observed proton[94, 95] so that, for example, a smaller T_1 value would result in a smaller STD response.

The variation of the normalized STD responses for different protons within a ligand also contains information on *how* the ligand binds to the target protein.[96, 97] The magnitude of a normalized STD response is related to the corresponding ligand proton proximity to target protons, a consequence of the distance dependence of intermolecular ^1H–^1H cross-relaxation. Thus, a stronger STD response implies closer intermolecular contact, which can be interpreted as information on which part of the ligand is responsible for most of the binding energy. To compare directly the normalized STD responses, the same assumptions as described above should be fulfilled. This 'ligand epitope mapping' approach is best suited for weakly binding ligands and/or the use of sufficiently short saturation times, in order to ascertain that the magnetizations of the ligand protons are not equalized due to spin diffusion during the lifetime of the ligand–target complex. This approach has been substantially extended to determine indirectly the conformation of a ligand when bound to a target protein with a known structure. In this method, called SOS-NMR,[98] STD is applied to a ligand complexed to a series of perdeuterated target protein samples, each with different specific amino acid types protonated. The relative STD intensities of the ligand peaks from the different samples contain quantitative information on which protons on the ligand are in the vicinity of which type of amino acid in the target protein. With a known target protein structure, the structure of the ligand–target complex can be deduced from this information provided that experiments have been performed on a sufficient number of differently labeled protein samples to define the binding site uniquely. An analysis of 272 unique crystal structures of ligand–protein complexes showed that 3–9 differently labeled protein samples would be enough to identify unambiguously the ligand binding site of more than 90 % of the ligand–protein complexes. Obviously, SOS-NMR is a very resource-intensive method, but could prove highly valuable in cases where it has not been possible to obtain any structural information on the ligand–target complex either by X-ray crystallography or by 'traditional' NMR methods.

Finally, STD has been demonstrated to be applicable for the detection of ligand binding in very demanding systems such as a virus[99] and an integral membrane protein either reconstituted in liposomes[100] or on living cells.[101]

WaterLOGSY. The WaterLOGSY experiment[73, 74] relies on water-mediated magnetization transfer to compounds that bind to the target protein. The most favored version of this experiment[102] is essentially an NOE experiment starting with selective inversion of the bulk water magnetization followed by a long mixing time (up to several seconds). The inverted bulk water magnetization is transferred to the target protein and binding compounds via several possible magnetization transfer pathways:[6] (i) direct ^1H–^1H cross-relaxation between compounds and tightly bound water molecules at the binding site, (ii) chemical exchange of inverted bulk water with protein hydroxyl and amine groups at the binding site, which in turn will transfer the inverted magnetization to the bound compound protons, and (iii) chemically exchanging hydroxyl and amine groups on the protein that are not situated in the binding site may contribute via spin diffusion through the protein. During the mixing time, the ^1H–^1H cross-relaxation will give rise to NOEs with opposite signs depending on whether the magnetization transfer takes place in the bound state (slow tumbling, long rotational correlation time) or in solution (fast tumbling, short rotational correlation time). Reversibly binding compounds will experience magnetization transfer in both the bound and free states, whereas nonbinding compounds will only experience magnetization transfer in the free state. For the binding compounds, the contribution from the bound state will dominate provided that the ligand excess is not extremely high and/or the interaction very weak. Consequently, the resulting WaterLOGSY spectrum will consist of positive signals from compounds binding to the target protein and negative signals from compounds not binding to the protein. To detect also very weak binders, WaterLOGSY spectra should be collected in both the presence and absence of target protein. The signals from a very weak binder may be negative, but less negative compared with when the target protein is absent. As with the STD experiment, efficient detection of reversibly binding compounds is based on the long T_1 values of the free ligand protons and the use of an excess of ligand leading to a build-up of the population of ligands that has experienced magnetization transfer during the mixing time.

Both STD and WaterLOGSY are very popular fragment screening techniques and share many features. However, STD is an incoherent technique, i.e. it relies on transfer of saturation (incoherence), whereas WaterLOGSY takes advantage of the same physical processes but in a coherent way, i.e. polarization transfer. Both techniques use a relatively high ligand:protein ratio and the sensitivity with respect to both protein consumption and detectable dissociation constant range is comparable. However, for targets with a low proton density, e.g. RNA, the use of WaterLOGSY is clearly advantageous.[103] Targets with a low proton density suffer from inefficient spin diffusion, leading to low STD sensitivity. WaterLOGSY, on the other hand, is much less dependent on spin diffusion efficiency since the method mainly relies on magnetization transfer from the water molecules surrounding the target. Nevertheless, the low spin diffusion efficiency in DNA can be exploited in STD experiments to obtain information on the binding site of ligands. By saturating at different frequencies (and therefore different DNA regions) and observing the relative STD responses, base-pair intercalators have been distinguished from minor groove binders.[104] For the best sensitivity, the WaterLOGSY experiment should be performed in H_2O with only small amounts of D_2O for field-frequency locking. The intense H_2O signal, however, raises the need for efficient water suppression schemes and is a source of potential spectral artifacts. Since signals from nonbinding compounds appear in the spectrum as negative peaks, the spectrum will be more complex than the corresponding STD spectrum.

Line broadening and relaxation filter experiments. The spin–spin relaxation times (T_2) of protons in a slowly tumbling entity such as a target protein are much shorter than those of protons from a faster tumbling small organic compound. Since the ^1H signal linewidth at half-height is proportional to $(\pi T_2)^{-1}$, a simple approach is to measure the linewidths of the ligand ^1H signals in the absence and presence of a target protein. The method requires that the magnet field homogeneity is identical for the samples. This is achieved by employing a gradient shimming step for each sample and by checking that the shimming quality is identical for all samples by observing the linewidth of an inert compound present in all samples, e.g. the calibration standard DSS (2,2-dimethyl-2-silapentane-5-sulfonic acid). Upon binding of the ligand to the target protein, the tumbling rate of the small molecule is decreased. This results in a decreased T_2, which is manifested by both a broadening and a concomitant decrease of ligand ^1H peak heights. Further, if fast exchange kinetics are assumed and possible exchange contributions to the linewidths are small or negligible, it is also straightforward to estimate the binding affinity.[105] However, since the primary NMR screening is often performed on cocktails of fragments, spectral overlap both from other compounds and from the target protein may render accurate lineshape analysis difficult. In order to detect weakly binding ligands reliably, the molar excess of ligand over target protein must be kept fairly low.

Rather than measuring the ligand linewidths, a more common setup is to apply a relaxation filter, i.e. a spinlock (e.g. a CPMG sequence)[106] directly after the 90 degree detection pulse and before the acquisition of the signal.[75] The relaxation filter experiments are performed on two samples that are identical except that the target protein is present in one and absent in the reference sample. The spectra collected on these two samples are compared for changes in NMR signal intensities of the small molecules; see Figure 4.4 for an example. A long enough relaxation filter will eliminate the target protein signals. With the assumptions that the exchange contributions to the linewidths are negligible and that the ligands bind with a similar on-rate (e.g. a diffusion-controlled on-rate) and to one site only, the signals for a higher affinity ligand will decrease more than for a lower affinity ligand with a given time length of the spinlock. Therefore, in general, it is possible to use the outcome of the relaxation-filtered experiment to rank ligands with respect to affinity.[24, 107] A convenient way to affinity rank ligands is to collect relaxation-filtered spectra with different durations of the spinlock. The duration of the spinlock for a given target determines the detection cut-off and, therefore, when longer spinlock times are used, weaker binding ligands are detected. For example, simulations have suggested that for a relatively small target protein of MW $\sim$15 kDa, a spinlock of 400 ms will eliminate the signals from ligands binding with an affinity of 500 μM or tighter for equimolar amounts of protein and ligand.[24] Here, it can be noted that it is often erroneously claimed that fast exchange is always a requirement for direct detection of binding by ligand-detected methods. That is not the case for the transverse relaxation filter technique provided that no molar excess of ligand over target protein is used. In the extreme case of a covalent binder, for example, all ligand molecules would then be bound to the target protein during the spinlock time and no NMR signal from the ligand would pass the relaxation filter.

It is also possible to utilize the differences in spin–lattice relaxation (T_1) between a large protein and small organic molecule to detect binding by applying an inversion–recovery pulse sequence.[76] However, detection of binding is only possible for *selective* T_1 measurements, i.e. the inversion of fragment ^{1}H signals must be performed by the use of frequency-selective pulses. As a consequence, it is not practical to use this technique for primary fragment screening if the screened fragment signals are to be observed, since a separate inversion pulse would have to be designed for every fragment in the library. Instead, a weakly binding 'spy' molecule should be employed in competition experiments where only selectively inversed signals from the 'spy' molecule are repeatedly monitored.[76]

Paramagnetic spin labels. A paramagnetic spin label is an atom with an unpaired electron, which will enhance the relaxation of nearby protons due to the strong electron-proton dipole–dipole interaction. By covalently attaching a paramagnetic spin label on selected target protein side-chains, the transverse ^{1}H relaxation times of any fragments binding in the vicinity of the spin label will decrease dramatically. For detection, a transverse relaxation filter experiment is usually applied. This technique has been dubbed SLAPSTIC (spin labels attached to protein side-chains as a tool to identify interacting compounds).[81, 82] The strong dipole–dipole interaction between an unpaired electron and a proton will dominate over all other relaxation mechanisms and is effective over much longer through-space distances than is typical for ^{1}H–^{1}H relaxation (up to 15–20 Å). Further, the strong transverse relaxation effect makes it possible to use low protein concentrations, on the order of a factor 10 lower than in a normal transverse relaxation filter experiment.[35] The main disadvantage of the method is the need to introduce the spin label at a suitable amino acid side-chain near the binding site of interest. There is a risk that the introduction of the spin label might alter the structure and/or the binding properties of the target protein. An attractive variant of the method is the use of second-site screening. A spin-label is then attached to a known ligand and used to detect fragments binding to a site adjacent to the binding site of the spin-labeled ligand. The quenching effect of the spin label on compounds is observed if, and only if, both the spin-labeled compound and the other compound bind at the same time and in the vicinity of each other. Here, it is important that the attachment of the spin label does not alter either the first or the second binding site.

Fluorine relaxation filter. The ^{19}F nucleus exhibits a number of features that make it attractive to use for detection in NMR binding experiments:[88] (i) the ^{19}F transverse relaxation rate and chemical shift are very sensitive to changes in the microenvironment (e.g. upon ligand–target complex formation), (ii) ^{19}F occurs at 100 % natural abundance and has a gyromagnetic ratio nearly as high as that for ^{1}H and (iii) the absence of ^{19}F in biological macromolecules, most organic compounds and buffer components results in very clean ^{19}F spectra. However, direct observation of ^{19}F in a primary fragment screen would require that all fragments contain at least one fluorine atom. In practice, therefore, ^{19}F detection in screening campaigns is applied in competition experiments where a fluorine containing 'spy' molecule is employed.[86, 87] The observed parameter of the spy molecule is typically the ^{19}F signal intensity after a spin-echo filter.

4.5.2 Protein-detected Techniques

Protein-detected techniques rely on detecting changes in the NMR observables of a protein upon exposure to a ligand. The best known approach is the pioneering SAR-by-NMR fragment-linking scheme introduced by Fesik and co-workers in 1996.[32] In the first step of the scheme, ligands are identified by monitoring alterations of target signals in a 2D ^{1}H–^{15}N correlation spectrum. The spectrum can be regarded as a residue-resolved map over the entire protein backbone (except prolines). The amide resonances in the binding site of the ligand usually show the largest changes as compared with a spectrum of the *apo*-protein. If a cocktail of fragments is tested, then deconvolution of the cocktail will be necessary to identify the active fragment. Protein-detected techniques require isotopically enriched samples, but have the benefit (provided that sequence-specific resonance assignments have been obtained) of being able to identify directly the binding epitope on the target, which also makes it possible to distinguish between specific and nonspecific binding events. It is also possible to assess whether any significant conformational changes of the target protein occur upon binding. Further, protein-detected methods do not rely on fast exchange to retrieve information from the bound state, making it possible to detect both low- and high-affinity hits. The major drawbacks of protein-detected methods are the limited applicability to typical pharmaceutical target proteins and the high demands on resources. Most therapeutically important target proteins have molecular weights above 30 kDa, and obtaining sequential resonance assignments for such proteins is a major task in terms of both resources and time, despite recent advances in data acquisition schemes. An indirect and more limited approach would be to assign only resonances of active site residues by the use of known substrates or ligands. There are, in any case, substantial demands on the protein production. Milligram amounts of isotopically labeled target protein (e.g. ^{13}C, ^{15}N, ^{2}H) are to be produced in a suitable expression host. The purified protein must be soluble, stable for at least a couple of weeks at room temperature and remain monodisperse at the high concentrations needed. Even if it were possible to obtain the sequence-specific resonance assignments for a target protein of 30–40 kDa, the overlap in the 2D spectrum would be considerable due to both severe line-broadening effects and the large number of peaks in the spectrum. Consequently, it would be very difficult to reliably re-assign the peaks that have moved upon ligand binding. Therefore, efforts have been dedicated to develop protein-detected methods that are more generally applicable to larger proteins.

A passable way would be to decrease the number of peaks in the 2D spectrum. One approach is to record 2D ^{1}H–^{13}C correlation spectra on target proteins with selective ^{13}C labeling of the methyl groups of valine, leucine and isoleucine.[108] In this approach, the sensitivity is increased almost threefold for small proteins (MW$<$20 kDa) compared with 2D ^{1}H–^{15}N correlation spectra on uniformly ^{15}N-labeled targets. For larger proteins, the ^{13}C labeling should be combined with ^{2}H labeling in order to decrease dipole–dipole relaxation. In another labeling strategy, termed site-selective screening,[109, 110] it is possible to screen selectively for binding to a selected epitope without the need for sequence-specific assignments. First, pairs of sequential amino acid residues that reside in the binding site of interest (e.g. the active site) and are unique, i.e. they only appear once in the protein sequence, are identified. Such a unique amino acid pair, XY, is then selectively labeled so that amino acid X is labeled with ^{13}C and amino acid Y with ^{15}N. Performing an HNCO-type correlation spectrum of a protein labeled in this way will result in only one signal. Thus, chemical shift perturbations upon addition of a ligand that binds in the vicinity to the labeled amino acid pair are easily detected even for large proteins due to the reduced spectral complexity. However, this simplicity of the resulting spectrum is both the strength and weakness of this method. With only one probe for the binding site, it is not possible to separate direct binding in the vicinity of the amino acid pair from small shifts due to indirect effects from locations other than the binding site. To minimize this risk, several unique amino acid pairs should be identified and selectively labeled.

4.5.3 Choice of NMR Technique for Primary Fragment Screening

From the survey of the NMR techniques above, the general recommendation would be to use a ligand-detected technique for the primary fragment screen. A protein-detected method could be considered in cases where the isotopically labeled target protein is readily produced in large quantities and with known 3D structure. Preferably also the sequential resonance assignments of the target protein should be known or it should at least be clear from inspection of the 2D ^{1}H–^{15}N correlation spectrum that the assignments will be straightforward to obtain. Assuming that these conditions are fulfilled, one case where a protein-detected

NMR technique would be the first choice is when one wants to find fragment binders to a specific site for which there are no known binders, e.g. a newly identified allosteric site. The introduction of a paramagnetic spin-label either at the site of interest or on a known binder binding at an adjacent site would be an alternative method. However, the long distances (15–20 Å) over which spin-labels exert their effect render the method less precise. Another case when a protein-detected method could be preferred is when the target protein is relatively small and the structural information needed later in the project is to come from the use of NMR spectroscopy and not X-ray crystallography. Even though protein-detected NMR techniques generally are not the first choice for the primary fragment screening, they could prove very valuable at later stages to provide more precise information regarding binding sites. In this context, an approach where careful analysis of the protein chemical shift perturbations leads to a more accurately characterized ligand-binding site should be mentioned.[111] Most drug-like molecules contain aromatic rings that have an effect on the protein chemical shifts upon binding. By quantifying the spatial dependence of the ligand ring current field, more precise information on the binding geometry can be obtained.

The ligand-detected NMR techniques for primary fragment screening described above can all be used when a suitable spy molecule is available. The screening will then detect fragments able to displace the spy molecule, most likely due to overlapping binding sites. If there is no suitable spy molecule available, the choice of a ligand-detected NMR technique for primary fragment screening is in practice limited to four techniques: transverse relaxation filter, STD, WaterLOGSY or SLAPSTIC. SLAPSTIC requires the covalent attachment of a spin label on a protein side-chain at or near the binding site of interest. This leads to several other requirements for the use of SLAPSTIC in primary fragment screening: (i) detailed knowledge of the 3D structure of the target protein; (ii) the presence of a protein side-chain in the vicinity of the binding site of interest amenable to spin labeling; (iii) the attachment of the spin-label should not affect the structural integrity or the binding properties of the target protein. At least for targets for which there is no detailed structural information, it is difficult to say if these requirements are fulfilled. Therefore, to select a screening technique where the target protein can be used without modification will be very tempting. The other three techniques are all straightforward to use for fragment screening of novel targets. The related STD and WaterLOGSY techniques appear to be currently the most popular techniques, one major reason being the relatively low protein consumption. Typical protein and fragment concentrations are 0.5–5 and 50–200 μM, respectively. This large ligand excess is a potential drawback since the ligands then may start to populate nonspecific, low-affinity binding sites. The transverse relaxation filter experiment, on the other hand, does not use a very high ligand molar excess. Compared with STD and WaterLOGSY, the protein concentrations used in fragment screening are generally higher, but the fragment concentrations are also significantly lower. Typical protein and fragment concentrations are 4–10 and 20–40 μM, respectively. With these concentrations, transverse relaxation-filtered spectra with a sufficient signal-to-noise ratio can be obtained in as short time as 10–15 min, which is very competitive compared with STD and WaterLOGSY. The throughput is, of course, hampered by the need to collect reference spectra in the absence of protein. However, if using the same ligand concentrations, the reference spectra may not have to be collected for every NMR screening campaign. Protein consumption for these techniques may or may not be a critical concern. As discussed above, if structure-based drug design by X-ray crystallography is going to be used to drive the chemistry after the fragment screening, large amounts of target protein will be necessary in any case. With a target protein that is readily produced in large quantities, the question of whether the NMR fragment screen will require 1 or 3 mg of target protein may not be a burning issue. When the availability of target protein is scarce, recycling of protein should be considered. After the fragment screen (or part of the fragment screen), the samples are collected, pooled and subjected to exhaustive dialysis in order to remove the fragments and DMSO. A prerequisite is high stability of the target protein. In this context, the lower fragment concentrations used with the transverse relaxation filter technique is advantageous compared with STD and WaterLOGSY.

4.5.4 Competition Experiments

The purpose of a competition experiment is to obtain information on the binding site of fragment hits and to rank them with respect to affinity. If a protein-detected method has been used, this information is already at hand. The same applies if displacement of a spy molecule has been monitored in the primary fragment screen. Otherwise, either a previously known high affinity binder with known binding site (e.g. staurosporine for many protein kinases) or a weakly binding spy molecule (e.g. ATP for protein kinases)[112] can be used. The fragment hits binding to the same binding site, can be ranked with respect to binding affinity (see Figure 4.3). If the K_d of the known binder, here denoted I, is known (e.g. from an isothermal titration calorimetry measurement), then it is possible to estimate the K_d for the primary hits, denoted L, by assessing the displacement caused by the addition of the known binder, by using the relation[6, 113] $K_d(L) = [L]K_I/(I_{50} - K_I)$, where K_I is the dissociation constant of the known binder used for competition and I_{50} is the concentration of the known binder that causes the bound population of L to decrease to half of what it was in the absence of the known binder. The concentration [L] can be approximated to the total ligand concentration in cases where L has been added to the sample in large molar excess compared with the target. Usually, however, the absolute values of primary hit dissociation constants are not very important, but rather the relative affinities.

The competition experiments also serve to find out whether the primary hits bind specifically or nonspecifically. Fragments for which it is not possible to compete out by an excess of a known high-affinity binder bind either specifically to another site or nonspecifically to many hydrophobic patches on the protein surface. Alternatively, nondenaturing ESI-MS may have to be applied to find out whether fragments that are not displaced by a known binder bind specifically to another site or if they are nonspecific binders. If there are fragment hits that bind specifically to nonoverlapping binding sites, it is worthwhile to test if it is possible to observe interligand NOEs, especially if it has not been possible to determine any structures of fragment–target complexes. The observed interligand NOEs show that the fragments bind simultaneously at adjacent sites and the observed NOEs should give hints on how to link the two fragments or expand one of them.[79, 80] Fast exchanging ligands with overlapping binding sites can also exhibit interligand NOEs that are mediated via protons in the target protein.[114] It is possible to exploit these interligand NOEs to determine the relative orientation of the two competing ligands in the binding pocket.

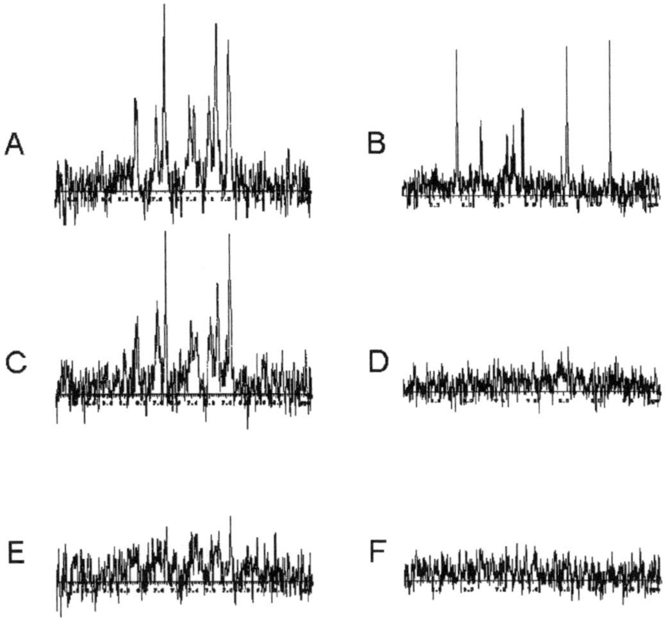

Figure 4.3 *STD competition experiments using a compound with known binding site to rank the binding affinity of fragment hits or fragment analogues to a target protein. (A) and (B) are the aromatic parts of STD spectra of two different fragment hits in the absence of competing compound. In the STD spectra (C) and (D), the competing compound is present in equimolar amounts compared to the fragment hits, and in (E) and (F), the competing compound is present in excess over the fragment hits. From this series of experiments, it is possible to conclude that the fragment hit in the STD spectra to the left (A, C and E) binds to the target with a higher affinity than the fragment hit in the STD spectra to the right. Further, since it is possible to displace fully both fragment hits with the competing compound, the fragment hits bind specifically to the target protein and, most probably, at a site overlapping with the competing compound.*

4.6 Cases

This section provides descriptions of two cases of fragment-based screening campaigns performed at Biovitrum. The first case, where the adipocyte fatty acid binding protein (A-FABP) was the target protein, is described in some detail,[24] whereas the second example is described in more general terms. Both target proteins (A-FABP and the Ser/Thr kinase) were subjected to both HTS and fragment screening by NMR. The fragment screening started at the same time as assay development for HTS was initiated, i.e. as soon as soluble target protein had been produced. Since no assay development or formatting is necessary to start fragment screening by NMR, the results presented here were available before the HTS was completed. The goal with the fragment screenings was to find soluble compounds with low micromolar or better potencies and high ligand efficiencies. The further development of these compounds was then performed together with the HTS hits by the respective project teams.

4.6.1 Expansion of Primary Fragment Hit – A-FABP

Fatty acid binding proteins (FABPs) constitute a family of homologous proteins, all of which reversibly bind long-chain fatty acids, bile acids and retinoic acid with high affinity.[115–117] Members of this protein family have MWs of 14–15 kDa and a prominent feature is a large and well-defined binding pocket that can accommodate these lipophilic ligands. Although the exact physiological role of the FABPs is still not clearly understood, they are thought to be involved in the storage, transport and targeting of fatty acids to appropriate locations within the cells.[118–121] A-FABP, for example, is known to interact physically with hormone-sensitive lipase,[122] possibly to enable A-FABP to bind and transport the product of the enzymatic reaction. In 1996, Hotamisligil *et al.* found that mice with a disruption in the gene encoding A-FABP develop dietary obesity comparable to wild-type animals but, unlike control mice, do not develop insulin resistance or diabetes.[123] This was further supported by data in a genetically obese mouse model.[124] Based on these results, it was hypothesized that blocking of A-FABP may represent a route to prevent obesity-induced insulin resistance.

In the A-FABP fragment screening campaign,[24] transverse relaxation-filtered ^{1}H 1D NMR was employed as the binding assay, a fluorescence polarization assay utilizing a fatty acid analogue with a fluorescent group was used to determine potency and X-ray crystallography was used to obtain structural information. The fragment library that was screened contained 531 diverse fragments divided into 57 cocktails. The relaxation filter spectra for one cocktail containing 10 fragments are shown in Figure 4.4, together with the identified hit BVT.1960. A total of 52 primary fragment hits were detected. The hit criterion used was $>80\,\%$ reduction in signal intensity in the presence of A-FABP as compared with the reference spectrum when applying a spinlock time of 400 ms. The primary hits were confirmed by repeating the relaxation filter experiment with two spinlock times (100 and 400 ms) on the individual compounds This allowed a classification of the hits into weaker (14) and stronger (38) binders. Most of the initial hits (43 out of 52, 83 %) comprised a CO_2^-, SO_3^- or PO_3^- group, indicating that they are fatty acid mimics. This feature was further accentuated in the strong binder category, where 95 % (36 out of 38) of the hits comprised a CO_2^-, SO_3^- or PO_3^- group. It was possible to obtain X-ray crystallographic structures of A-FABP in complex with several of the primary fragment hits.

The left panel of Figure 4.5 shows a region of the crystal structure of A-FABP in complex with the primary screening hit BVT.1960. As in previously published structures of ligand–target complexes between FAPB and fatty acids,[115–117] an interaction between the carboxylate group of the NMR hit and a conserved tyrosine side-chain (Tyr128) is clearly observed. In addition, a novel interaction, not found in complexes between FABPs and fatty acids, was found for a distal hydroxyl group, which forms a hydrogen bond to the side-chain of Asp76. More than 30 structures of ligand–target complexes of human A-FABP were determined and an outstanding feature of these structures is how the ligands cluster in one region of the binding pocket (see the right panel of Figure 4.5). Noteworthy is that there are a large number of conserved water molecules in the binding pocket and only a few of these are displaced by ligands.

A-FABP belongs to a large family of fatty acid binding proteins and selectivity against particularly the heart and muscle isoform (H-FABP) is considered important. Mice with a disruption of the gene encoding H-FABP have been reported to suffer from stress

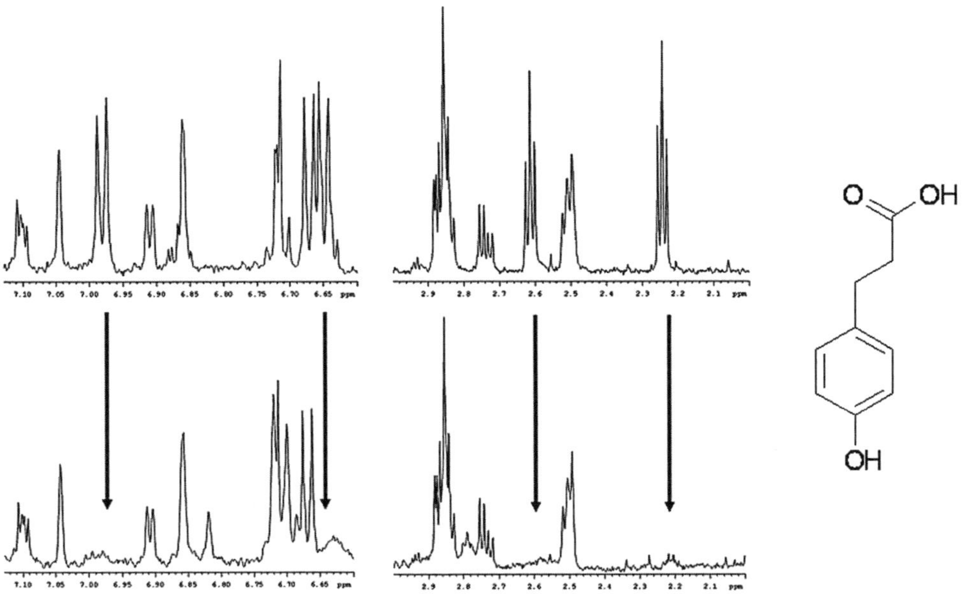

Figure 4.4 *^{1}H 1D relaxation filter spectra of part of the aromatic region (left) and the aliphatic region (right) for a cocktail containing 10 compounds in the absence (top) and presence (bottom) of A-FABP. The two doublet signals at 6.64 and 6.98 ppm and the two triplets at 2.25 and 2.62 ppm in the reference spectrum (top) originate from BVT.1960 (structure shown to the right). These signals disappear when A-FABP is present in the sample, as pointed out by the arrows. The signals from the nonbinding compounds in the mixture remain unaffected. The spinlock time in this experiment was 400 ms. Adapted with permission from van Dongen* et al., J. Am. Chem. Soc., ***124****, 11874. Copyright 2002 American Chemical Society.*

intolerability, in some cases even leading to death.[125] The potencies of the highest ranked NMR hits were therefore measured and compared between A-FABP and H-FABP using the fluorescence polarization assay. BVT.1960 gave an IC_{50} value of 0.59 mM for A-FABP and it was shown to be at least 25 times selective over H-FABP (IC_{50} = 30 mM). This selectivity is remarkable for such a small compound with a molecular weight of only 166 g mol^{-1}. The amino acid sequences of A-FABP and H-FABP are 65 % identical[126] with only a few substitutions in the ligand-binding pocket. With the structure of the complex between A-FABP and BVT.1960 at hand, a comparison of the amino acid sequences of the two proteins reveals that the selectivity of BVT.1960 against H-FABP can be attributed to the effect of Val115–Leu and Cys117–Leu substitutions (illustrated in Figure 4.5). Presumably, the additional bulkiness introduced by these substitutions hampers the beneficial van der Waals interactions experienced by one of the flanks of BVT.1960.

Because of the >25 times selectivity for A-FABP over H-FABP and the availability of a structure of a ligand–target complex, the BVT.1960 scaffold was chosen as a starting point for testing of compound analogues. Based on the structure of the ligand–target complex, it can be concluded that there should be room for extension of BVT.1960 on the opposite side to the one interacting with Val115 and Cys117 (Figure 4.5). Substructure searches in

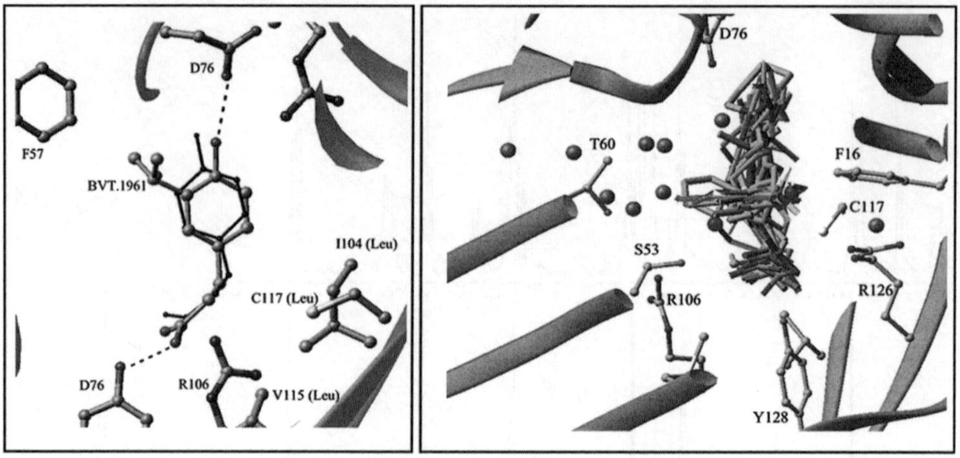

Figure 4.5 *Left: crystal structure of the ligand-binding site of human A-FABP. The structure of one of the primary hits from the NMR screen (BVT.1960 shown in thin black lines) was used to design a higher affinity ligand (BVT.1961 shown in thicker lines) with a similar binding mode and retained selectivity against H-FABP. There are three amino acid residues (Ile104, Val115 and Cys117) in the vicinity of these ligands that are different between A-FABP and H-FABP. The bulkier leucine side-chains of H-FABP in the corresponding positions could account for its lower affinity for these ligands. Adapted with permission from van Dongen* et al., J. Am. Chem. Soc., ***124****, 11874. Copyright 2002 American Chemical Society. Right: a compounded figure of several ligands in complex with human A-FABP. The ligands all cluster in one region of the binding pocket, leaving most of the solvent molecules (spheres) unperturbed. Approximately 10 water molecules in the binding pocket are conserved in all structures of the ligand–target complexes.*

the Available Chemicals Directory (ACD-3D, MDL Information Systems) and the in-house compound collection were conducted to find compounds with the BVT.1960 scaffold and small extensions at different positions. Eleven compounds were identified and tested in the fluorescence polarization assay against both A-FABP and H-FABP. As illustrated in Figure 4.6, it was discovered that substitutions at the ethylene linker carbons resulted in a loss of selectivity, whereas replacing the *p*-hydroxyl group on the aromatic ring resulted in decreased potency. Substitutions at the *ortho* and *meta* positions with respect to the ethylene linker, on the other hand, appeared to have the capacity to combine improved potency with retained selectivity.

On the basis of these findings, 12 additional analogues to BVT.1960, mainly differing in substitutions on the phenyl ring, were synthesized. Binding studies by NMR and potency ranking using the fluorescence polarization assay for these compounds distinctly pointed to a beneficial effect of substitutions at the *meta* (R_2) position. As is shown in Figure 4.7, an isopropyl substitution to produce BVT.1961 combined a potency of 10 μM with a completely retained selectivity for A-FABP over H-FABP. An X-ray structure confirmed that BVT.1961 adopts an equivalent binding mode as BVT.1960 (Figure 4.5). In addition, two compounds, with a benzyl or phenone group at the *meta* position, showed potencies in the low μM range.

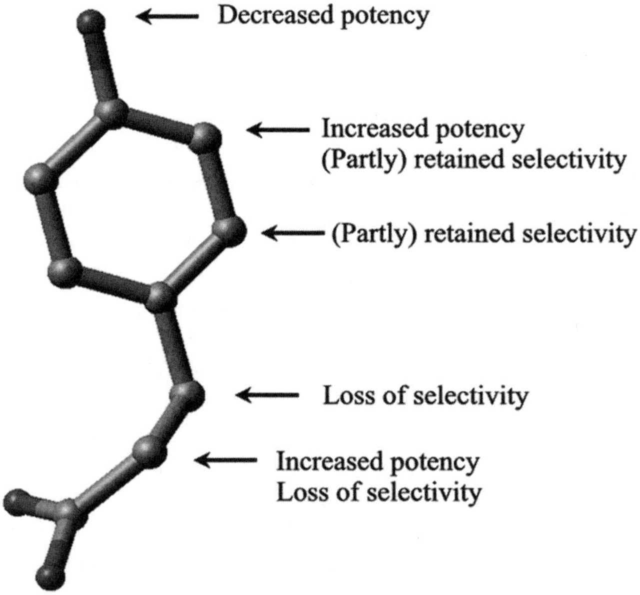

Figure 4.6 *A schematic illustration of what impact various substitutions on BVT.1960 had on the potency on A-FABP and selectivity against H-FABP. These findings were based on 11 close analogues to BVT.1960. Adapted with permission from van Dongen* et al., J. Am. Chem. Soc., ***124****, 11874. Copyright 2002 American Chemical Society.*

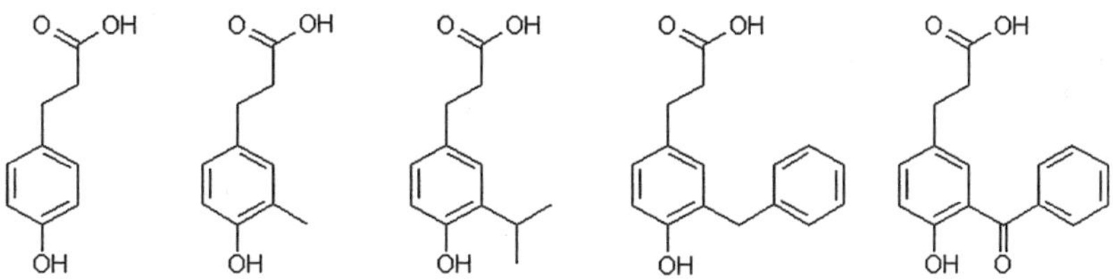

IC_{50}(μM):	590	260	10	5.5	4.6
Selectivity:	>25	9	26	10	17
MW:	166	184	208	256	270
LE (kcal/mol):	0.37	0.41	0.46	0.38	0.36
BVT #:	1960	4096	1961	1963	1962

Figure 4.7 *Summary of the measured potencies for A-FABP and selectivities against H-FABP for analogues of BVT.1960. The numbers are based on measurements using the fluorescence polarization assay. Adapted with permission from van Dongen* et al., J. Am. Chem. Soc., ***124****, 11874. Copyright 2002 American Chemical Society.*

In summary, the fragment screening campaign very efficiently generated a series of three compounds for A-FABP which fulfill the hit criteria that are normally set in an HTS, i.e. at least 10 μM potency and reasonable selectivity. In addition, these compounds are small, have high ligand efficiencies and are highly soluble. Hence they allow for further expansion, in order to generate molecules with improved affinity and selectivity, without the immediate risk of obtaining drug candidates too large and lipophilic to become optimal drugs. The generation of highly soluble hits makes it possible to obtain structures of ligand–target complexes very early in the drug discovery process (pre-HTS in this case) and learn more about key interactions in the binding pocket, which promises to streamline the upcoming optimization efforts of compounds originating both from fragment screens and HTS. Furthermore, the compounds generated from fragment screening could constitute very different classes of compounds to what is obtained from screening of traditional HTS libraries. In this context, it is interesting to note that Vertex also has published results of an NMR screen using the SHAPES library of A-FABP.[6] The primary screen yielded 13 hits with affinities ranging from 0.3 to 800 μM and two crystal structures of ligand–target complexes were obtained. Based on the structural information, 134 commercially available analogues were purchased and tested in a calorimetric screen. Nine compounds with low micromolar to nanomolar potencies were identified and they were different from those presented here. As was noted by Lepre *et al.*,[6] although both groups applied very similar approaches to exactly the same target protein, the resulting lead series were very different. This illustrates that fragment-based methods are particularly effective for discovering *novel* leads.

4.6.2 Expansion of Primary Fragment Hit – Ser/Thr Kinase

In general, it is difficult to develop binding fragments into high potency lead molecules without structural information such as crystal structures of fragment–target complexes. This example, however, serves to illustrate that in favorable cases it is possible to obtain promising hit and lead series by fragment-based screening despite a lack of both structural information and synthetic chemistry resources.

Protein kinases have become the second most important class of drug targets, after G-protein-coupled receptors.[127] The main therapeutic areas include oncology, inflammation and metabolic diseases. As most protein kinase inhibitors target the ATP binding site, not only specificity but also intellectual property is an important issue. There are several compound libraries targeted at the ATP binding site of protein kinases on the market, but there are also many pharmaceutical companies using these libraries, leading to a crowded patent situation. With this in mind, fragment-based screening should have an important role to play in kinase inhibitor discovery, especially when it comes to finding novel and patentable scaffolds.

The Ser/Thr kinase target was a novel target: no small molecule inhibitors had been published and no crystal structure was available. The fragment screening by NMR took place before the HTS campaign but there was a low to medium throughput kinase activity assay (a nonhomogeneous scintillation assay using ATP-γ-^{33}P) available that was not compatible with HTS. Using this assay, it was shown that the nonselective protein kinase inhibitor staurosporine is a highly potent inhibitor of this Ser/Thr kinase and could therefore be used for competition experiments. The NMR screen using STD identified 32 binding fragments which were tested in the kinase activity assay at two concentrations, 250 and 500 μM. Six of the binding fragments showed an inhibition of >50 % at 250 μM concentration and an additional five fragments showed >50 % inhibition at 500 μM. All the binding fragments showed at least some inhibition at 500 μM, with the exception of one that appeared to be an activator. Dose–response curves were collected for the most potent inhibitor fragments and the two most potent fragment inhibitors showed remarkable 0.8 and 13 μM potencies. This translated to ligand efficiencies of 0.55 and 0.51 kcal mol^{-1}, respectively, indicating very favorable binding interactions. The most potent fragment hits were shown to compete for binding with the known ATP-pocket binder staurosporine. Despite considerable efforts, no crystal structure of this target kinase was obtained.

In order to check if it was possible to optimize the binding scaffolds, close fragment analogues to the two most potent fragment hits were tested in the activity assay. Several of these analogues showed inhibition but none was more potent than the original fragment hits, which was hardly surprising considering their high ligand efficiency. The in-house compound collection and commercial sources were then searched for larger compounds in which the most potent fragment hits are present as a substructure (exact or very similar to the fragment hit). Seven and 12 compounds were found for the 0.8 and 13 μM fragments, respectively. Among the seven analogues for the 0.8 μM fragment, one was active (IC_{50} = 4 μM; LE = 0.28 kcal mol^{-1}), but this compound was fairly large and would probably require cumbersome chemistry to develop, so this track was dropped. The substructure analogues of the 13 μM fragment hit looked more promising. Of the 12 analogues tested, three showed good potency (IC_{50} = 1, 2 and 6 μM; LE = 0.33, 0.35 and 0.34 kcal mol^{-1}, respectively) and competed for binding with staurosporine, as seen by STD (see Figure 4.8). These three compounds were diverse, highly soluble and would constitute the first compounds in three series. Especially the 1 μM analogue was considered very interesting since there was no mentioning in the literature of this type of compound. Further, the compound showed significant inhibition (>50 % inhibition at 10 μM) to only one other kinase in a selectivity panel consisting of 30 diverse kinases, which compared favorably with nearly all of the compounds found by the HTS campaign. In conclusion, fragment screening coupled with a standard biochemical kinase assay succeeded in identifying several lead series, including a truly novel lead series, to a Ser/Thr kinase target despite the fact that no structural information and no synthetic chemistry resources were used. Only a small number of in-house and commercially available analogues had been tested up to this stage. This gave a significant boost to the outcome of the HTS from which three promising hit series had been identified, all from kinase targeted libraries.

4.6.3 Difficult Targets for Fragment-based Screening by NMR

As discussed above, fragment-based screening does in general succeed in finding good lead series with high solubility and ligand efficiency as well as novel binding scaffolds. However, there are types of targets with associated difficulties.

Target proteins forming large multimeres. It is important to ascertain that the target protein is monodisperse at the protein concentration and in the buffer that is going to be used for the screening. In one case with a large protease target, this was not properly checked before the fragment screen. STD was used to screen fragments at a protein and fragment concentration of 1 and 100 μM, respectively. The hit rate turned out to be extremely high,

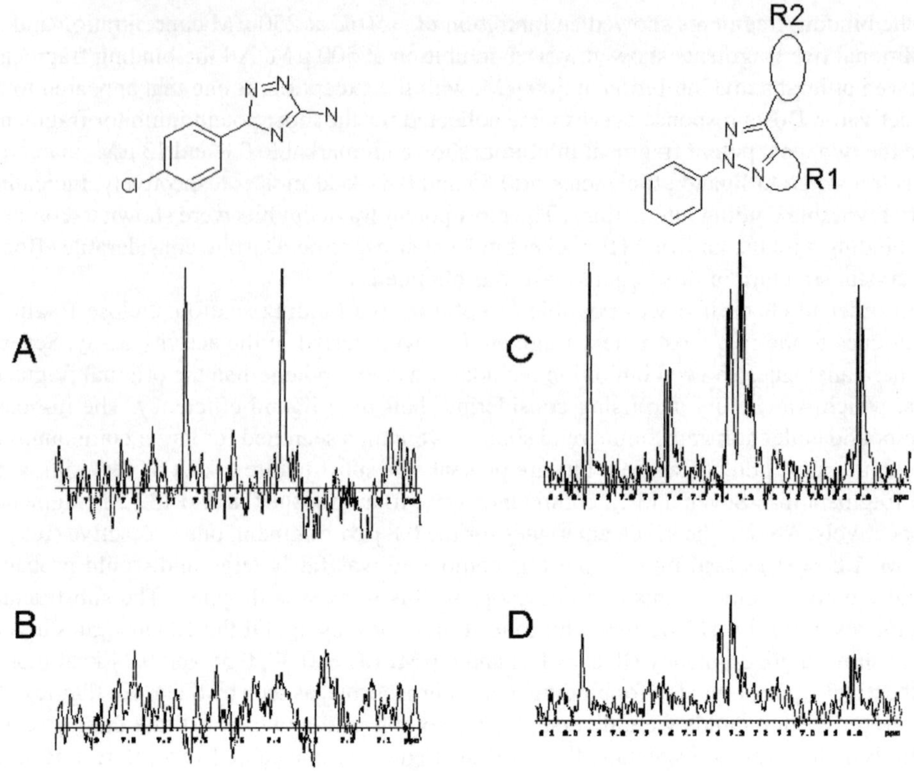

Figure 4.8 *The aromatic part of STD-NMR spectra of (A) 2 μM kinase + 50 μM fragment hit; (B) 2 μM kinase + 50 μM fragment hit + staurosporine; (C) 2 μM kinase + 50 μM analogue hit; (D) 2 μM kinase + 50 μM analogue hit + staurosporine. The fragment hit (IC_{50} = 13 μM) is shown at top left and a sketch of the fragment hit analogue (IC_{50} = 1 μM), containing a substructure very similar to the fragment hit, is shown at top right. As seen by the decrease in the STD signals on addition of staurosporine, both the fragment hit and the analogue compete for binding with staurosporine.*

well over 20 %, and none of the hits were completely competed out using a previously known active site binder. A high degree of nonspecific binding was naturally suspected and analytical ultracentrifugation indirectly confirmed the suspicion, showing that the protease was present in solution as a mixture of monomers, dimers, tetramers and octamers, the last approaching 1 MDa in weight. Since spin diffusion becomes more efficient with slower tumbling times, STD will in general detect increasingly weaker binders as the target molecular weight increases. Vigilance should in general be exercised when screening very large proteins by NMR, even when they are monodisperse, since large proteins offer many more hydrophobic patches that can serve as low-affinity binding sites for organic molecules than smaller proteins. In this case, it was not possible to select rationally fragment hits to develop since no crystal structure was at hand and all fragment hits showed very low and similar potencies in an activity assay. To proceed, we would have to optimize the buffer conditions with the goal of obtaining monomeric protein and then rerun the fragment screen.

Unfortunately, it was very difficult to produce larger amounts of the target protein and the fragment-based screening project was stopped at this point.

It is also important to test the temperature stability of the target protein before the fragment screening. Partly unfolded proteins will transiently expose hydrophobic patches to the solvent (and the fragments) to a much higher degree than fully folded proteins and will therefore pick up more nonspecific binders. Partly unfolded proteins may also form large, but soluble, aggregates.

Target proteins with large solvent-exposed binding pockets. Large amounts of bound water in the targeted binding pocket will pose a special problem for fragment binding. The binding affinity of fragments may not be sufficient to displace the bound water in a solution where the water concentration is ~55 M. This will result in a very low hit rate. Our experiences include a phosphatase and a protease, both with large solvent-exposed active sites, and hit rates below 1%. The low hit rate can be counteracted to some extent by using higher protein and fragment concentrations and also by performing the NMR screen at low temperature in order to increase the tumbling time of the protein.

4.7 Future Directions

4.7.1 Multitarget Drugs

Due to the redundancy in biological networks, modulating the function of multiple protein targets simultaneously can be beneficial for treating complex diseases.[128, 129] Many currently marketed drugs act via multiple targets, but the discovery of their multiple mechanism of action was usually serendipitous and retrospective. Compounds that are designed to modulate functions of several targets are in general larger and more lipophilic than compounds in the clinic or in the market.[130, 131] The reason for this is the current 'framework combination strategy', where two selective ligands are combined into one dual ligand, which most likely will contain features that are important for one of the target proteins only. The result of this strategy is contradictory to the notion that larger compounds are more selective than smaller compounds,[130, 132] i.e. multitarget compounds should typically be smaller than target-selective compounds. Fragment-based screening is the ideal approach to find a core scaffold capable of binding to two or several target proteins. The core scaffold could then be optimized to a compound with appropriately balanced affinities between the target proteins. Naturally, the greatest likelihood of success would be when the target proteins share a conserved binding site, e.g. kinases. The RAMPED-UP NMR method,[133] where several differently labeled target proteins can be screened for binders simultaneously, will probably prove to be useful in finding multitarget core scaffolds.

4.7.2 Membrane Protein Targets

Currently published examples of fragment-based screening by NMR are only applied to soluble target proteins. There are, however, NMR techniques that have been used to detect the binding of small molecules to integral membrane proteins reconstituted in liposomes. STD has been used to detect the binding of peptides to integrin embedded in liposomes[100] and the bound conformation of a GPCR-bound peptide was determined by transferred NOE experiments.[134] The problem is rather the difficulty in producing and successfully reconstituting milligram amounts of therapeutically interesting integral membrane proteins, e.g. GPCRs, membrane-bound enzymes (e.g. tyrosine kinases) and ion channels. In recent years, the number of crystal structures of integral membrane proteins has increased dramatically; especially notable is a second high-resolution GPCR structure.[135] This is mainly a consequence of the impressive progress in the protein production process[136] and leaves good hope that integral membrane protein targets will be subjected to fragment-based screening by NMR and structure-based drug design in the very near future.

The integral membrane proteins that will prove most suitable for fragment screening by NMR can be predicted (i) to be possible to produce in milligram quantities and reconstitute in liposomes and (ii) not to contain large amounts of detergents in the binding pocket of interest, i.e. the binding pocket should be hydrophilic. If the binding pocket is lipophilic, the fragments will probably not be able to displace the detergent molecules that could bind with relatively high affinity. A method that appears to be very promising for target proteins that are difficult to produce or that are insoluble, such as membrane proteins, is called TINS (target-immobilized NMR screening).[137, 138] Binding is detected by comparing ^{1}H 1D spectra in the presence and absence of the target protein that is immobilized on a solid support. The compounds are pumped through a dual flow cell and binding is detected as a simple reduction in ligand peak height in the cell with target protein present. The binders are then washed off and the experiment is repeated with the next fragment cocktail using the same immobilized protein. Hence only a single protein sample is required to screen the fragment library. The use of paramagnetic spin labels, attached either to the target protein in the vicinity of the binding pocket of interest or to a known ligand, can also be predicted to play a role in fragment screening of membrane proteins.

4.8 Acknowledgements

I am grateful to many colleagues, past and present, for providing a stimulating and fun environment for the daily work. Thanks are due to everyone at iNovacia and all those from the former Structural Chemistry Department at Biovitrum. I would, however, like to especially acknowledge the following individuals: the NMRers, Maria van Dongen, Johan Weigelt, Tomas Åkerud, Mats Wikström and Toshiaki Nishida; mass spectrometry, Agneta Tjernberg; dynamic light scattering, analytical ultracentrifugation, microcalorimetry and other biophysical techniques, Natalia Markova, Carina Norström and Dan Hallén; X-ray crystallography, Jonas Uppenberg, Stefan Svensson and Derek Ogg; medicinal chemistry, Jan Vågberg, Styrbjörn Byström, Wei Berts, Annika Jenmalm and Katarina Roos; computational chemistry, Micael Jacobsson, Mats Kihlén, Anna-Lena Gustavsson, Evert Homan, Jerk Vallgårda and René Avontuur; and biochemical assays, Thomas Lundbäck and Eva-Maria Axén.

References

[1] Rees, D. C., *et al.*, Fragment-based lead discovery. *Nat Rev Drug Discov*, 2004, **3**, 660–672.
[2] Erlanson, D. A., *et al.*, Fragment-based drug discovery. *J Med Chem*, 2004, **47**, 3463–3482.
[3] Carr, R. A., *et al.*, Fragment-based lead discovery: leads by design. *Drug Discov Today*, 2005, **10**, 987–992.

[4] Hajduk, P. J. and Greer, J., A decade of fragment-based drug design: strategic advances and lessons learned. *Nat Rev Drug Discov*, 2007, **6**, 211–219.

[5] Zartler, E. R. and Shapiro, M. J., Fragonomics: fragment-based drug discovery. *Curr Opin Chem Biol*, 2005, **9**, 366–370.

[6] Lepre, C. A., *et al.*, Theory and applications of NMR-based screening in pharmaceutical research. *Chem Rev*, 2004, **104**, 3641–3676.

[7] Hann, M. M., *et al.*, Molecular complexity and its impact on the probability of finding leads for drug discovery. *J Chem Inf Comput Sci*, 2001, **41**, 856–864.

[8] Lipinski, C. and Hopkins, A., Navigating chemical space for biology and medicine. *Nature*, 2004, **432**, 855–861.

[9] Andrews, P. R., *et al.*, Functional group contributions to drug-receptor interactions. *J Med Chem*, 1984, **27**, 1648–1657.

[10] Hopkins, A. L., *et al.*, Ligand efficiency: a useful metric for lead selection. *Drug Discov Today*, 2004, **9**, 430–431.

[11] Lipinski, C. A., *et al.*, Experimental and computational approaches to estimate solubility and permeability in drug discovery and development settings. *Adv Drug Deliv Rev*, 1997, **46**, 3–26.

[12] Kuntz, I. D., *et al.*, The maximal affinity of ligands. *Proc Natl Acad Sci USA*, 1999, **96**, 9997–10002.

[13] Abad-Zapatero, C. and Metz, J. T., Ligand efficiency indices as guideposts for drug discovery. *Drug Discov Today*, 2005, **10**, 464–469.

[14] Dalvit, C., *et al.*, NMR-based quality control approach for the identification of false positives and false negatives in high throughput screening. *Curr Drug Discov Technol*, 2006, **3**, 115–124.

[15] Teague, S. J., *et al.*, The design of leadlike combinatorial libraries. *Angew Chem Int Ed*, 1999, **38**, 3743–3748.

[16] Wenlock, M. C., *et al.*, A comparison of physiochemical property profiles of development and marketed oral drugs. *J Med Chem*, 2003, **46**, 1250–1256.

[17] Tjernberg, A., *et al.*, Mechanism of action of pyridazine analogues on protein tyrosine phosphatase 1B (PTP1B). *Bioorg Med Chem Lett*, 2004, **14**, 891–895.

[18] Lundqvist, T., The devil is still in the details – driving early drug discovery forward with biophysical experimental methods. *Curr Opin Drug Discov Devel*, 2005, **8**, 513–519.

[19] Dalvit, C., *et al.*, A general NMR method for rapid, efficient and reliable biochemical screening. *J Am Chem Soc*, 2003, **125**, 14620–14625.

[20] Hajduk, P. J., *et al.*, Novel inhibitors of Erm methyltransferases from NMR and parallel synthesis. *J Med Chem*, 1999, **42**, 3852–3859.

[21] Boehm, H. J., *et al.*, Novel inhibitors of DNA gyrase: 3D structure based biased needle screening, hit validation by biophysical methods and 3D guided optimization. A promising alternative to random screening. *J Med Chem*, 2000, **43**, 2664–2674.

[22] Hajduk, P. J., *et al.*, Identification of novel inhibitors of urokinase via NMR-based screening. *J Med Chem*, 2000, **43**, 3862–3866.

[23] Erlanson, D. A., *et al.*, Site-directed ligand discovery. *Proc Natl Acad Sci USA*, 2000, **97**, 9367–9372.

[24] van Dongen, M. J., *et al.*, Structure-based screening as applied to human FABP**4**, a highly efficient alternative to HTS for hit generation. *J Am Chem Soc*, 2002, **124**, 11874–11880.

[25] Wendt, M. D., *et al.*, Identification of novel binding interactions in the development of potent, selective 2-naphthamidine inhibitors of urokinase. Synthesis, structural analysis and SAR of *N*-phenyl amide 6-substitution. *J Med Chem*, 2004, **47**, 303–324.

[26] Gill, A. L., *et al.*, Identification of novel p38alpha MAP kinase inhibitors using fragment-based lead generation. *J Med Chem*, 2005, **48**, 414–426.

[27] Joshi, M., *et al.*, Discovery of low-molecular-weight ligands for the AF6 PDZ domain. *Angew Chem Int Ed*, 2006, **45**, 3790–3795.

[28] Congreve, M., *et al.*, Application of fragment screening by X-ray crystallography to the discovery of aminopyridines as inhibitors of beta-secretase. *J Med Chem*, 2007, **50**, 1124–1132.

[29] Huth, J. R., *et al.*, Discovery and design of novel hsp90 inhibitors using multiple fragment-based design strategies. *Chem Biol Drug Des*, 2007, **70**, 1–12.

[30] Geschwindner, S., *et al.*, Discovery of a novel warhead against beta-secretase through fragment-based lead generation. *J Med Chem*, 2007, **50**, 5903–5911.

[31] Edwards, P. D., *et al.*, Application of fragment-based lead generation to the discovery of novel, cyclic amidine beta-secretase inhibitors with nanomolar potency, cellular activity and high ligand efficiency. *J Med Chem*, 2007, **50**, 5912–5925.

[32] Shuker, S. B., *et al.*, Discovering high-affinity ligands for proteins: SAR by NMR. *Science*, 1996, **274**, 1531–1534.

[33] Hajduk, P. J., *et al.*, Discovery of potent nonpeptide inhibitors of stromelysin using SAR by NMR. *J Am Chem Soc*, 1997, **119**, 5818–5827.

[34] Pellecchia, M., *et al.*, NMR-based structural characterization of large protein–ligand interactions. *J Biomol NMR*, 2002, **22**, 165–173.

[35] Jahnke, W., *et al.*, Second-site NMR screening and linker design. *Curr Top Med Chem*, 2003, **3**, 69–80.

[36] Liu, G., *et al.*, Fragment screening and assembly: a highly efficient approach to a selective and cell active protein tyrosine phosphatase 1B inhibitor. *J Med Chem*, 2003, **46**, 4232–4235.

[37] Wyss, D. F., *et al.*, Non-peptidic small-molecule inhibitors of the single-chain hepatitis C virus NS3 protease/NS4A cofactor complex discovered by structure-based NMR screening. *J Med Chem*, 2004, **47**, 2486–2498.

[38] Raimundo, B. C., *et al.*, Integrating fragment assembly and biophysical methods in the chemical advancement of small-molecule antagonists of IL-2, an approach for inhibiting protein–protein interactions. *J Med Chem*, 2004, **47**, 3111–3130.

[39] Howard, N., *et al.*, Application of fragment screening and fragment linking to the discovery of novel thrombin inhibitors. *J Med Chem*, 2006, **49**, 1346–1355.

[40] Huc, I. and Lehn, J. M., Virtual combinatorial libraries: dynamic generation of molecular and supramolecular diversity by self-assembly. *Proc Natl Acad Sci USA*, 1997, **94**, 2106–2110.

[41] Congreve, M. S., *et al.*, Detection of ligands from a dynamic combinatorial library by X-ray crystallography. *Angew Chem Int Ed*, 2003, **42**, 4479–4482.

[42] Hajduk, P. J., *et al.*, NMR-based discovery of lead inhibitors that block DNA binding of the human papillomavirus E2 protein. *J Med Chem*, 1997, **40**, 3144–3150.

[43] Fejzo, J., *et al.*, The SHAPES strategy: an NMR-based approach for lead generation in drug discovery. *Chem Biol*, 1999, **6**, 755–769.

[44] Lepre, C. A., *et al.*, Applications of SHAPES screening in drug discovery. *Comb Chem High Throughput Screen*, 2002, **5**, 583–590.

[45] Jahnke, W., *et al.*, NMR reporter screening for the detection of high-affinity ligands. *Angew Chem Int Ed*, 2002, **41**, 3420–3423.

[46] Kamionka, M., *et al.*, *In silico* and NMR identification of inhibitors of the IGF-I and IGF-binding protein-5 interaction. *J Med Chem*, 2002, **45**, 5655–5660.

[47] Fattorusso, R., *et al.*, Discovery of a novel class of reversible non-peptide caspase inhibitors via a structure-based approach. *J Med Chem*, 2005, **48**, 1649–1656.

[48] Hajduk, P. J., *et al.*, NMR-based modification of matrix metalloprotcinase inhibitors with improved bioavailability. *J Med Chem*, 2002, **45**, 5628–5639.

[49] Huth, J. R. and Sun, C., Utility of NMR in lead optimization: fragment-based approaches. *Comb Chem High Throughput Screen*, 2002, **5**, 631–643.

[50] Sun, C., *et al.*, NMR in pharmacokinetic and pharmacodynamic profiling. *Chembiochem*, 2005, **6**, 1592–1600.
[51] Tjernberg, A., *et al.*, Determination of dissociation constants for protein–ligand complexes by electrospray ionization mass spectrometry. *Anal Chem*, 2004, **76**, 4325–4331.
[52] He, Y., *et al.*, Synthesis and evaluation of novel bacterial rRNA-binding benzimidazoles by mass spectrometry. *Bioorg Med Chem Lett*, 2004, **14**, 695–699.
[53] Hofstadler, S. A. and Griffey, R. H., Analysis of noncovalent complexes of DNA and RNA by mass spectrometry. *Chem Rev*, 2001, **101**, 377–390.
[54] Swayze, E. E., *et al.*, SAR by MS: a ligand based technique for drug lead discovery against structured RNA targets. *J Med Chem*, 2002, **45**, 3816–3819.
[55] Graffinity. Graffinity's fragment based drug discovery process – RAISE. http://www. graffinity.com/t_elements.php
[56] Huber, W., A new strategy for improved secondary screening and lead optimization using high-resolution SPR characterization of compound-target interactions. *J Mol Recognit*, 2005, **18**, 273–281.
[57] Karlsson, R., *et al.*, Biosensor analysis of drug-target interactions: direct and competitive binding assays for investigation of interactions between thrombin and thrombin inhibitors. *Anal Biochem*, 2000, **278**, 1–13.
[58] Maly, D. J., *et al.*, Combinatorial target-guided ligand assembly: identification of potent subtype-selective c-Src inhibitors. *Proc Natl Acad Sci USA*, 2000, **97**, 2419–2424.
[59] Barker, J., Fragment screening by biochemical assay. *Expert Opin Drug Discov*, 2006, **1**, 225–236.
[60] Blundell, T. L., *et al.*, High-throughput crystallography for lead discovery in drug design. *Nat Rev Drug Discov*, 2002, **1**, 45–54.
[61] Bemis, G. W. and Murcko, M. A., The properties of known drugs. 1. Molecular frameworks. *J Med Chem*, 1996, **39**, 2887–2893.
[62] Bemis, G. W. and Murcko, M. A., Properties of known drugs. 2. Side chains. *J Med Chem*, 1999, **42**, 5095–5099.
[63] Lepre, C. A., Library design for NMR-based screening. *Drug Discov Today*, 2001, **6**, 133–140.
[64] Aronov, A. M. and Bemis, G. W., A minimalist approach to fragment-based ligand design using common rings and linkers: application to kinase inhibitors. *Proteins*, 2004, **57**, 36–50.
[65] Baurin, N., *et al.*, Design and characterization of libraries of molecular fragments for use in NMR screening against protein targets. *J Chem Inf Comput Sci*, 2004, **44**, 2157–2166.
[66] Schuffenhauer, A., *et al.*, Library design for fragment-based screening. *Curr Top Med Chem*, 2005, **5**, 751–762.
[67] Siegal, G., *et al.*, Integration of fragment screening and library design. *Drug Discov Today*, 2007, **12**, 1032–1039.
[68] Babaoglu, K. and Shoichet, B. K., Deconstructing fragment-based inhibitor discovery. *Nat Chem Biol*, 2006, **2**, 720–723.
[69] Markova, N. DynaPro DLS plate reader for buffer optimization, in *Wyatt Application Note 16*, 2005. http://www.mals.com/literature/appnotes/DynaProAN.cfm
[70] Meyer, B. and Peters, T., NMR spectroscopy techniques for screening and identifying ligand binding to protein receptors. *Angew Chem Int Ed*, 2003, **42**, 864–890.
[71] Wyss, D. F., *et al.*, NMR-based approaches for lead discovery. *Curr Opin Drug Discov Dev*, 2002, **5**, 630–647.
[72] Mayer, M. and Meyer, B., Characterization of ligand binding by saturation transfer difference NMR spectroscopy. *Angew Chem Int Ed*, 1999, **38**, 1784–1788.
[73] Dalvit, C., *et al.*, Identification of compounds with binding affinity to proteins via magnetization transfer from bulk water. *J Biomol NMR*, 2000, **18**, 65–68.

[74] Dalvit, C., *et al.*, WaterLOGSY as a method for primary NMR screening: practical aspects and range of applicability. *J Biomol NMR*, 2001, **21**, 349–359.

[75] Hajduk, P. J., *et al.*, One-dimensional relaxation- and diffusion-edited NMR methods for screening compounds that bind to macromolecules. *J Am Chem Soc*, 1997, **119**, 12257–12261.

[76] Dalvit, C., *et al.*, High-throughput NMR-based screening with competition binding experiments. *J Am Chem Soc*, 2002, **124**, 7702–7709.

[77] Lin, M., *et al.*, Diffusion-edited NMR – affinity NMR for direct observation of molecular interactions. *J Am Chem Soc*, 1997, **119**, 5249–5250.

[78] Ni, F., Recent developments in transferred NOE methods. *Prog. Nucl. Magn. Res.*, 1994, **26**, 517–606.

[79] Li, D., *et al.*, The inter-ligand Overhauser effect: a powerful new NMR approach for mapping structural relationships of macromolecular ligands. *J Biomol NMR*, 1999, **15**, 71–76.

[80] Becattini, B. and Pellecchia, M., SAR by ILOEs: an NMR-based approach to reverse chemical genetics. *Chemistry*, 2006, **12**, 2658–2662.

[81] Jahnke, W., *et al.*, Second-site NMR screening with a spin-labeled first ligand. *J Am Chem Soc*, 2000, **122**, 7394–7395.

[82] Jahnke, W., *et al.*, Spin label enhanced NMR screening. *J Am Chem Soc*, 2001, **123**, 3149–3150.

[83] Clore, G. M. and Gronenborn, A. M., Theory and application of the transferred nuclear Overhauser effect to the study of the conformations of small ligands bound to proteins. *J. Magn. Reson.*, 1982, **48**, 402–417.

[84] Jahnke, W., Spin labels as a tool to identify and characterize protein–ligand interactions by NMR spectroscopy. *ChemBioChem*, 2002, **3**, 167–173.

[85] Dalvit, C., *et al.*, Sensitivity improvement in 19F NMR-based screening experiments: theoretical considerations and experimental applications. *J Am Chem Soc*, 2005, **127**, 13380–13385.

[86] Dalvit, C., *et al.*, Fluorine-NMR competition binding experiments for high-throughput screening of large compound mixtures. *Comb Chem High Throughput Screen*, 2002, **5**, 605–611.

[87] Dalvit, C., *et al.*, Fluorine-NMR experiments for high-throughput screening: theoretical aspects, practical considerations and range of applicability. *J Am Chem Soc*, 2003, **125**, 7696–7703.

[88] Peng, J. W., Cross-correlated ^{19}F relaxation measurements for the study of fluorinated ligand–receptor interactions. *J Magn Reson*, 2001, **153**, 32–47.

[89] Dalvit, C., *et al.*, NMR-based screening with competition water-ligand observed via gradient spectroscopy experiments: detection of high-affinity ligands. *J Med Chem*, 2002, **45**, 2610–2614.

[90] Wang, Y. S., *et al.*, Competition STD NMR for the detection of high-affinity ligands and NMR-based screening. *Magn Reson Chem*, 2004, **42**, 485–489.

[91] Hoffman, R. A. and Forsén, S., High resolution nuclear magnetic double and multiple resonance *Prog. Nucl Magn Reson Spectrosc*, 1966, **1**, 15–204.

[92] Cutting, B., *et al.*, Sensitivity enhancement in saturation transfer difference (STD) experiments through optimized excitation schemes. *Magn Reson Chem*, 2007, **45**, 720–724.

[93] Mayer, M. and James, T. L., Detecting ligand binding to a small RNA target via saturation transfer difference NMR experiments in D_2O and H_2O. *J Am Chem Soc*, 2002, **124**, 13376–13377.

[94] Jayalakshmi, V. and Krishna, N. R., Complete relaxation and conformational exchange matrix (CORCEMA) analysis of intermolecular saturation transfer effects in reversibly forming ligand–receptor complexes. *J Magn Reson*, 2002, **155**, 106–118.

[95] Yan, J., *et al.*, The effect of relaxation on the epitope mapping by saturation transfer difference NMR. *J Magn Reson*, 2003, **163**, 270–276.

[96] Mayer, M. and Meyer, B., Group epitope mapping by saturation transfer difference NMR to identify segments of a ligand in direct contact with a protein receptor. *J Am Chem Soc*, 2001, **123**, 6108–6117.
[97] Sandstrom, C., *et al.*, Atomic mapping of the interactions between the antiviral agent cyanovirin-N and oligomannosides by saturation-transfer difference NMR. *Biochemistry*, 2004, **43**, 13926–13931.
[98] Hajduk, P. J., *et al.*, SOS-NMR: a saturation transfer NMR-based method for determining the structures of protein–ligand complexes. *J Am Chem Soc*, 2004, **126**, 2390–2398.
[99] Benie, A. J., *et al.*, Virus–ligand interactions: identification and characterization of ligand binding by NMR spectroscopy. *J Am Chem Soc*, 2003, **125**, 14–15.
[100] Meinecke, R. and Meyer, B., Determination of the binding specificity of an integral membrane protein by saturation transfer difference NMR: RGD peptide ligands binding to integrin alphaIIbbeta3. *J Med Chem*, 2001, **44**, 3059–3065.
[101] Claasen, B., *et al.*, Direct observation of ligand binding to membrane proteins in living cells by a saturation transfer double difference (STDD) NMR spectroscopy method shows a significantly higher affinity of integrin alpha(IIb)beta3 in native platelets than in liposomes. *J Am Chem Soc*, 2005, **127**, 916–919.
[102] Dalvit, C., Homonuclear 1D and 2D NMR experiments for the observation of solvent–solute interactions. *J Magn Reson B*, 1996, **112**, 282–288.
[103] Johnson, E. C., *et al.*, Application of NMR SHAPES screening to an RNA target. *J Am Chem Soc*, 2003, **125**, 15724–15725.
[104] Di Micco, S., *et al.*, Differential-frequency saturation transfer difference NMR spectroscopy allows the detection of different ligand–DNA binding modes. *Angew Chem Int Ed*, 2006, **45**, 224–228.
[105] Feeney, J., *et al.*, The effects of intermediate exchange processes on the estimation of equilibrium constants by NMR. *J Magn Reson* 1979, **33**, 519–529.
[106] Meiboom, S. and Gill, D., Modified spin-echo method for measuring nuclear relaxation times. *Rev. Sci. Instrum.*, 1958, **29**, 688–691.
[107] van Dongen, M., *et al.*, Structure-based screening and design in drug discovery. *Drug Discov Today*, 2002, **7**, 471–478.
[108] Hajduk, P. J., *et al.*, NMR-based screening of proteins containing ^{13}C-labeled methyl groups. *J Am Chem Soc*, 2000, **122**, 7898–7904.
[109] Weigelt, J., *et al.*, Site-selective screening by NMR spectroscopy with labeled amino acid pairs. *J Am Chem Soc*, 2002, **124**, 2446–2447.
[110] Weigelt, J., *et al.*, Site-selective labeling strategies for screening by NMR. *Comb Chem High Throughput Screen*, 2002, **5**, 623–630.
[111] McCoy, M. A. and Wyss, D. F., Spatial localization of ligand binding sites from electron current density surfaces calculated from NMR chemical shift perturbations. *J Am Chem Soc*, 2002, **124**, 11758–11763.
[112] McCoy, M. A., *et al.*, Screening of protein kinases by ATP-STD NMR spectroscopy. *J Am Chem Soc*, 2005, **127**, 7978–7979.
[113] Cheng, Y. and Prusoff, W. H., Relationship between the inhibition constant (K_1) and the concentration of inhibitor which causes 50 per cent inhibition (I_{50}) of an enzymatic reaction. *Biochem Pharmacol*, 1973, **22**, 3099–3108.
[114] Sanchez-Pedregal, V. M., *et al.*, The INPHARMA method: protein-mediated interligand NOEs for pharmacophore mapping. *Angew Chem Int Ed*, 2005, **44**, 4172–4175.
[115] Banaszak, L., *et al.*, Lipid-binding proteins: a family of fatty acid and retinoid transport proteins. *Adv Protein Chem*, 1994, **45**, 89–151.

[116] Reese-Wagoner, A., *et al.*, Structural properties of the adipocyte lipid binding protein. *Biochim Biophys Acta*, 1999, **1441**, 106–116.
[117] Xu, Z., *et al.*, Crystal structure of recombinant murine adipocyte lipid-binding protein. *Biochemistry*, 1992, **31**, 3484–3492.
[118] Hanhoff, T., *et al.*, Insights into binding of fatty acids by fatty acid binding proteins. *Mol Cell Biochem*, 2002, **239**, 45–54.
[119] Zimmerman, A. W. and Veerkamp, J. H., New insights into the structure and function of fatty acid-binding proteins. *Cell Mol Life Sci*, 2002, **59**, 1096–1116.
[120] Boord, J. B., *et al.*, Cytoplasmic fatty acid-binding proteins: emerging roles in metabolism and atherosclerosis. *Curr Opin Lipidol*, 2002, **13**, 141–147.
[121] Haunerland, N. H. and Spener, F., Fatty acid-binding proteins – insights from genetic manipulations. *Prog Lipid Res*, 2004, **43**, 328–349.
[122] Jenkins-Kruchten, A. E., *et al.*, Fatty acid-binding protein–hormone-sensitive lipase interaction. Fatty acid dependence on binding. *J Biol Chem*, 2003, **278**, 47636–47643.
[123] Hotamisligil, G. S., *et al.*, Uncoupling of obesity from insulin resistance through a targeted mutation in aP2, the adipocyte fatty acid binding protein. *Science*, 1996, **274**, 1377–1379.
[124] Uysal, K. T., *et al.*, Improved glucose and lipid metabolism in genetically obese mice lacking aP2. *Endocrinology*, 2000, **141**, 3388–3396.
[125] Binas, B., *et al.*, Requirement for the heart-type fatty acid binding protein in cardiac fatty acid utilization. *FASEB J*, 1999, **13**, 805–812.
[126] Veerkamp, J. H., *et al.*, Structural and functional features of different types of cytoplasmic fatty acid-binding proteins. *Biochim Biophys Acta*, 1991, **1081**, 1–24.
[127] Cohen, P., Protein kinases – the major drug targets of the twenty-first century?, *Nat Rev Drug Discov*, 2002, **1**, 309–315.
[128] Csermely, P., *et al.*, The efficiency of multi-target drugs: the network approach might help drug design. *Trends Pharmacol Sci*, 2005, **26**, 178–182.
[129] Morphy, R. and Rankovic, Z., Fragments, network biology and designing multiple ligands. *Drug Discov Today*, 2007, **12**, 156–160.
[130] Morphy, R. and Rankovic, Z., The physicochemical challenges of designing multiple ligands. *J Med Chem*, 2006, **49**, 4961–4970.
[131] Morphy, R., The influence of target family and functional activity on the physicochemical properties of pre-clinical compounds. *J Med Chem*, 2006, **49**, 2969–2978.
[132] Hopkins, A. L., *et al.*, Can we rationally design promiscuous drugs?, *Curr Opin Struct Biol*, 2006, **16**, 127–136.
[133] Zartler, E. R., *et al.*, RAMPED-UP NMR: multiplexed NMR-based screening for drug discovery. *J Am Chem Soc*, 2003, **125**, 10941–10946.
[134] Inooka, H., *et al.*, Conformation of a peptide ligand bound to its G-protein coupled receptor. *Nat Struct Biol*, 2001, **8**, 161–165.
[135] Cherezov, V., *et al.*, High-resolution crystal structure of an engineered human beta2-adrenergic G protein-coupled receptor. *Science*, 2007, **318**, 1258–1265.
[136] Lundstrom, K., Structural genomics and drug discovery. *J Cell Mol Med*, 2007, **11**, 224–238.
[137] Vanwetswinkel, S., *et al.*, TINS, target immobilized NMR screening: an efficient and sensitive method for ligand discovery. *Chem Biol*, 2005, **12**, 207–216.
[138] Marquardsen, T., *et al.*, Development of a dual cell, flow-injection sample holder and NMR probe for comparative ligand-binding studies. *J Magn Reson*, 2006, **182**, 55–65.

5

Application of Protein–Ligand NOE Matching to the Rapid Evaluation of Fragment Binding Poses

William J. Metzler, Brian L. Claus, Patricia A. McDonnell, Stephen R. Johnson, Valentina Goldfarb, Malcolm E. Davis, Luciano Mueller and Keith L. Constantine

5.1 Introduction

'Fragments', or low molecular weight compounds, are increasingly being recognized as attractive starting points for deriving novel, potent inhibitors against an expanding array of therapeutic targets. Since the first reports of fragment-based screening by the Abbott biomolecular NMR laboratory,[1] where it was demonstrated that a structure-guided combination of fragments can yield a progressable lead series, numerous examples detailing the development of fragments into mature drug candidates have been documented (see ref. 2 for a review). The benefits of fragments are numerous. The low complexity of fragments results in a statistically high probability of their pharmacophores matching those in the target protein's binding site.[3] Hence fewer fragments need to be screened in order to identify a 'hit'. The result of this is that fragment screening libraries can be composed of a relatively small number of compounds, typically on the order of only several hundred to several thousand. Although their absolute binding affinities tend to be low, fragment hits are generally efficient binders, that is, they tend to have high binding affinities relative to their heavy atom counts. This is because a high proportion of their atoms usually make favorable interactions with the receptor. Furthermore, the simplicity of fragments makes them highly amenable to optimization, since additional functionality can be added

Fragment-Based Drug Discovery: A Practical Approach Edited by Edward R. Zartler and Michael J. Shapiro

without loss of drug-like character.[4] A recent retrospective analysis indicates that the success of fragment optimization and progression can be accurately predicted.[2] The main factor for successful fragment progression is the availability of structural data to guide the optimization process.

Optimization of the potency and specificity of lead-like and drug-like compounds using experimental structural information on protein–ligand complexes is a well-established component of the drug discovery process.[5] As noted above, the availability of experimental structural information on fragments bound to their target is crucial to driving the progression from relatively simple, minimally functionalized weak hits to potent, elaborated leads.[2] There are primarily three options for generating molecular structure-based information: molecular modeling, X-ray crystallography and NMR spectroscopy. While various computational tools can often generate predictive binding models of larger inhibitors, binding mode prediction for fragments using computational methods remains problematic. Although fragments generally have fewer internal degrees of freedom than larger compounds, their location and orientation within a binding pocket can be more difficult to predict, since alternative binding poses are more likely to yield similar scores or predicted binding energies, i.e. multiple minima problems are more common. Whereas progress has been made in the computational prediction of fragment binding poses (e.g. see ref. 6), experimental binding pose characterization remains critically important.[7, 8] When feasible, X-ray crystallography can be used simultaneously to screen a fragment library for hits and to provide high-resolution structural information on the binding poses of the hits obtained.[9–12] A challenge in the crystallographic determination of protein–fragment complexes is the need for high-resolution structures, typically <2.0 Å, to place correctly pseudo-symmetry-related groups with similar electron densities, e.g. a methyl versus an amino group. Traditional NMR methods, based on observed and assigned protein–ligand NOEs, have been used to characterize the structures of fragments bound to proteins [1, 13, 14] and to guide the design of potent 'linked compounds', as exemplified in the original report of the 'SAR by NMR' method.[1] The main challenges for the NMR-based approaches to structure determination are feasibility and time. Regarding the latter, both the acquisition and analysis of the multiple data sets necessary to define the bound structure of the ligand can greatly constrain the number of protein–ligand complexes that can be characterized, limiting the information that can be generated for the numerous hits identified in a fragment-based screen.

In an effort to accelerate the structure determination of protein–ligand complexes by NMR, we have developed an approach, termed 'NOE matching',[15] that allows one to determine the binding pose of a ligand without first having to determine resonance assignments for the target protein. Briefly, NOE matching is a pattern-matching procedure that derives 'cost' values for trial binding poses based on how well each trial pose predicts an experimental 3D ^{13}C-edited, ^{13}C/^{15}N-filtered HSQC-NOESY spectrum.[16] The key advantage of the NOE matching approach is that it can be applied without NMR assignments for the protein in the protein–ligand complex. We have previously shown that NOE matching can identify the correct binding pose with reasonable accuracy. In this chapter, we briefly summarize the NOE matching protocol and its previous application to lead-like compounds bound to small proteins.[15] We then describe the application of NOE matching to fragments and discuss some of the unique challenges encountered with these small compounds. Finally, because the majority of pharmaceutically relevant targets are relatively large and often poorly behaved proteins, we investigate the application of NOE matching to these challenging systems. We describe how additional information can be obtained and incorporated into the NOE matching protocol. With these enhancements, we expect that NOE matching will enable the binding poses of ligands to be determined for a variety of systems, including large complexes.

5.2 Summary of the NOE matching Protocol

The NOE matching protocol is described pictorially in Figure 5.1. Two files are needed for input: an experimental NOE peak list with ligand protons assigned and a set of trial binding poses to be evaluated or scored. The list of experimental peaks is typically derived from a 3D ^{13}C-edited, ^{13}C/^{15}N-filtered HSQC-NOESY spectrum.[16–20] (Hereafter, this type of spectrum will be referred to as a 3D X-filtered NOESY spectrum.)

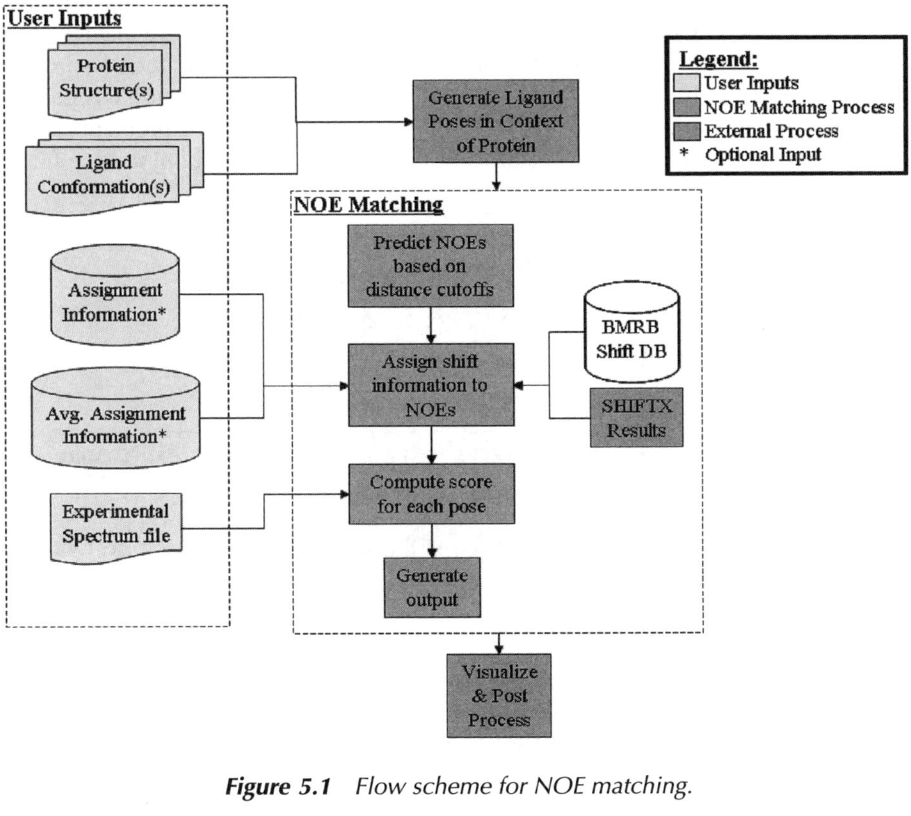

Figure 5.1 Flow scheme for NOE matching.

These peaks are grouped based on the protein ^{1}H and ^{13}C chemical shift values. This procedure *identifies* (but does not *assign*) ^{1}H^{13}C groups on the protein that give rise to protein–ligand NOEs. Isotope-filtered NMR methods[19, 20] are applied to assign the bound ^{1}H resonances of the ligand. (We note that for complex ligands, such as peptides and natural products, isotope labeling of the ligand may also be required to obtain bound-state assignments.) A set of possible binding poses is generated by any suitable method or combination of methods. In general, the sampling of binding poses must be extensive, as poses that are similar to the true pose must be sampled if they are to be identified by the NOE matching procedure. Each pose is used to predict a 3D X-filtered NOESY spectrum. The chemical shifts used for these predicted spectra can be derived from actual experimental assignments, from chemical shift prediction algorithms (as described later in this chapter) or most simply by using the average chemical values for diamagnetic proteins available from the Biological Magnetic Resonance Data Bank (BMRB).[21]

The central idea of NOE matching is to determine how well the predicted 3D X-filtered NOESY spectrum matches to the experimental 3D X-filtered NOESY spectra. Poses which produce a good match between predicted and experimental spectrum are expected to resemble more closely the true binding pose. This expectation depends critically on having enough information in the observed pattern of NOEs to define the binding pose in a nondegenerate fashion. NOE matching casts the problem of measuring the similarity between predicted and experimental spectra as an equally partitioned bipartite graph weighted matching problem,[22] wherein experimentally identified $^1H^{13}C$ groups are matched to HC atom groups predicted to give rise to NOEs by the given pose. In addition to providing a COST value for each pose, potential assignments for the experimental NOE peaks are generated. A hypothetical example of a 3D X-filtered NOESY matching problem cast as an equally partitioned bipartite graph is shown in Figure 5.2. The algorithm used[22] finds an

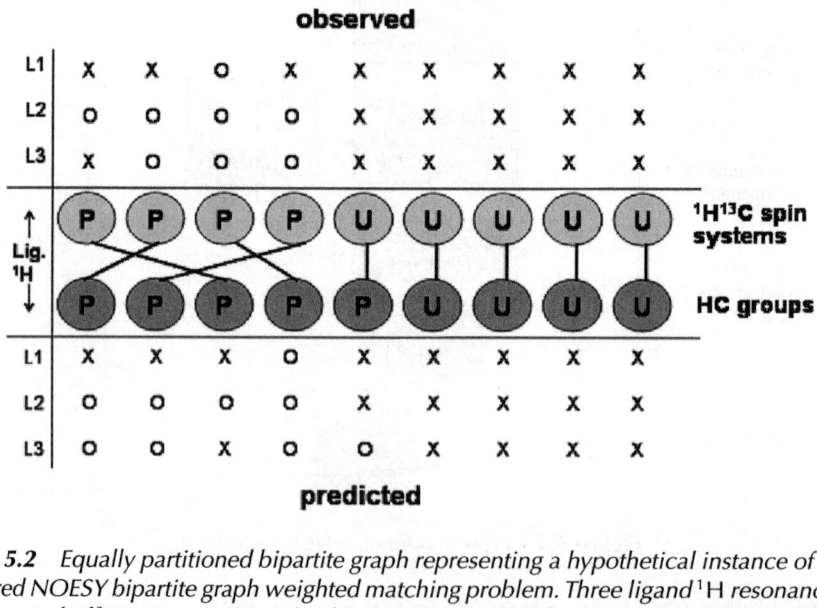

Figure 5.2 *Equally partitioned bipartite graph representing a hypothetical instance of the 3D X-filtered NOESY bipartite graph weighted matching problem. Three ligand ^{1}H resonances and four protein $^1H^{13}C$ groups are involved in experimentally observed NOEs. The binding pose predicts NOEs involving the same three ligand ^{1}H atoms and five protein HC groups. Each edge between the observed and predicted nodes has an edge weight value associated with it. The sum of these edge weights defines the COST of the pose. See ref. 15 for additional details.*

optimal (minimum COST) complete matching of the bipartite graph in polynomial [$O(N^3)$] time. Below we briefly describe the implementation of the NOE matching COST function.

The function C(*k*,*q*) defines the edge weight values (Figure 5.2):

$$C\,(k,q) \sum_i M_i\,(k,q) \quad i = 1, N_L \tag{5.1}$$

where N_L is the number of resolved, assigned ligand ^{1}H groups (e.g. $N_L = 3$ in Figure 5.2). Equation (5.1) has been modified to ensure that obviously incorrect group-to-group matchings (e.g. an aliphatic group to an aromatic group) do not occur. This will be described in detail elsewhere. Referring to Figure 5.2, the matching cost *M* between an experimental peak and a predicted peak is defined by the following expressions:

$$M_i(\mathrm{X},\mathrm{X}) = 0 \qquad \text{(no experimental peak, no predicted peak)} \tag{5.2}$$

$$M_i(\mathrm{O},\mathrm{X}) = K_1(\mathrm{IE_i})^2 \qquad \text{(experimental peak present, no predicted peak)} \tag{5.3}$$

$$M_i(\mathrm{X},\mathrm{O}) = K_2(\mathrm{IP}_i)^2 \qquad \text{(no experimental peak, predicted peak present)} \tag{5.4}$$

$$M_i(O,O): \qquad \text{(experimental peak present, predicted peak present)}$$

If $\mathrm{IE}_i > \mathrm{IP}_i$

$$M_i(\mathrm{O},\mathrm{O}) = K_\mathrm{H}[f(\mathrm{H})/\sigma\mathrm{H}]^2 + \mathrm{K_C}[\mathrm{f}(\mathrm{C})/\sigma\mathrm{C}]^2 + \mathrm{K_3}[\mathrm{f}'(\mathrm{I})]^2$$

Else

$$M_i(\mathrm{O},\mathrm{O}) = K_\mathrm{H}[f(\mathrm{H})/\sigma\mathrm{H}]^2 + K_\mathrm{C}[f(\mathrm{C})/\sigma\mathrm{C}]^2 + K_4[f'(\mathrm{I})]^2 \tag{5.5}$$

End If

See ref. 15 for a detailed description of the terms and parameters in Equations (5.2)–(5.5). Equation (5.5) has been modified to incorporate data from nonuniformly labeled samples. These modifications will be described elsewhere.

The total cost of a given pose corresponds to an optimal solution of the complete matching problem, which is a permutation ε of $\{1, 2, \ldots, N\}$ that minimizes

$$\mathrm{COST_{pose}} = \sum_j C\,[j, \varepsilon\,(j)] \quad j = 1, N \tag{5.6}$$

5.3 Applications to Lead-like Compounds Bound to Small Proteins

The ability of NOE matching to identify the correct pose from a collection of decoy poses was demonstrated on three test cases involving lead-like compounds bound to small proteins. One of these test cases involves muscle fatty acid-binding protein (mFABP) and the other two involve the leukocyte function-associated antigen 1 I-domain (LFA-1). As these results have been presented in detail elsewhere,[15] here we only summarize the results obtained when the BMRB average chemical shifts (diamagnetic protein statistics) were used for chemical shift predictions. The compounds used for these test cases are shown in Scheme 5.1 (compound **3** contains a proprietary core, which is represented by an ellipse).

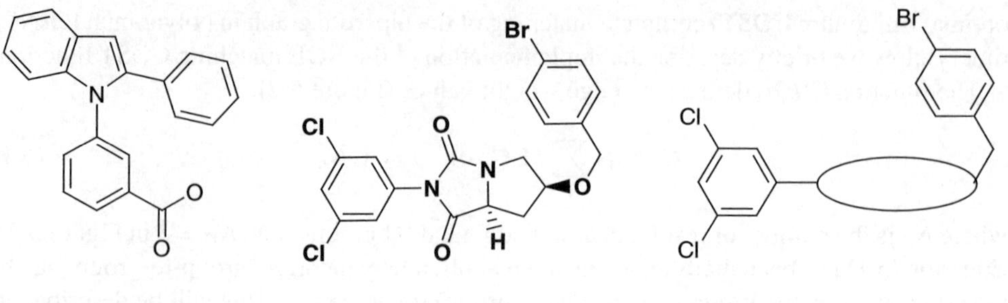

Scheme 5.1 Structures of compounds used for test cases.

5.3.1 mFABP

In the case of mFABP bound to compound **1** (Scheme 5.1), the 'target pose' was derived from a high-resolution NMR ensemble (Supporting Information for ref. 15). A total of 461 trial binding poses were generated, with RMSDs to the target pose ranging from 0.20 to 6.66 Å. The experimental 3D X-filtered NOESY spectrum ($\tau_m = 60$ ms) for this complex yielded 140 peaks, which were clustered into 54 protein $^1H^{13}C$ groups. (Although not discussed further here, we have shown that NOE matching performed well on this system when up to 80 % of the peaks were deleted; see Table S2 in the Supporting Information for ref. 15).

The NOE matching protocol contains adjustable parameters that define weights for the observed and predicted NOE intensities (or for functions of differences between the observed and predicted intensities) and parameters that define weights for functions of the differences between the observed and predicted chemical shifts. Another parameter (the standard deviation multiplier, SDM) is a scaling factor which modulates the tolerances associated with the predicted chemical shifts. For the tests reported, the weighting parameters were held constant and the SDM parameter was coarsely varied. For mFABP/**1** using BMRB-predicted chemical shifts, an SDM value of 0.50 was found to be near optimal (SDM values of 0.25, 0.50 and 0.75 gave very similar results, in general). Some of the results obtained from applying NOE matching to this system using SDM = 0.50 are shown in Figure 5.3. The correlation coefficient (r) between the COST and RMSD is 0.912.

The pose with the minimum COST value has an RMSD of 0.75 Å to the target pose. The target pose itself ranks 14 out of 462 poses. In addition, a majority of the experimentally identified $^1H^{13}C$ groups are correctly assigned for low COST poses. For mFABP/**1**, NOE matching worked very well using no protein NMR assignments and the simplest chemical shift prediction scheme (predictions set to BMRB average shifts).

5.3.2 LFA-1

NOE matching was applied to two different complexes involving LFA-1 (see Scheme 5.1, compounds **2** and **3**). For LFA-1/**2**, an X-ray structure of the complex (Supporting Information for ref. 15; S. Sheriff, unpublished work) was used as the target pose. A total

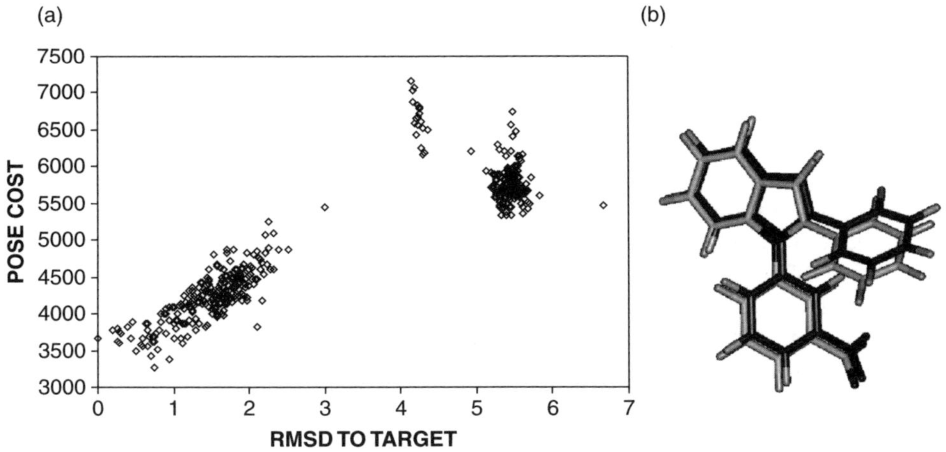

Figure 5.3 (A) COST versus the RMSD (Å) to the target pose for mFABP/**1** obtained with SDM = 0.50 and with the predicted protein chemical shifts set to the corresponding BMRB average values. (B) Superposition of target pose and the minimum cost pose (dark gray) from (A).

of 1500 trial poses were generated for LFA-1/**2**, with minimum and maximum RMSDs to the trial pose of 0.22 and 8.52 Å, respectively. Some limited sampling of the protein conformation was used in this case. The experimental 3D X-filtered NOESY spectrum ($\tau_m = 100$ ms) for this complex yielded 69 peaks, which were clustered into 51 protein $^1H^{13}C$ groups. It is important to note that the hydantoin core of **2** did not yield any protein–ligand NOEs. Results for LFA-1/**2** (SDM = 0.25) are shown in Figure 5.4 (r = 0.858 between COST and RMSD).

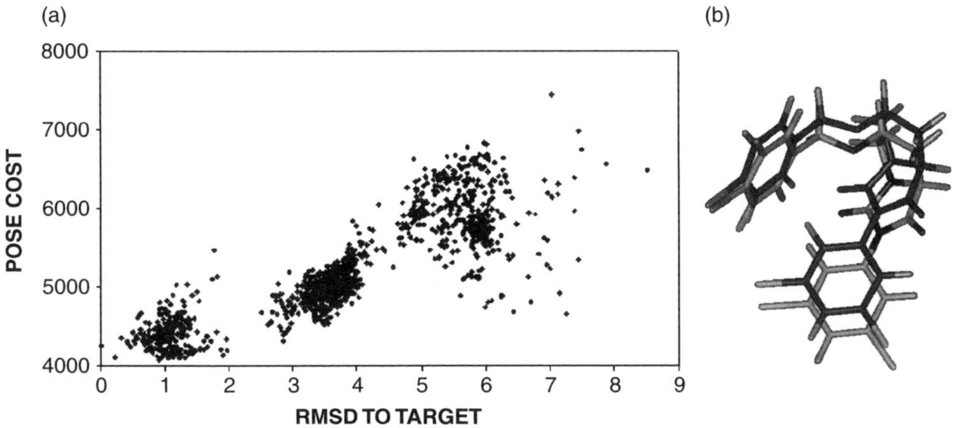

Figure 5.4 (A) $COST_{pose}$ versus the RMSD (Å) to the target pose for LFA-1/**2**, with SDM = 0.25 and with the predicted protein chemical shifts set to the corresponding BMRB average values. (B) Superposition of the target pose and the minimum cost pose (dark gray) from (A).

For LFA-1/**2**, the pose with the minimum COST value has an RMSD of 0.91 Å to the target pose. The target pose itself ranks 73 out of 1501 poses. LFA-1 was not experimentally assigned for this complex, so statistics regarding correctly assigned $^{1}H^{13}C$ groups are not available for this case. While NOE matching was successful overall for LFA-1/**2**, the results were not as good as those obtained with mFABP/**1**. Although the lowest COST poses cluster around the known target pose, there are few poses with relatively low COST that are quite different from known poses. Discriminating among substantially different poses with similar low COST, in the absence of a known pose, is an issue that can be addressed, in part, by developments discussed later in this chapter.

Although the exact composition of compound **3** (Scheme 5.1) cannot be revealed at this time, it is more favorable than compound **2** for NOE matching. Compound **3** has a core that contains two methyl groups, both of which give rise to protein–ligand NOEs. For LFA-1/**3**, the target pose was derived from a well-resolved experimental ensemble, generated as described (Supporting Information for ref. 15) from NMR and X-ray data. A total of 350 trial poses were generated for LFA-1/**3**, with minimum and maximum RMSDs to the trial pose of 0.18 and 7.63 Å, respectively. A single protein conformation was used in this case. The experimental 3D X-filtered NOESY spectrum ($\tau_m = 100$ ms) for this complex yielded 74 peaks, which were clustered into 44 protein $^{1}H^{13}C$ groups. Results for LFA-1/**3** (SDM = 0.25) are shown in Figure 5.5 ($r = 0.973$ between COST and RMSD).

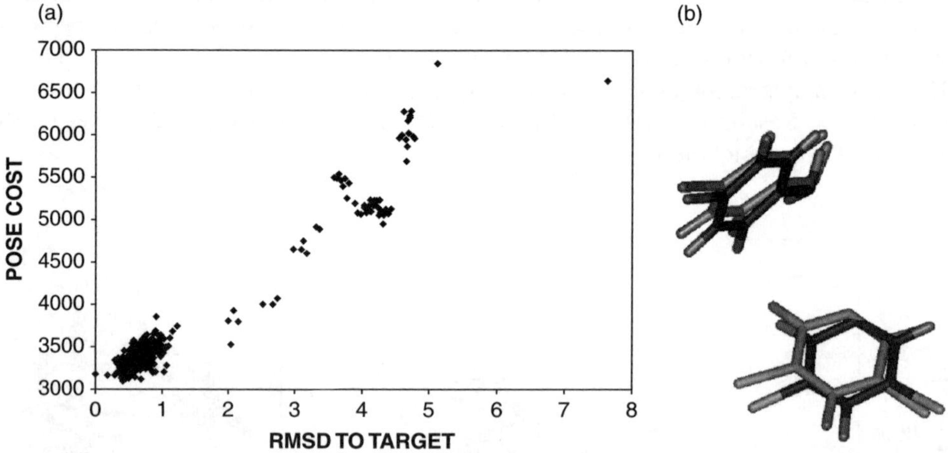

Figure 5.5 *(A) $COST_{pose}$ versus the RMSD (Å) to the target pose for LFA-1/**3** with SDM = 0.25 and with predicted protein chemical shifts set to the corresponding BMRB average values. (B) Superposition of the target pose and the minimum cost pose (dark gray) from (A), showing only the non-proprietary moieties.*

The pose with the minimum COST value has an RMSD of 0.41 Å to the target pose. The target pose itself ranks 11 out of 351 poses. As with mFABP/**1**, a majority of the experimentally identified $^{1}H^{13}C$ groups are correctly assigned for low COST poses for LFA-1/**3**.

The three examples described in this section all involve relatively small, well-behaved proteins in complex with lead-like compounds. The results obtained indicate that, for lead-like or drug-like compounds bound to suitable targets, NOE matching can yield relatively accurate binding poses without using any protein NMR assignments or accurate chemical shift predictions. Later in this chapter, we will focus on challenges presented by simpler compounds and/or large proteins for NOE matching in particular and for detecting and analyzing protein–ligand NOE interactions in general.

5.4 Enhanced Pose Generation and Pose Scoring

For NOE matching to identify the correct pose accurately, it is essential that the ensemble of poses contains one or more poses that are very similar to the true pose. Generation of a broad sampling of poses with existing software, however, is not straightforward. Most pose generation algorithms, such as Glide,[23] are designed to identify the optimal pose, discarding false poses. Identification of the true pose by NOE matching, on the other hand, relies on the fact the COST of correct poses should be significantly lower than the COST of decoy poses. Therefore, it is important that a wide distribution of poses is generated and evaluated, as this increases the probability that the lowest COST poses obtained during an NOE matching run reflect a global minimum of the COST function and not a local minimum. To address this concern, we have adapted an internally developed posing engine, *Poser*. For predefined ligand and protein conformations, *Poser* provides a systematic and exhaustive sampling of poses in a binding site.

Poser generates all possible poses of user-supplied ligand conformations within a specified resolution. The only limitation imposed is the absence of significant steric clashes between the ligand and protein. A regularly spaced grid is centered on a user-specified binding site. By default, the binding site consists of the entire protein, but can be specified by defining the coordinate boundaries of the binding site. A mask is created by labeling each grid point as either inside the protein or outside the protein. Further, the shortest distance of each grid point to the protein surface is also calculated. For these purposes, the atomic radii for the protein are set to 90 % of the Bondi atomic radii, as specified in OEChem, v1.4.[24] This softening of the protein radii is performed to account for small displacements that may occur, but that are not explicitly modeled.

The geometric centroid of each conformation of each ligand is placed iteratively on every grid point that is not labeled as being inside the protein. The molecule is then rotated about the grid point by a user-defined number of degrees. Each resulting pose is then checked for bumps by making sure the pose does not overlap a grid point labeled as being inside the protein. The user may specify a scaling factor to decrease the Bondi radii of the ligand atoms for these purposes. If the pose does not exceed the user-specified limit of allowable bumps, the pose is written out and saved.

If the user requires contact between the ligand and the protein, the maximum distance from the geometric centroid of each conformation to its molecular surface is calculated. The molecule is only placed on those grid points that are labeled as being within this distance plus 1 Å.

Examples of samplings obtained by *Poser* are provided in Table 5.1. To generate the input conformers for *Poser*, we used the program Omega[25] with standard defaults. This

Table 5.1 *Statistics on pose generation with* Poser

Target	Box size (Å)	No. of conformers[a]	Poses evaluated[b] (millions)	Poses saved[c]	Best conformer found[d]	Best pose found[e]
FKBP-12	11 × 12 × 16	5	>629	45 693	0.51	0.64
PDF	12 × 14 × 10	3	>485	50 792	0.52	0.96
CDK2 (1)	11 × 12 × 16	5	>632	157 380	0.30	0.45
CDK2 (2)	16 × 15 × 12	13	>808	10 579	0.49	0.69
Bcl-x_L	19 × 16 × 17	1	>191	9 556	0.25	0.71

[a] Number of conformers generated by Omega and used as input into *Poser* run.
[b] Poses were generated with a grid spacing of 1 Å, a rotational sampling of 5° and a radii scaling of 0.9. No steric clashes between the target and ligand were allowed. For Bcl-x_L, a 10° rotational sampling was used.
[c] Total number of poses that fit into the binding site.
[d] The RMSD of the Omega conformer for the ligand with the lowest RMSD to the target conformer. Compounds were aligned for best fit before calculating RMSD.
[e] The RMSD of the *Poser*-generated binding pose with the lowest RMSD to the target binding pose. RMSDs were calculated with ligand molecules in the context of protein, that is, no alignment was performed.

typically led to three or more low-energy conformers of each compound. The conformer with the lowest RMSD to the target conformer was generally ~0.5 Å, reflecting that the experimentally determined conformation of a compound is often different from the computationally defined local and global energy minima that exist in the absence of the target protein. This RMSD value sets the limit for what we could expect our pose sampling with *Poser* to achieve. For each of the NOE matching cases with fragments described below, poses were generated with a 'rotational sampling' of 5°. This led to hundreds of millions of poses being evaluated for each input conformer of the compound and tens of thousands of poses being saved for later evaluation by NOE matching. In general, the best pose generated using *Poser* was within 1 Å of the target pose and, as Table 5.1 indicates, was often closer.

5.5 Applications to Fragment-like Compounds

Although the boundaries between drug-like, lead-like and fragment-like compounds can be somewhat fuzzy, fragment-like compounds are generally smaller and less functionalized than lead-like/drug-like compounds. This distinguishing feature carries with it a significant consequence; namely, the binding of a fragment to its receptor is often much more difficult to characterize structurally than that of a lead-like molecule. There are several reasons for this. First, the binding affinity of fragments tends to be weaker than what one might typically observe for a more complex molecule, leading to the requirement of higher compound concentrations to attain receptor saturation. Second, the lack of structural complexity of fragments provides fewer distinguishing features that can be used to guide structural refinement. Third, binding of a fragment to its receptor may not be limited to a single binding pose. The reduced potency and structural simplicity of fragments presents challenges for both X-ray and NMR structural determinations.

In applying NOE matching to fragment pose determination, we were very concerned that the fragments might not be large enough to contact enough of the binding pocket (i.e. multiple residue types) to give rise to sufficient information content in the observed NOE patterns (a requirement of NOE matching) to permit discrimination between true and decoy poses. To determine whether the 3D X-filtered NOESY experiment contains enough information to enable NOE matching to identify the correct binding pose, we ran tests using simulated data derived from a CDK2/**4** complex, an FKBP-12/**5** complex and a peptide deformylase (PDF)/**6** complex. The compounds used for these test cases are shown in Scheme 5.2.

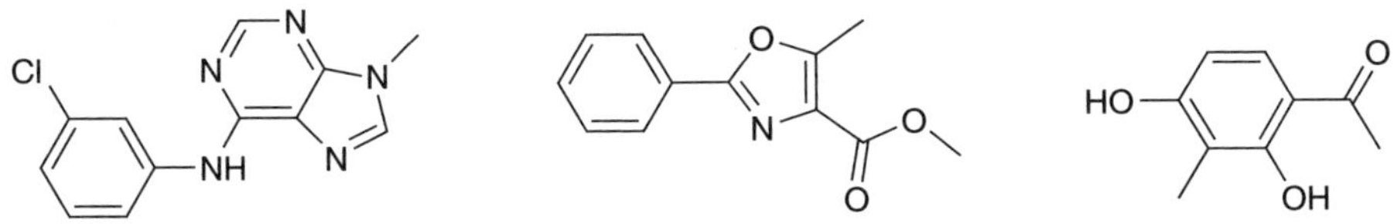

4 **5** **6**

Scheme 5.2 *Structures of compounds used for test cases.*

5.5.1 CDK2

Compound **4** is an ATP-mimic that binds in the active site of the catalytic domain of CDK2 kinase. The crystal structure of CDK2 in complex with **4** (1ckp[26]) served as the 'target pose' for the NOE matching simulations. To generate the required input files for NOE matching, a list of CDK2/**4** NOEs was derived from distances calculated using the CDK2/**4** complex using a distance cutoff of 5 Å and using the BMRB average chemical shifts for the simulated 'experimental' chemical shifts. The simulated NOE list for this complex contained a total of 69 peaks, which were clustered into 43 protein $^{1}H^{13}C$ groups. Trial binding poses were generated with *Poser*. The compound binding site was defined as the active site of the protein; the 'posing box' was expanded by 1 Å in all coordinate axes. For each of the 13 ligand conformers (generated with Omega), over 808 million poses were generated and evaluated by *Poser*, with 10 579 poses being retained. The RMSDs of the retained trial poses to the target pose ranged from 0.69 to 7.85 Å. The NOE matching protocol was run using BMRB-predicted chemical shifts.

The results obtained from applying NOE matching to CDK2/**4** are shown in Figure 5.6. The pose with the minimum COST value has an RMSD of 0.74 Å to the target pose. The pose with the closest RMSD to the target pose itself ranks 14 out of 10 579 poses.

5.5.2 FKBP-12

Compound **5** is one of numerous fragments identified in an in-house NMR screen of FKBP-12. The solution structure of FKBP-12 in complex with **5** was determined by restrained simulated annealing[27] using data derived from standard three-dimensional NMR techniques. The average structure of the resultant ensemble of NMR structures was calculated and subjected to unrestrained energy minimization; this structure served as the 'target pose' for the NOE matching simulations. The experimentally determined resonance assignments were used for the ligand resonance assignments. To generate the required input files for NOE matching, a list of FKBP-12/**5** NOEs was derived from distances calculated using the FKBP-12/**5** complex using a distance cutoff of 5.0 Å (similar to what we observed in our experimental NOESY spectrum) and using the BMRB[21] average chemical shifts for

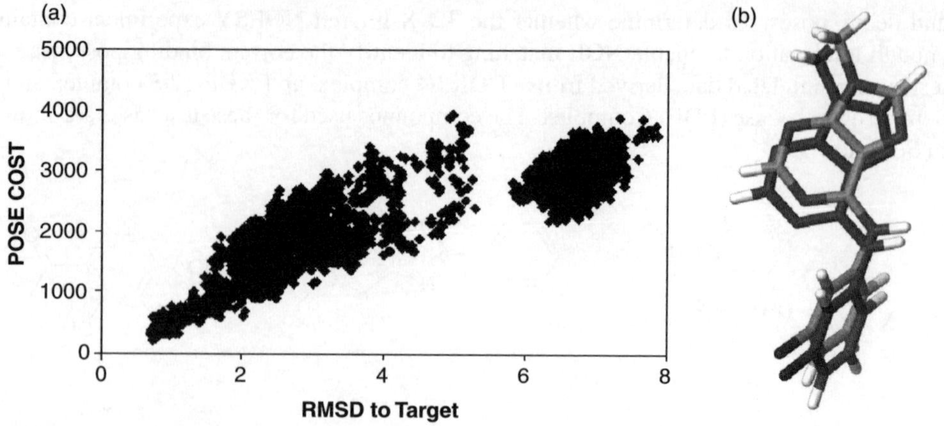

*Figure 5.6 (A) COST versus the RMSD (Å) to the target pose for CDK2-/**4**. The predicted protein chemical shifts were set to the corresponding BMRB average values. The 3D X-filtered NOESY spectrum used as input for NOE matching was simulated from the target structure. (B) Superposition of target pose and the minimum cost pose (dark gray) from (A).*

the simulated 'experimental' chemical shifts. The simulated NOE list for this complex contained a total of 110 peaks, which were clustered into 53 protein $^{1}H^{13}C$ groups. Trial binding poses were generated with *Poser*. The compound binding site was defined as the active site of the protein; the 'posing box' was expanded by 1 Å in all coordinate axes. For each of the five ligand conformers (generated with Omega), 629 669 376 poses were generated and evaluated by *Poser*, with 45 693 poses being retained. The RMSDs of the retained trial poses to the target pose ranged from 0.64 to 7.56 Å. The NOE matching protocol was run using BMRB-predicted chemical shifts.

The results obtained from applying NOE matching to FKBP-12/**5** are shown in Figure 5.7. The pose with the minimum COST value has an RMSD of 0.64 Å to the target pose; this pose also had the lowest RMSD of all 45 693 with respect to the target pose.

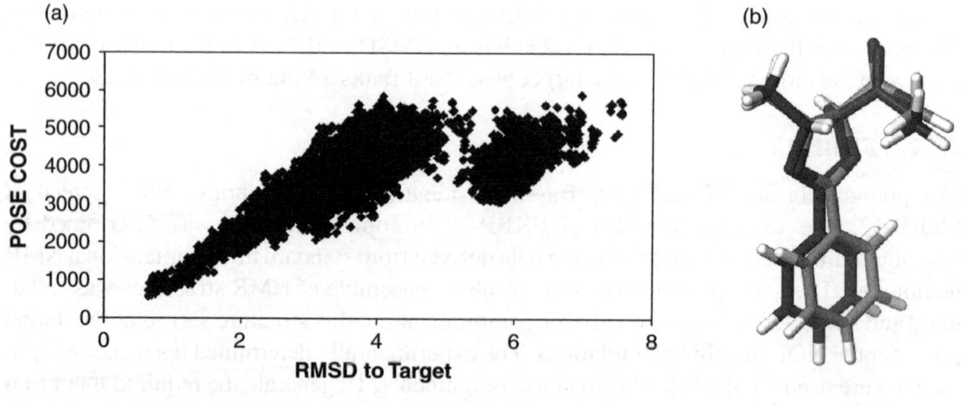

*Figure 5.7 (A) COST versus the RMSD (Å) to the target pose for FKBP-12/**5**. The predicted protein chemical shifts were set to the corresponding BMRB average values. (B) Superposition of target pose and the minimum cost pose (dark gray) from (A).*

5.5.3 PDF

Compound **6** is one of numerous fragments identified in an in-house NMR screen of PDF. The solution structure of PDF in complex with **6** was determined by restrained simulated annealing[27] using data derived from standard three-dimensional NMR techniques. The average structure of the resultant ensemble of NMR structures was calculated and subjected to unrestrained energy minimization; this structure served as the 'target pose' for the NOE matching simulations. To generate the required input files for NOE matching, a list of PDF/**6** NOEs was derived from distances calculated using the PDF/**6** complex using a distance cutoff of 4.5 Å (a conservative upper bound estimate compared to our real experimental NOESY data) and using the BMRB average chemical shifts for the simulated 'experimental' chemical shifts. The simulated NOE list for this complex contained a total of 62 peaks, which were clustered into 48 protein $^1H^{13}C$ groups. Trial binding poses were generated with *Poser*. The compound binding site was defined as the active site of the protein; the 'posing box' was expanded by 1 Å in all coordinate axes. For each of the three ligand conformers (generated with Omega), 485 968 896 poses were generated and evaluated by *Poser*, with 50 792 poses being retained. From this set of poses, 1000 poses were selected by random sampling for scoring by NOE matching. The RMSDs of the retained trial poses to the target pose ranged from 0.96 to 5.75 Å. The NOE matching protocol was run using BMRB-predicted chemical shifts.

The results obtained from applying NOE matching to PDF/**6** are shown in Figure 5.8. The pose with the minimum COST value has an RMSD of 1.02 Å to the target pose. The pose with the closest RMSD to the target pose itself ranks 6 out of 1000 poses.

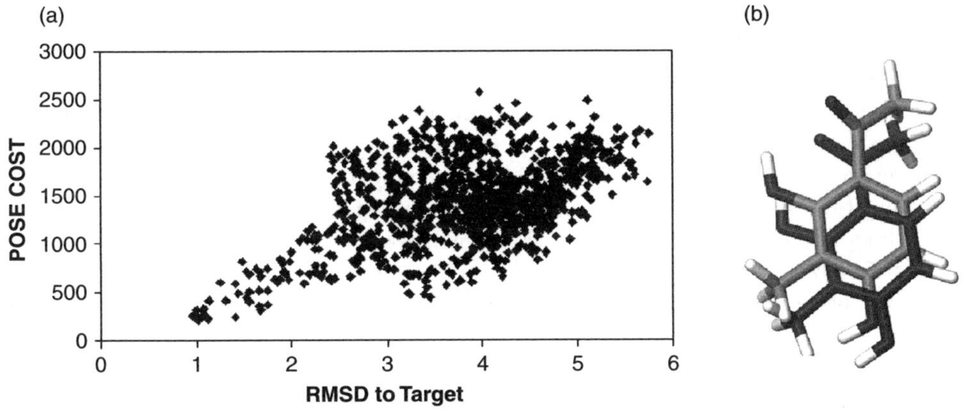

Figure 5.8 *(A) COST versus the RMSD (Å) to the target pose for PDF/**6**. The predicted protein chemical shifts were set to the corresponding BMRB average values. The 3D X-filtered NOESY spectrum used as input for NOE matching was simulated from the target structure. (B) Superposition of target pose and the minimum cost pose (dark gray) from (A).*

Two predominant binding modes were scored as having a low COST by NOE matching. These binding modes can be observed in Figure 5.8A as the lowest COST poses (around 1 Å from the target pose) and a second binding mode whose members are ~3.4 Å from the target pose. These two binding modes represent poses that are ~180° flipped with respect to each other. The target pose for PDF represents a particularly challenging case for NOE matching. An examination of the distribution of the predicted 3D X-filtered NOEs reveals that, although most predicted protein–ligand NOEs are rich in their information content in placing the ligand in the correct region of the pocket, they contain little discrimination power between the two predominant binding modes. These NOEs, which arise from the ligand's central ring and are to PDF methyl groups that lie directly above the ring, are readily satisfied in both the correct and the decoy pose that has the ligand flipped by 180° in the binding pocket. NOEs from the ligand methyl groups at opposite ends of the compound contain the only true information to distinguish between the poses, and residue types at both ends of the pocket are similar – each end of the binding pocket contains isoleucines, leucines and valines. A unique residue in one end of the pocket is a histidine. It is predominantly this residue that allows the NOE matching to score the correct binding pose with a lower COST than the decoy pose.

For the three cases shown above using simulated data on protein/fragment complexes, NOE matching worked with varying degrees of success. As the fragment becomes structurally less complex, the differences in the COST between correct or close to correct poses and decoy poses becomes smaller. Whereas for the CDK2 case NOE matching readily identified the correct pose, for the PDF case, the gradation in COST as structures became more dissimilar to the target pose was very shallow (Figure 5.8A). Nevertheless, the COST for the poses dissimilar to the target pose (observed at approximately 3.5 Å from the target pose in Figure 5.8A) is over a factor of two higher than that for the correct poses. Hence it is evident that, given high-quality data, NOE matching can identify the correct pose even for fragments.

5.5.4 PDF with Experimental Data

In order to determine whether NOE matching will work on small proteins with fragments using 'typical' NMR data, we repeated the calculations, but this time using the experimental NOE cross peak list. A 3D X-filtered NOESY spectrum ($\tau_m = 150$ ms) for the PDF/**6** was acquired on a 1.5 mM sample of PDF with a room temperature probe. The experimentally determined resonance assignments were used for the ligand resonance assignments. The 3D X-filtered NOESY yielded 109 peaks, which were clustered into 78 protein $^1H^{13}C$ groups. Trial binding poses were generated with *Poser* as described above for the simulated example. The NOE matching protocol was run using BMRB-predicted chemical shifts. This test case represents a practical application of NOE matching on this small protein–fragment complex.

The results obtained from applying NOE matching to PDF/**6** are shown in Figure 5.9. The pose with the minimum COST value has an RMSD of 1.12 Å to the target pose. The pose with the closest RMSD to the target pose itself ranks 13 out of 1000 poses. The results obtained with experimental data are similar to the results obtained with simulated data.

5.5.5 PDF with Experimental Data and SHIFTX Chemical Shifts

In the absence of sequence-specific protein NMR resonance assignments, the data used for NOE matching are limited to the unassigned experimental protein ^1H and ^{13}C chemical shifts, the predicted protein ^1H and ^{13}C chemical shifts, and the experimental and predicted

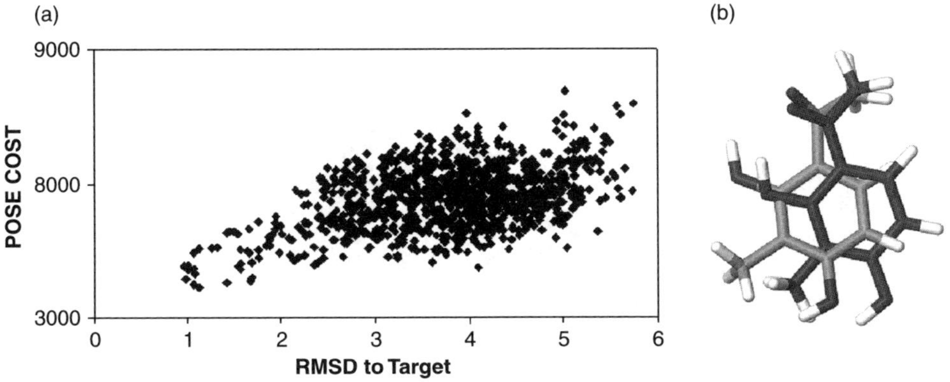

Figure 5.9 *(A) COST versus the RMSD (Å) to the target pose for PDF/**6**. The predicted protein chemical shifts set to the corresponding BMRB average values. Cross peaks from the experimental 3D X-filtered NOESY spectrum were used as input for NOE matching. (B) Superposition of target pose and the minimum cost pose (dark gray) from (A).*

NOE intensities. Due to the relatively large uncertainties associated with predicted chemical shifts, there are often several $^{1}H^{13}C$ groups within the binding pocket that yield predicted chemical shifts that match the experimental chemical shifts within the defined tolerances. NOE matching evaluations have been typically carried out with protein chemical shifts set to those corresponding to BMRB average values. To begin to assess whether predicted chemical shifts might improve the overall ranking of poses, we performed an initial NOE matching evaluation of the PDF poses in which the protein chemical shifts were assigned values predicted with the program SHIFTX.[28] SHIFTX is a computer program (developed by Wishart and co-workers) which predicts ^{1}H, ^{13}C and ^{15}N chemical shifts using a hybrid prediction approach that employs precalculated, empirically derived chemical shift hypersurfaces in combination with classical or semiempirical equations (for ring current, electric field, hydrogen bonds and solvent effect). The hyper-surfaces in SHIFTX are generated using a database of IUPAC-referenced protein chemical shifts (RefDB)[29] and corresponding high-resolution (<2.1 Å) crystal structures. Although SHIFTX generally improves the overall prediction of the chemical shifts relative to the BMRB values for residues that have low mobility in the target structure, we believe that it can be misleading to attempt to predict dynamic residues using a single static structure as a template. For example, if an aromatic side-chain in a binding pocket can adopt multiple conformations and the target pdb structure we have selected has the side-chain aromatic ring in an incorrect conformation (e.g. rotated 90° around χ^2), then the SHIFTX-predicted chemical shifts of many nearby atoms would move upfield (downfield), rather than downfield (upfield), due to reorientation of the local magnetic field predicted for the ring currents. Such a case would adversely affect the scoring in NOE matching. To minimize this possibility, we used the SHIFTX chemical shift prediction only for those residues whose buried surface area was greater than 65 % of their accessible surface area;[30] for all other residues, which were deemed surface exposed and thus potentially mobile, we used the average BMRB value for the chemical shift. Using these chemical shifts, predicted as just described, we re-ran the NOE matching protocol on the poses for the PDF case with experimental data. The results are shown in Figure 5.10.

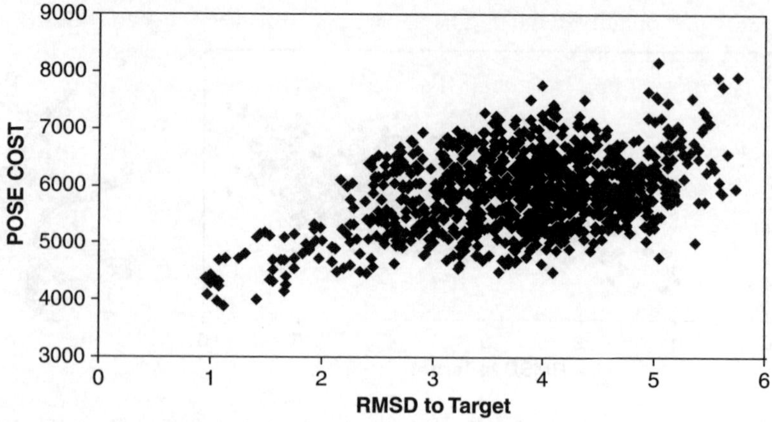

Figure 5.10 *COST versus the RMSD (Å) to the target pose for PDF/**6**. The predicted protein proton chemical shifts were set to the values determined using SHIFTX.*

SHIFTX chemical shift prediction provided a small improvement in the NOE matching scoring. In general, for lower COST poses, poses that were more dissimilar to the target pose (that is, decoy poses) were now scored with a higher relative COST than poses that were more similar to the target pose. For example, the COST of poses with RMSDs of 0.96, 1.12, 2.35 and 4.09 Å from the target pose increased by 5.7, 5.4, 7.6 and 8.5%, respectively. By improving the COST difference between correct and decoy poses, incorporation of SHIFTX-predicted shifts into the NOE matching protocol leads to higher confidence results. We expect that, in general, as chemical shift prediction tools improve, so will the results obtained from NOE matching. Rapid empirical predictions of ligand-induced chemical shift changes for protein resonances could benefit NOE matching (e.g. see ref. 31). Also, accurate quantum mechanics-based predictions of binding-induced chemical shift changes for ligand[32] and/or protein resonances could be applied to evaluate further a small number of selected poses.

5.5.6 Bcl-x_L with Experimental Data

Recently, a potent inhibitor of Bcl-x_L was discovered after applying the 'SAR by NMR' technique to an initial hit from an NMR-based screen.[14] We applied NOE matching to one of the commercially available initial hits [4-fluorobiphenyl-4-carboxylic acid 7, $K_d = 300 \pm 30$ μM].

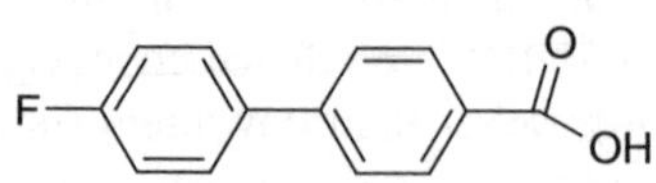

7

Compound 7

The deposited coordinates of the NMR structure of this fragment in complex with Bcl-x_L (PDB entry 1YSG)[14] were used as the target pose. Trial poses were generated using

Poser. Due to the simplicity of the fragment, only one low-energy ligand conformation was used as input. Trial binding poses were generated using *Poser* allowing a 10° rotational sampling. The 'posing box' was defined as the compound binding site plus an additional 3 Å in all coordinate axes. In total, over 191 million poses were generated and evaluated by *Poser*, with 9556 poses being retained. The NOE matching protocol was run using BMRB-predicted chemical shifts. The experimental NOE list for NOE matching on Bcl-x_L/**7** was derived from a 3D ^{13}C-edited NOESY spectrum (τ_m = 100 ms) recorded on a sample consisting of 3.0 mM **7**–1.2 mM ^{13}C/^{15}N Bcl-x_L. The data were recorded on a Varian INOVA spectrometer at 303 K equipped with a Cold Probe. We identified a total of 57 NOEs. (We note that the published NOE list, which was unavailable for our use, contained 75 NOEs.[12]) The 57 experimental NOEs were clustered into 30 protein ^{1}H^{13}C groups for NOE matching.

The results obtained from applying NOE matching to Bcl-x_L/**7** are shown in Figure 5.11. The pose with the minimum COST value has an RMSD of 0.71 Å to the target pose; this pose had the lowest RMSD of all 9556 with respect to the target pose. As shown in Figure 5.11B, the pose with the lowest COST overlays nicely with the target pose.

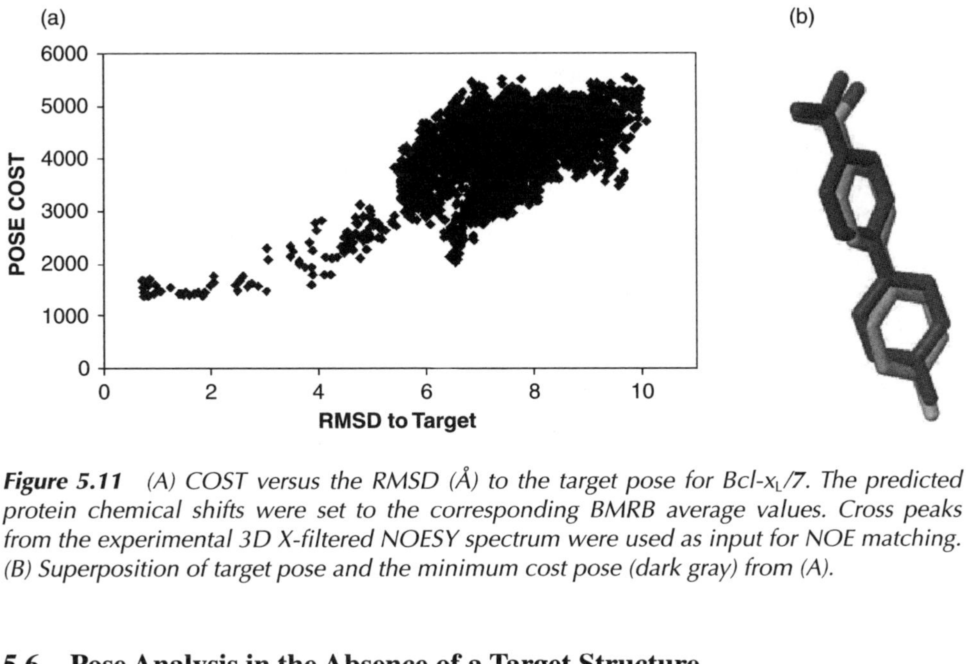

Figure 5.11 *(A) COST versus the RMSD (Å) to the target pose for Bcl-x_L/7. The predicted protein chemical shifts were set to the corresponding BMRB average values. Cross peaks from the experimental 3D X-filtered NOESY spectrum were used as input for NOE matching. (B) Superposition of target pose and the minimum cost pose (dark gray) from (A).*

5.6 Pose Analysis in the Absence of a Target Structure

In all of the cases examined above, the results of NOE matching were evaluated with respect to the target pose, that is, trial poses were compared directly with a known structure. Poses with the lowest COST were bound similarly to the target pose (low RMSD), whereas poses with higher COST tended to be bound more dissimilarly to the target pose (higher RMSD). Although this is important to demonstrate in test cases, one will not be able to rely on such a measure in practice (as the target pose will be unknown). An alternative way to look at the test cases is look at the COST of the poses ranked from lowest COST to highest COST, as has been done in Figure 5.12. In this figure are indicated both the RMSD of the lowest COST pose to the target pose and the RMSD and rank of the pose in the ensemble with the lowest RMSD to the target pose.

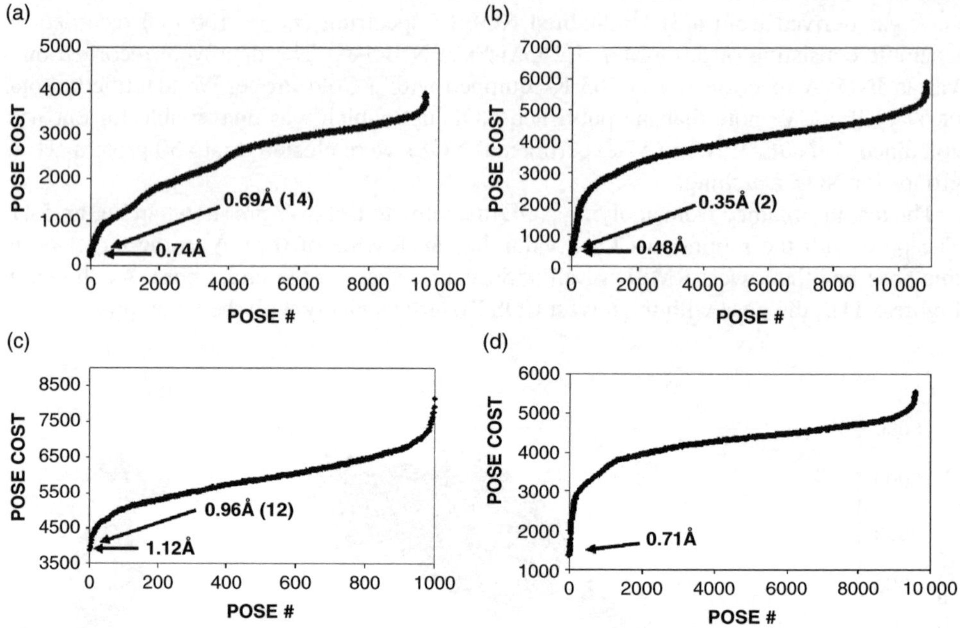

Figure 5.12 *The poses for the NOE matching runs are ordered based on COST and plotted from lowest COST to highest COST. In each panel, the RMSD of the lowest COST pose to the target pose is listed. Also indicated are the RMSD and rank of the pose in the ensemble with the lowest RMSD to the target pose. (A) CDK2/**4** complex using simulated NOE data; (B) FKBP-12/**5** complex using simulated NOE data; (C) PDF/**6** using experimental NOEs and predicted assignments set to SHIFTX values; (D) Bcl-x_L/**7** using experimental NOEs.*

Figure 5.12 clearly indicates that in all cases, for both simulated and experimental data, the correct pose is within the set of poses having the lowest COST, as expected. One only needs to ensure that a correct pose is a member of the ensemble of trial poses; a robust way to achieve this is to do extensive, systematic sampling, as we have done with *Poser*. (As discussed later in this chapter, highly flexible binding sites present the additional challenge of adequately sampling the conformational space of the protein prior to using *Poser*.) In the absence of very unusual protein chemical shifts, the correct pose (or a pose very similar to it) will always be scored with low COST; therefore, one need not consider any of the poses determined to be of higher COST by NOE matching, facilitating the analysis of the results. Typically, we have found that any pose with a COST more than 50 % greater than the lowest COST pose can be discarded from subsequent analysis. As will be seen below, this becomes important when evaluating hundreds of thousands or even millions of trial poses.

Even after discarding higher COST poses, the number of poses to consider can still be very large. The challenge becomes how to assess whether these poses are similar or distinct. To determine how many 'unique' poses exist, we have initially resorted to clustering the poses according to their pairwise RMSDs and then evaluated the binding poses. Although the nature of the systematic sampling done by *Poser* does not lend itself to standard clustering algorithms, we have done so anyway on the premise that the pose distribution will not be uniform once the pose set is filtered by the bump check, contact criteria and NOE matching score.

As an example, we show the results for clustering the poses of the Bcl-x_L/**7** test case. There were 9556 poses generated by *Poser* that were evaluated by NOE matching. The COST values from NOE matching ranged from 1383 to 5541, as shown in Figure 5.12D. The target pose had a COST of 2015. Poses were clustered by RMSD using an internally developed average-linkage method. The intra-group average pairwise similarity was required to be below a threshold of 0.8 Å. For these poses, the result was 1077 clusters, of which 221 were singletons. This was still too many to visualize. By eliminating any pose with a COST greater than two times that of the lowest COST pose, the number of poses to be considered was reduced from 9556 to 156. These poses were clustered as before, this time resulting in 44 clusters, of which 21 were singletons. The largest and second-largest clusters contained 22 and 12 members, respectively, and differed significantly in structure. A plot of the 10 most populated clusters versus their NOE matching COST (Figure 5.13A) indicates that two of the clusters (clusters 2 and 7) contain members that have significantly lower COST than all of the other clusters. The RMSD of the poses representing the centroids of these two clusters indicates that the members of the clusters are structurally similar; the RMSD is 1.14 Å. Moreover, if one plots the

(a)

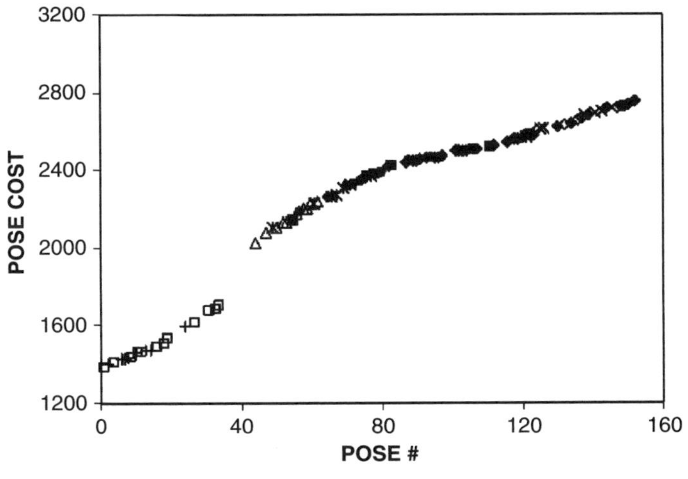

Figure 5.13 *(A) The lowest COST poses for the NOE matching run on Bcl-x_L/**7** were clustered based on RMSD to each other. The individual poses for each of the 10 lowest COST clusters were ordered based on COST and plotted from lowest COST to highest COST. Members of the different clusters are indicated with distinct symbols. All of the lowest COST poses belonged to one of two clusters, as shown in (B). (B) COST versus the RMSD (Å) to the cluster 2 centroid for Bcl-x_L/**7**. Member poses of the 10 lowest scoring clusters are indicated as in (A).*

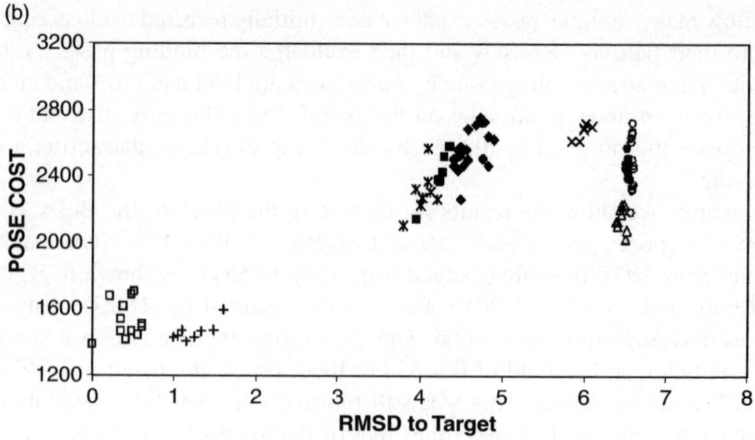

Figure 5.13 (Continued).

RMSD of each of the members of the 10 clusters versus their RMSD to the centroid of cluster 2 (Figure 5.13B), the two clusters with the lowest COST are clearly similar to each other, whereas the remaining (higher COST) clusters are structurally distinct from clusters 7 and 2.

The poses representing the centroids of clusters 2, 7 and 1 (the largest cluster with 22 members) are superimposed in Figure 5.14. Pose members of the lower COST clusters

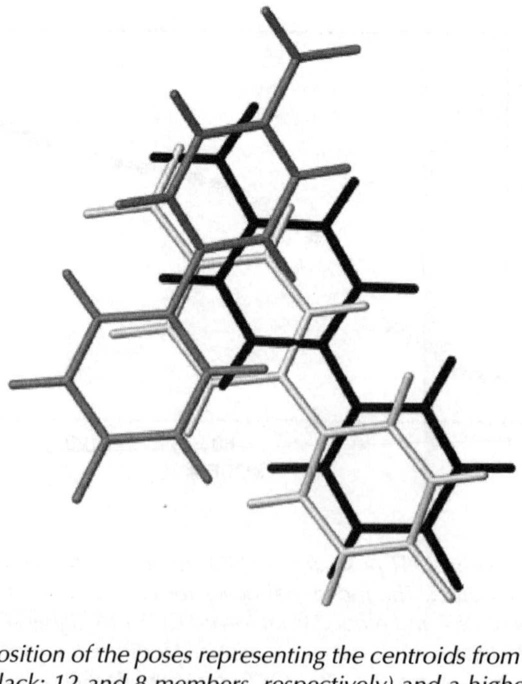

Figure 5.14 Superposition of the poses representing the centroids from the two lowest COST clusters (white and black; 12 and 8 members, respectively) and a higher COST cluster (gray; 22 members).

(clusters 2 and 7) are similar to each other, exhibiting a translation and small tilt in the binding pocket. The members of the higher COST cluster (cluster 1) show a much larger translation and are rotated by ~60° in the binding pocket with respect to the poses in clusters 2 and 7.

5.7 Consequences of Protein Structure Selection on NOE Matching

As mentioned, protein conformational variability presents a significant challenge for NOE matching and also for all other computational docking/scoring procedures. Although not attempting to address this issue thoroughly in this chapter, we demonstrate the effects of using a limited number of different protein conformations for the Bcl-x_L/**7** complex. In the previous Bcl-x_L/**7** example, we had used the protein coordinates from the NMR structure of Bcl-x_L taken from the Bcl-x_L/**7** complex as our starting structure. Since this structure itself was modeled primarily by using assigned protein–ligand NOEs[14] in conjunction with a (mostly fixed) starting structure, the possibility of a significantly different protein conformation with **7** bound cannot be excluded. Because numerous structures have been determined for Bcl-x_L, both in the absence and presence of bound ligands, we had the opportunity to address the consequences of initial protein structure selection on the NOE matching outcome. We ran several more test cases, using protein coordinates generated by different methods. These included apo Bcl-x_L structures[33] determined by either NMR (PDB entry 1LXL) or X-ray crystallography (PDB entry 1MAZ) and the structure of a 16-residue BAK peptide–Bcl-x_L complex determined by NMR (PDB entry 1BXL).[34]

In the first test case, we used the Bcl-x_L coordinates from the apo NMR structure (1LXL). The RMSD of backbone atoms involved in secondary structure of 1LXL is 2.03 Å from 1YSG. Trial binding poses of **7** were generated with *Poser* using a binding site box set to ligand binding site plus an additional 1 Å in all coordinate axes and a 5° rotational sampling. Over 728 000 poses were evaluated, with only 184 poses being retained by *Poser*. The low number of poses retained by *Poser* reflects both the compressed binding site typically found in NMR structures due to the force fields used during refinement and real structural differences between apo and bound structures that partially occlude the binding site. NOE matching was applied using our 57 experimental intermolecular NOEs and with the predicted protein resonance assignments set to the BMRB average values. The COST values from NOE matching ranged from 1745 to 4562, as shown in Figure 5.15A. Clustering of the poses based on RMSD and a similarity of 0.8 Å resulted in 18 clusters, of which four were singletons. Comparison with the poses in each of the clusters to the pose for **7** found in the Bcl-x_L/**7** complex is complicated by the structural differences between the apo and bound protein structures. However, if one superimposes the protein backbone atoms involved in secondary structure and then calculates the RMSD for resultant positions of **7** in each of the structures, the pose with the lowest COST has an RMSD of 3.32 Å with the target pose. For the ensemble of *Poser* trial poses, RMSDs to the target pose ranged from 3.16 to 7.50 Å. As mentioned above and shown in Figure 5.15B, the addition of **7** causes some residues in the binding pocket of apo Bcl-x_L to adopt different conformations in bound Bcl-x_L. For example, F97, Y101 and R139 rearrange upon complex formation with the largest movement occurring with Y101.

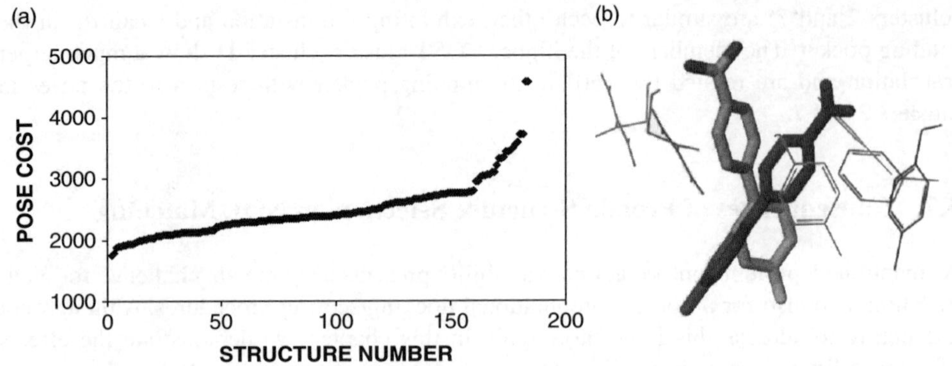

Figure 5.15 *(A) COST versus rank for the NOE matching run on Bcl-x_L/7 poses using the protein coordinates from 1LXL. The poses of the NOE matching run are ordered based on COST and plotted from lowest COST to highest COST. (B) Superposition of target pose and the minimum cost pose (dark gray) from (A). The side-chains of F97, Y101 and R139 from 1YSG (light gray) and 1LXL (dark gray) are also displayed.*

In the second test case, we used the Bcl-x_L coordinates from the apo X-ray structure (1MAZ). The RMSD of backbone atoms involved in secondary structure of 1MAZ is 1.39 Å from 1YSG and 1.54 Å from 1LXL, the NMR apo structure. Trial binding poses of **7** were generated with *Poser* using a binding site box set to ligand binding site plus an additional 1 Å in all coordinate axes and a 5° rotational sampling. Over 734 000 poses were evaluated, with 53 797 poses being retained by *Poser*. To reduce the calculation time while we optimize the code for dealing with larger number of poses, 10 000 poses were randomly selected from the *Poser* ensemble for evaluation with NOE matching. NOE matching was applied using our 57 experimental intermolecular NOEs and with the predicted protein resonance assignments set to the corresponding BMRB average values. The COST values from NOE matching ranged from 1429 to 4429 (Figure 5.16A). Clustering of the poses based on RMSD and a similarity of 0.8 Å resulted in 522 clusters, of which 122 were singletons. As indicated above, comparison with the poses in each of the clusters to the pose for **7** found in the Bcl-x_L/**7** complex is complicated by the structural differences between the apo and bound protein structures. Superimposing the protein backbone atoms involved in secondary structure and then calculating the RMSD for resultant positions of **7** in each of the structures, the pose with the lowest COST has an RMSD of 3.19 Å with the target pose. For the ensemble of *Poser* trial poses, RMSDs to the target pose ranged from 1.40 to 8.33 Å. As mentioned above, several residues in the binding pocket of apo Bcl-x_L to adopt different conformations upon **7** binding (Figure 5.16B).

In the third test case, we used the Bcl-x_L coordinates from the NMR structure of the complex BAK–Bcl-x_L (1BXL). The RMSD of backbone atoms involved in secondary structure of 1BXL is 1.53 Å from 1YSG. Trial binding poses of **7** were generated with *Poser* using a binding site box set to ligand binding site plus an additional 1 Å in all coordinate axes and a 5° rotational sampling. Over 795 000 poses were evaluated, with 127 289 poses being retained by *Poser*. To reduce the calculation time while we optimized

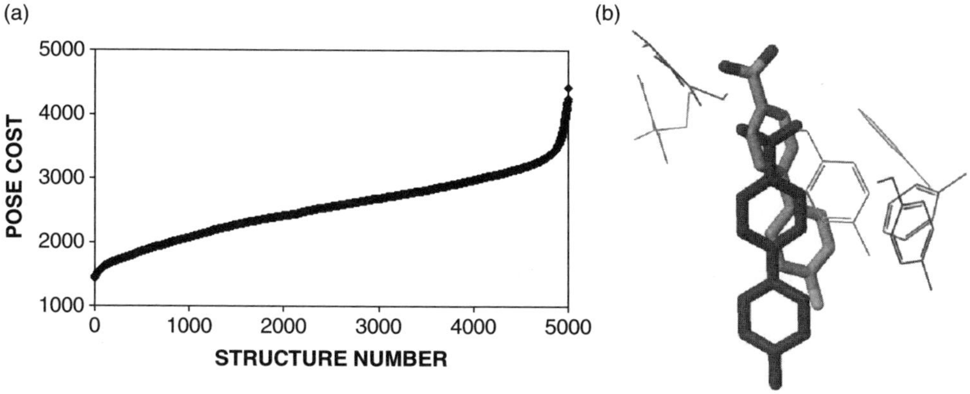

Figure 5.16 (A) COST versus rank for the NOE matching run on Bcl-x_L/7 poses using the protein coordinates from 1MAZ. The poses of the NOE matching run are ordered based on COST and plotted from lowest COST to highest COST. (B) Superposition of target pose and the minimum cost pose (dark gray) from (A). The side-chains of F97, Y101 and R139 from 1YSG (thin light gray) and 1MAZ (thin dark gray) are also displayed.

the code for dealing with larger numbers of poses, 10 000 poses were selected from the *Poser* ensemble based on an RMSD sampling for evaluation with NOE matching. NOE matching was applied using our 57 experimental intermolecular NOEs and with the predicted protein resonance assignments set to the corresponding BMRB average values. The COST values from NOE matching ranged from 1246 to 5023 (Figure 5.17A). Clustering of the poses based on RMSD and a similarity of 0.8 Å resulted in 1222 clusters, of which 260 were

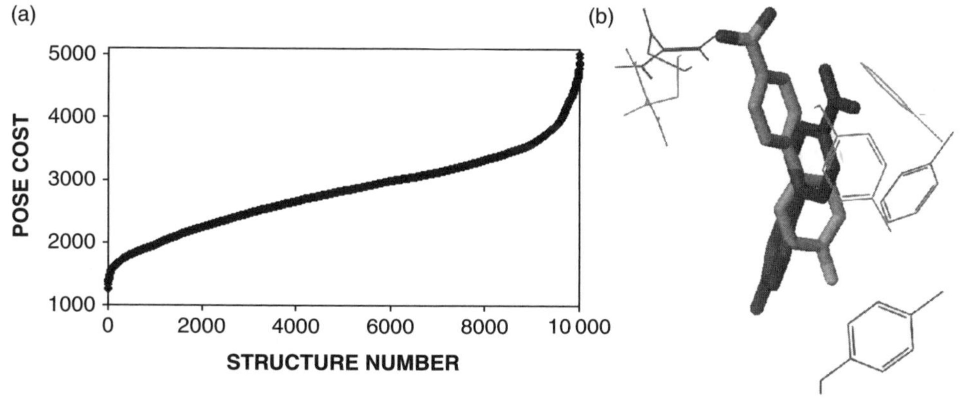

Figure 5.17 (A) COST versus rank for the NOE matching run on Bcl-x_L/7 poses using the protein coordinates from 1BXL. The poses of the NOE matching run are ordered based on COST and plotted from lowest COST to highest COST. (B) Superposition of target pose and the minimum cost pose (dark gray) from (A). The side-chains of F97, Y101 and R139 from 1YSG (thin light gray) and 1BXL (thin dark gray) are also displayed.

singletons. Superimposing the protein backbone atoms involved in secondary structure and then calculating the RMSD for resultant positions of **7** in each of the structures, the pose with the lowest COST has an RMSD of 2.98 Å with respect to the target pose. For the ensemble of *Poser* trial poses, RMSDs to the target pose ranged from 1.08 to 7.93 Å. As mentioned above, several residues in the binding pocket of apo Bcl-x_L to adopt different conformations upon **7** binding (Figure 5.17B).

The test cases above demonstrate the importance of coordinate selection for use during the NOE matching runs. None of the test cases using alternative Bcl-x_L protein coordinates have satisfactorily reproduced the experimentally determined target pose for Bcl-x_L/**7**. This is because of significant backbone and side-chain rearrangement in the binding pocket. NOE matching is predicated on the fact that somewhere in the ensemble of poses to be evaluated is a pose that adequately reflects the 'correct' pose. We are evaluating how often significant movement of side-chains occurs and how to identify in advance when such movements will complicate NOE matching analysis.

It is interesting to note the significant difference in the number of poses accepted by *Poser* for the three test cases using different Bcl-x_L starting coordinates. Whereas the apo NMR structure accepted only 184 poses, the apo X-ray structure accepted >57 000 poses and the NMR structure taken from the BAK complex accepted >127 000 poses. This is because the binding site in the BAK peptide-bound Bcl-x_L structure is more open than in the other structures. The more poses that can be sampled, the better the chance one has of finding a pose close to the correct pose. Moreover, nine poses of **7** generated from the BAK–Bcl-x_L structure had COSTs lower than any COSTs from poses using the other structures of Bcl-x_L. The best scoring pose obtained using the BAK–Bcl-x_L structure has a COST of 1246, whereas the best scoring pose obtained using the 1YSG structure had a COST of 1383. These results suggest that, of the four protein conformations used, the BAK–Bcl-x_L conformation may be the most similar to the true protein conformation in the Bcl-x_L/**7** complex. (Rigorous proof of this suggestion could only be obtained by from an X-ray or high-resolution full NMR structure of the Bcl-x_L/**7** complex.) These results indicate that it is best to use all available experimental protein conformations for NOE matching. Additional, computationally derived protein conformations may also be useful, provided that one can be confident that these conformations are realistic. Because of the importance of sampling the correct protein conformation in addition to the correct ligand conformation, location and orientation, we are in the process of evaluating the use of protein ensembles as input target structures for NOE matching.

The use of RMSD to evaluate the success of NOE matching has been used, in part, for convenience. An alternative criterion for pose evaluation is to gauge how well the lowest COST poses explain any known SAR data and to gauge how predictive they are. In other words, if the lowest COST poses display the correct interactions with the binding site and if they *correctly predict potential new interactions*, then they should explain any previously known SAR data and should greatly facilitate structure-based lead optimization. After all, the primary goal of this work is to be able to identify rapidly poses that provide information to the chemists to guide the next round of synthesis. In all of the test cases for Bcl-x_L/**7**, NOE matching was successful at one level. It was able to distinguish between the two predominant poses for the complex: the low COST poses shown in Figures 5.11B, 5.14B, 5.15B, 5.16B and 5.17B and a second predominant pose in which **7** is flipped 180° in the binding pocket.

In all cases, the flipped poses had significantly higher COST, typically greater than 50 % more than the lowest COST poses. We are in the process of examining whether local interactions observed between Bcl-x_L and **7** are reproduced by the low COST structures.

5.8 Applications to Fragment-like Compounds Bound to Large Proteins

5.8.1 Nonuniform Protein Labeling

In its initial embodiment,[15] NOE matching was designed for unlabeled compounds bound to uniformly protonated, $^{13}C/^{15}N$-labeled protein samples. The initial success of NOE matching is due to the fact that, even in the absence of protein assignments, the *pattern* of NOEs contains enough information to limit the number of residue types that need to be considered as the potential partner giving rise to the NOE. As the size of a protein becomes larger, sensitivity is dramatically reduced due to increased relaxation rates, yielding many fewer NOEs and limiting the information content that can be extracted. However, protein–ligand NOE interactions with high sensitivity and information content can be obtained by nonuniform isotopic labeling schemes. For example, specific types of amino acid residues that are isotope labeled in a particular manner (e.g. protonated or protonated and $^{13}C/^{15}N$ labeled) can be incorporated into an otherwise uniformly perdeuterated (or perdeuterated and ^{15}N labeled) protein background.[35–38] These procedures produce protein samples that are labeled by residue type (residue type specific labeling). Residue type and residue type/atom type specific labeling schemes yield enhanced NOE sensitivities and, by reducing spin diffusion, more accurate distance restraints. These labeling schemes have been used to observe protein–ligand NOEs for complexes involving large proteins.[39, 40]

With regard to NOE matching, selective labeling schemes are important in that they can provide the identities of the residue types involved in protein–ligand NOEs. Furthermore, when a particular residue type occurs only once in a binding pocket, residue-type specific labeling combined with protein–ligand NOE experiments directly provide sequence-specific resonance assignments. An approach using a series of residue-type specific labeled samples in conjunction with saturation transfer difference (STD) NMR experiments has been described for characterizing ligand binding poses (SOS NMR[41]). Compared with an STD spectrum, a 2D or 3D NOESY spectrum using a residue type specifically labeled sample contains significantly more information. For example, given a valine-type specific labeled sample, observation of an STD only indicates that *one or more* protons on *one or more* valines is/are close to a given ligand proton. The NOESY data provide information on *how many valines* and *which specific valine atoms* (methyl, HB, HA) are close to a given ligand proton. NOE matching utilizes the additional information obtained in a NOESY spectrum.

In an effort to extend the applicability of NOE matching to the larger proteins of pharmaceutical interest, for example kinases and phosphatases, we have adapted NOE matching to be able to utilize data from any residue (and atom)-specific labeling scheme. An illustrative example is provided below.

5.8.2 CDK2 with Residue-specific Labeling

To assess how NOE matching will work using data from a residue-specific labeled protein, the test case on the CDK2/**4** complex was re-run with modification of the NOE input list; only those NOEs that could be observed in a residue-specific labeled sample were included. While the development of cell free protein expression systems opens up all residue types to selective labeling, we included only those residue types existing in the active site that are routinely labeled in proteins expressed in *E. coli*. For the active site of CDK2, these residues included isoleucine, valine, leucine, lysine, phenylalanine and alanine. NOEs from residues such as aspartate and glutamate were removed from the list for this simulation. All other parameters for the NOE run were as described above. The simulated NOE list for this complex contained a total of 62 peaks, which were clustered into 40 protein $^{1}H^{13}C$ groups. The results obtained from applying NOE matching to CDK2/**4** are shown in Figure 5.18. The pose with the minimum COST value has an RMSD of 0.92 Å to the target pose. The pose with the closest RMSD to the target pose itself ranks 19 out of 10 579 poses. In comparison, if all NOEs are included as input for the NOE matching calculation, the pose with the minimum COST value has an RMSD of 0.74 Å to the target (see the previous section).

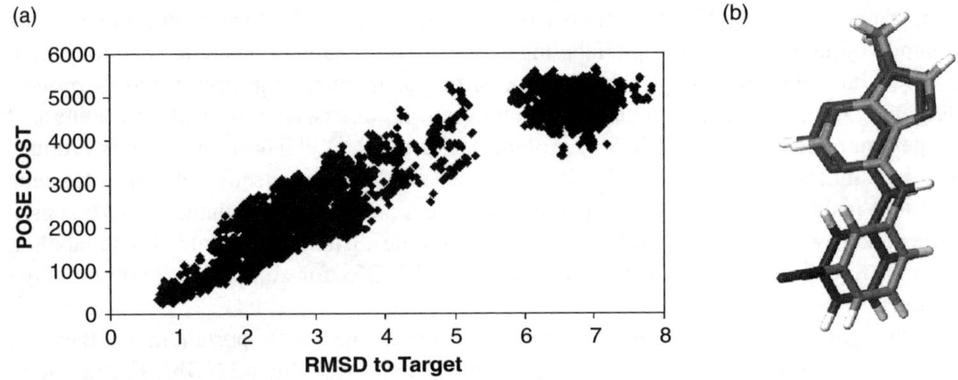

*Figure 5.18 (A) COST versus the RMSD (Å) to the target pose for CDK2/**4**. The predicted protein chemical shifts were set to the corresponding BMRB average values. The 3D X-filtered NOESY spectrum was filtered to simulate data that could be extracted from residue-specific labeled protein. (B) Superposition of target pose and the minimum cost pose (dark gray) from (A).*

5.9 Towards Larger Proteins by Nonuniform Labeling and Stability Enhancement

In this section, some of the approaches described above for enhancing the sensitivity and information content of protein–ligand NOEs are demonstrated for relatively large protein–inhibitor complexes. In addition, we demonstrate that a medium-quality 3D X-filtered NOESY spectrum can be obtained for a large protein–inhibitor complex by using a stabilized, uniformly $^{13}C/^{15}N$-labeled protein sample in conjunction with an elevated experimental temperature to increase the rotational correlation time of the protein–ligand complex.

These studies lay the groundwork for applying NOE matching to large proteins of high therapeutic interest.

We have recently undertaken NMR studies of several kinase–inhibitor complexes. Neither the kinase (which we subsequently refer to as 'kinaseX') nor the exact chemical structures of these inhibitors can be revealed at this time. The inhibitors all contain a heterocyclic core that is expected to bind to the 'hinge' region of kinaseX by accepting a hydrogen bond from a backbone HN, a basic aliphatic moiety that is expected to bind in the general location where the ribose and phosphates of ADP/ATP bind, and an aromatic substituent linked to the heterocyclic core. High-sensitivity standard 2D ^{1}H–^{1}H NOESY and TOCSY spectra were obtained in an inhibitor complexed to a uniformly ^{2}H-labeled kinaseX using ^{2}H-labeled buffer components, as predicted previously.[42] These spectra afforded the ^{1}H NMR assignments for kinaseX inhibitors and allowed the identification of several intermolecular NOE contacts as outlined below.

The ADP binding pocket of kinaseX contains two valines and three leucines. We produced a sample of kinaseX that incorporated [^{1}H]Leu into an otherwise fully deuterated protein and a second sample that incorporated [^{1}H]Val into an otherwise fully deuterated protein. {To prevent unwanted labeling of an amino acid in the biosynthetic pathway of desired amino acid, we supplement the growth media during induction with the undesired amino acid(s) that are ^{2}H-labeled; e.g. [^{2}H]Val was added to samples incorporating [^{1}H]Leu.} NOESY spectra of an inhibitor ('kinaseX inhibitor 1') in complex with kinaseX were recorded at 15 °C using these kinaseX samples (Figure 5.19). The heterocyclic core of kinaseX inhibitor 1 has aromatic ^{1}H resonances at 8.91 and 6.44 ppm and the aromatic substituent has aromatic ^{1}H resonances at 7.01 and 6.78 ppm. The inhibitor also has aliphatic ^{1}H resonances at 1.82 and 1.48 ppm. The heterocyclic core aromatic resonances give rise to NOEs of varying intensities to at least two Leu residues, whereas the aromatic substituent yields only one very weak (tentative) NOE involving a leucine at $F1 = 0.82$ ppm, $F2 = 7.01$ ppm (Figure 5.19A). The heterocyclic core has intense NOEs to valine resonances at 1.65 and 1.41 ppm and weak NOEs to a valine resonance at 2.10 ppm (Figure 5.19B). The resonances of the aromatic substituent at 7.01 and 6.78 ppm give rise to medium- and strong-intensity NOEs, respectively, involving a Val resonance at 0.14 ppm (Figure 5.19B).

KinaseX samples that are residue type-specifically labeled with [^{1}H]Thr, [^{1}H]Lys and [^{1}H]Met have been also produced and NOEs between these residues and inhibitors have been observed (data not shown). Since there is one threonine, one lysine and one methionine in the ADP binding pocket of kinaseX, sequence-specific assignments for these residues can be obtained directly by the observation of protein–inhibitor NOEs.

A cautionary note must be provided for using peaks from type-specifically labeled samples and merging peak lists from different spectra. An implicit assumption made when calibrating peaks from *uniformly* labeled samples is that the strongest NOE cross peaks correspond to distances approximating the van der Waals radii and the weakest NOE cross peaks correspond to distances in the range 5–5.5 Å. This assumption no longer holds true for spectra acquired in type-specific labeled proteins. In such spectra, distances corresponding to the strongest observed NOEs can be in excess of 5 Å and those for the weakest observed NOEs can be in excess of 9 Å. Hence it is essential for the NOEs from type-specifically labeled samples to be properly scaled before translating them into distances. This can be achieved by comparison with intra-ligand NOEs (2D NOEs or double reverse filtered

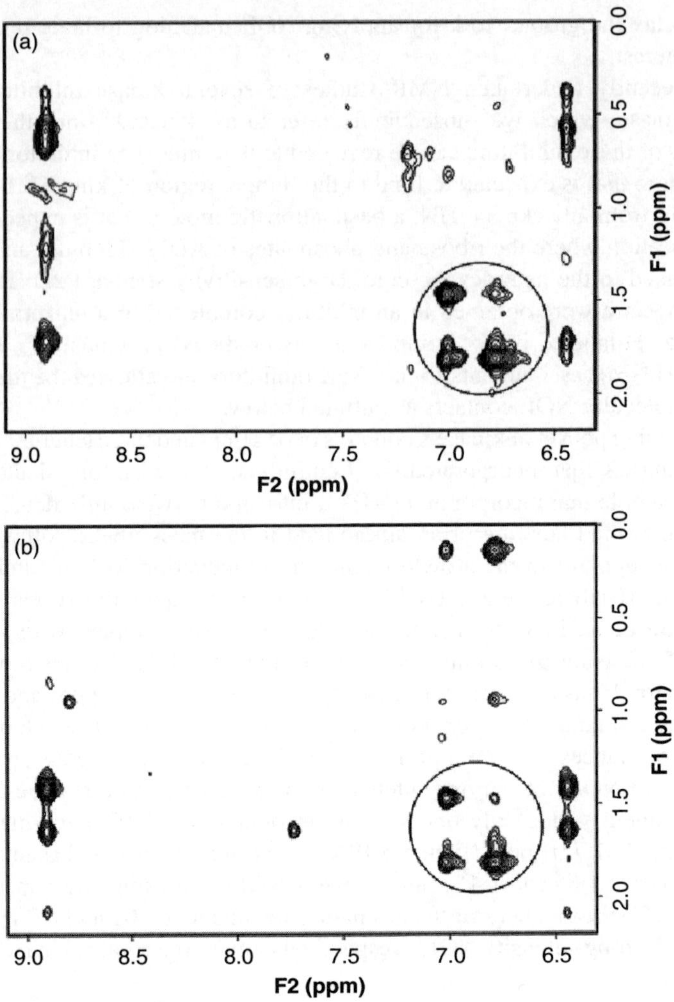

Figure 5.19 *Portions of NOESY spectra of kinaseX inhibitor 1 in complex with residue type specifically protonated samples of kinaseX. Intra-ligand cross peaks are circled in both spectra. (A) KinaseX inhibitor 1 in complex with [^{1}H]Leu (otherwise ^{2}H-labeled) kinaseX. Concentrations of both the protein and the inhibitor used were 140 μM. (B) KinaseX inhibitor 1 in complex with [^{1}H]Val (otherwise ^{2}H-labeled) kinaseX. Concentrations of both the protein and the inhibitor were 90 μM. Both spectra (A) and (B) were recorded at 15 °C, 600 MHz ^{1}H frequency, using a NOESY mixing time of 60 ms.*

NOEs). Failure to do so can result in the generation of highly inaccurate structures. Regardless of the labeling scheme, one must take care to choose a mixing time, or range of mixing times, that yields an adequate number of NOEs without suffering from severe spin diffusion effects. Although we are still gaining experience with the number of NOEs required to ensure reliable pose identification, for the systems that we have studied to date we have had from 9 to 22 NOEs per ligand proton group.

In addition to nonuniform labeling schemes, another approach for observing protein–ligand NOEs in larger systems involves collecting 3D X-filtered NOESY spectra on a uniformly labeled sample at elevated temperatures. The decreased rotational correlation time of the system at a higher temperature is generally expected to improve the sensitivity of the 3D X-filtered NOESY experiment. Exceptions can occur if exchange broadening increases with increased temperature. For this approach to be applied, it will often be necessary to increase the thermal stability of the target protein. This can be accomplished in a number of ways, including the rational design of point mutants,[43, 44] combinatorial mutagenesis in conjunction with stability screening,[45] deletion of flexible loops[46] and through the use of osmolytes.[47, 48] Although not applied here, another method that could potentially afford high-sensitivity 3D X-filtered NOESY data on large, uniformly labeled protein–ligand complexes is encapsulation in reverse micelles dissolved in low-viscosity fluids;[49, 50] this can greatly reduce the rotational correlation time.

We have produced kinaseX constructs with significantly enhanced thermal stability (J. Newitt *et al.*, unpublished work). Using one of these stability-enhanced constructs, a uniformly $^{13}C/^{15}N$-labeled kinaseX sample was prepared and complexed with an inhibitor ('kinaseX inhibitor 2'). This inhibitor has a heterocyclic core different from that of kinaseX inhibitor 1. Figure 5.20 shows a portion of a 3D X-filtered NOESY spectrum recorded at 35 °C for the complex with kinaseX inhibitor 2. The spectral region in Figure 5.20 displays NOE interactions between the protein and a resonance at 7.83 ppm (*F*3 position) arising from the heterocyclic core. Protein assignments for some of these peaks are based

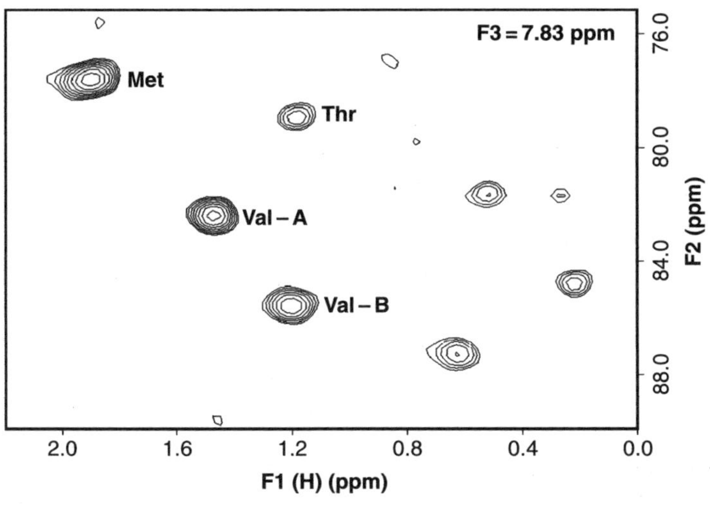

Figure 5.20 *Portion of a 3D X-filtered NOESY spectrum of uniformly* $^{13}C/^{15}N$*-labeled, stability-enhanced kinaseX in complex with kinaseX inhibitor 2. The protein and inhibitor concentrations used were 300 μM. The* F3 *(inhibitor* ^{1}H*) plane is at 7.83 ppm. Peaks with protein resonance assignments are labeled. (Note: Val-A and Val-B refer to the* γ_1 *and* γ *methyl, respectively, of the same valine residue.) The spectrum was recorded at 35 °C, 600 MHz* ^{1}H *frequency, using a NOESY mixing time of 100 ms on a Varian Inova spectrometer equipped with a Cold Probe. The spectrum is aliased in the* ^{13}C *(*F2*) dimension.*

on our previous studies with residue type-specifically ^{1}H-labeled kinaseX samples. In total, 20 ligand–protein peaks have been observed in this spectrum.

In the examples above, lead-like inhibitors of kinaseX with sub-micromolar affinities were studied. We have recently combined residue type-specific labeling ($^{1}H/^{13}C/^{15}N$-labeled amino acids incorporated into a $^{2}H/^{15}N$ background) with 3D X-filtered NOESY studies at elevated temperatures to characterize a fragment-like compound bound to kinaseX. Protein–ligand NOEs have also been observed in this case (data not shown). Due to the extreme conformational plasticity of protein kinases,[51] adequate sampling of protein conformational space is crucial for applying pose generation and NOE matching to kinase–inhibitor complexes, including those involving kinaseX. As described earlier in this chapter, pose sampling (including protein conformational sampling) is an ongoing area of research in our group and elsewhere. Flexibility is known to be a significant challenge for kinase–inhibitor docking procedures.[52] As we have shown for kinaseX, NOEs between kinases and inhibitors can readily be observed. Protein NMR assignments can be obtained for some of these interactions without undertaking a full sequential assignment of the protein. NOE data can provide detailed information on the location and orientation of inhibitor moieties (e.g. hinge-binding cores) that interact with the relatively rigid regions of kinases. In general, an accurate pose should be consistent with all of the observed NOE data; therefore, we expect enhanced NOE measurements and NOE matching to play significant roles in evaluating models of fragments and leads bound to large, flexible proteins.

5.10 Conclusion

In the pharamaceutical industry, NMR spectroscopy has demonstrated itself to be a powerful, highly versatile tool that has impact throughout the drug discovery process. NMR is frequently used as an assay to screen compound collections, to facilitate the assessment of hits, and to provide detailed structural and dynamical characterization of protein-ligand complexes. Because NMR can provide information in discrete units, the spectroscopist can "fine tune" data collection strategies.

The application of NMR to the characterization of biomolecular structures has, in most cases, followed bottom-up approaches[53] wherein discrete pieces of information (resonance assignments, NOE contacts, specific dihedral angle restraints, inter-atomic vector orientations within some reference frame, etc.) are gathered and finally used to define a consistent ensemble of structures. In some situations, this aspect of biomolecular NMR is a great advantage, since minimal information may be all that is needed to answer the specific question at hand, e.g. one may want to know if a particular aromatic ring on the ligand interacts with an aromatic ring from the protein. In other situations (complete structure determination, binding pose determination, etc.), the piecewise aspect of NMR is a disadvantage since, even with automation (reviewed in ref. 53), the bottom-up process of NMR-based structure determination is very time and resource consuming.

There have been efforts to utilize NMR data in top-down approaches for structure determinations. Perhaps the most ambitious protocol, and the one that is most closely analogous to X-ray crystallography, is the CLOUDS method.[54–56] In this approach, an unassigned 2D NOESY spectrum is transformed into a 'proton density' via relaxation matrix approaches and an atomic model is subsequently fitted to this 'proton density'. So far, this approach has only been demonstrated for very small proteins for which high-resolution, high-sensitivity data can be obtained. Another example of a top-down approach is the program AUREMOL,[53] wherein a trial structure is iteratively refined until a good match to the experimental data is obtained. Both CLOUDS and AUREMOL are focused on protein structure determination, not on ligand binding pose determinations.

As discussed elsewhere,[15] NOE matching is primarily a specialized top-down approach, focused on ligand binding pose evaluation, that can also readily incorporate information derived from bottom-up approaches. The results presented in this chapter demonstrate that NOE matching is applicable to fragments and lead-like/drug-like compounds bound to relatively small proteins. The main limitation to applying the method to larger proteins is the increased difficulty of observing protein–ligand NOEs in such systems. Our initial forays into approaches aimed at dealing with larger systems have been described in this chapter and our initial results are very promising. In addition to the protein stability enhancement and selective labeling strategies that we have utilized, other technologies that have yet to be explored have the potential to be major, permitting breakthroughs with respect to the application of NOE matching to large systems. These include the use of SAIL (stereo-arrayed isotope labeling) amino acids[57, 58] and the use of reverse-micelle encapsulation technologies.[49, 50]

Two key points regarding NOE matching are worth reiterating: (1) if the ensemble of trial poses contains some that are very similar to the true pose, NOE matching will generally score these poses with a low COST relative to most of the decoy poses; and (2) to ensure that one obtains a correct pose in the ensemble of trial poses, one needs to do extensive, systematic sampling of 'pose space'. Even with extensive sampling, one still may detect gaps in the RMSD space of the sampled poses with respect to the target pose (e.g. see Figures 5.4, 5.6 and 5.13); these gaps likely result from RMSD ranges for which no acceptable poses could be found, presumably due to steric hindrance, etc. As we have discussed, many additional improvements to NOE matching are possible. Some methods for evaluating the results of NOE matching in the absence of a known pose have been demonstrated in this chapter. Other ways to evaluate the results are also being explored. For the correct pose, most experimental NOE peaks should be assigned and the assignments should be plausible. As the bipartite graph matching algorithm requires predefined edge costs that cannot be adjusted during the search for an optimal match, it is difficult to incorporate explicitly connectivity information into the matching procedure. However, one could check the resulting assignments from NOE matching for consistency with known connectivity information. For example, we may know from TOCSY or COSY data that several experimental groups arise from the same (unassigned) residue – the assignments produced by NOE matching could be checked for consistency with this information. Another potential area of improvement involves ranking poses with low NOE matching COSTs by molecular mechanics energies and/or knowledge-based scoring potentials. Pose scoring based on observed and predicted ligand ^{1}H chemical shift changes[32] could also be used to rank a small subset of poses. More generally, NOE matching could be readily combined with other pose ranking procedures such as MM-GBSA[59] or MM-PBSA[60] as part of a consensus scoring approach (e.g. see ref. 61). Finally, as mentioned previously,[15] much of the NOE matching procedure may be recast in terms of Bayesian probabilities, e.g. a Bayesian analysis of chemical shifts can be used to predict the probability of a spin system arising from a specific amino acid type.[62]

This opens up the possibility of rigorously assigning likelihoods to the NOE assignments obtained from NOE matching, which in turn will facilitate the development iterative pose refinement strategies.

In conclusion, we expect that NOE matching will contribute significantly to our future drug discovery efforts and that the continued development of NOE matching and associated algorithms and technologies will keep us busy for some considerable time to come.

References

[1] Shuker, S. B., Hajduk, P. J., Meadows, R. P. and Fesik, S. W. (1996). Discovering high-affinity ligands for proteins: SAR by NMR. *Science* **274**, 1531–1534.

[2] Hajduk, P. J. and Greer, J. (2007). A decade of fragment-based drug design: strategic advances and lessons learned. *Nat. Rev. Drug Discov.* **6**, 211–219.

[3] Hann, M. M., Leach, A. R. and Harper, G. (2001). Molecular complexity and its impact on the probability of finding leads for drug discovery. *J. Chem. Inf. Comput. Sci.* **41**, 856–864.

[4] Oprea, T. I., Davis, A. M., Teague, S. J. and Leeson, P. D. (2001). Is there a difference between leads and drugs? A historical perspective. *J. Chem. Inf. Comput. Sci.* **41**, 1308–1315.

[5] Anderson, A. C. (2003). The process of structure-based drug design. *Chem. Biol.* **10**, 787–797.

[6] Vajda, S. and Guarnieri, F. (2006). Characterization of protein–ligand interaction sites using experimental and computational methods. *Curr. Opin. Drug Discov. Dev.* **9**, 354–362.

[7] Muchmore, S. W. and Hajduk, P. J. (2003). Crystallography, NMR and virtual screening: Integrated tools for drug discovery. *Curr. Opin. Drug Discov. Dev.* **6**, 544–549.

[8] Villar, H. O., Yan, J. and Hansen, M. R. (2004). Using NMR for ligand discovery and optimization. *Curr. Opin. Chem. Biol.* **8**, 387–391.

[9] Nienaber, V. L., Richardson, P. L., Klighofer, V., Bouska, J. J., Giranda, V. L. and Greer, J. (2000). Discovering novel ligands for macromolecules using X-ray crystallographic screening. *Nat. Biotechnol.* **18**, 1105–1108.

[10] Lesuisse, D., Lange, G., Deprez, P., Benard, D., Schoot, B., Delettre, G., Marquette, J.-P., Broto, P., Jean-Baptiste, V., Bichet, P., Sarubbi, E. and Mandine, E. (2002). SAR and X-ray. A new approach combining fragment-based screening and rational drug design: application to the discovery of nanomolar inhibitors of Src SH2. *J. Med. Chem.* **45**, 2379–2387.

[11] Hartshorn, M. J., Murray, C. W., Cleasby, A., Frederickson, M., Tickle, I. J. and Jhoti, H. (2005). Fragment-based lead discovery using X-ray crystallography. *J. Med. Chem.* **48**, 403–413.

[12] Petros, A. M., Dinges, J., Augeri, D. J., Baumeister, S. A., Betebenner, D. A., Bures, M. G., Elmore, S. W., Hajduk, P. J., Joseph, M. K., Landis, S. K., Nettesheim, D. G., Rosenberg, S. H., Shen, W., Thomas, S., Wang, X., Zanze, I., Zhang, H. and Fesik, S. W. (2006). Discovery of a potent inhibitor of the antiapoptotic protein Bcl-x_L from NMR and parallel synthesis. *J. Med. Chem.* **49**, 656–663.

[13] Hajduk, P. J., Sheppard, G., Nettesheim, D. G., Olejniczak, E. T., Shuker, S. B., Meadows, R. P., Steinman, D. H., Carrera, G. M., Jr, Marcotte, P. A., Severin, J., Walter, K., Smith, H., Gubbins, E., Simmer, R., Holzman, T. F., Morgan, D. W., Davidsen, S. K., Summers, J. B. and Fesik, S. W. (1997). Discovery of potent nonpeptide inhibitors of stromelysin using SAR by NMR. *J. Am. Chem. Soc.* **119**, 5818–5827.

[14] Oltersdorf, T., Elmore, S. W., Shoemaker, A. R., Armstrong, R. C., Augeri, D. J., Belli, B. A., Bruncko, M., Deckwerth, T. L., Dinges, J., Hajduk, P. J., Joseph, M. K., Kitada, S., Korsmeyer, S. J., Kunzer, A. R., Letai, A., Li, C., Mitten, M. J., Nettesheim, D. G., Ng, S. C., Nimmer, P. M., O'Connor, J. M., Oleksijew, A., Petros, A. M., Reed, J. C., Shen, W., Tahir, S. K., Thompson,

C. B., Tomaselli, K. J., Wang, B., Wendt, M. D., Zhang, H., Fesik, S. W. and Rosenberg, S. H. (2005). An inhibitor of Bcl-2 family proteins induces regression of solid tumours. *Nature* **435**, 677–681.

[15] Constantine, K. L., Davis, M. E., Metzler, W. J., Mueller, L. and Claus, B. L. (2006). Protein–ligand NOE matching: a high-throughput method for binding pose evaluation that does not require protein NMR resonance assignments. *J. Am. Chem. Soc.* **128**, 7252–7263.

[16] Fesik, S. W. and Zuiderweg, E. R. P. (1988). Heteronuclear three-dimensional NMR spectroscopy. A strategy for the simplification of homonuclear two-dimensional NMR spectra. *J. Magn. Reson.* **78**, 588–593.

[17] Petros, A. M., Kawai, M., Luly, J. R. and Fesik, S. W. (1992). Conformation of two non-immunosuppressive FK506 analogs when bound to FKBP by isotope-filtered NMR. *FEBS Lett.* **308**, 309–314.

[18] Lee, W., Revington, M. J., Arrowsmith, C. and Kay, L. E. (1994). A pulsed field gradient isotope-filtered 3D ^{13}C HMQC-NOESY experiment for extracting intermolecular NOE contacts in molecular complexes. *FEBS Lett.* **350**, 87–90.

[19] Zwahlen, C., Legault, P., Vincent, S. J. F., Greenblatt, J., Konrat, R. and Kay, L. E. (1997). Methods for measurement of intermolecular NOEs by multinuclear NMR spectroscopy: Application to a bacteriophage lambda N-peptide/boxB RNA complex. *J. Am. Chem. Soc.* **119**, 6711–6721.

[20] Breeze, A. L. (2000). Isotope-filtered NMR methods for the study of biomolecular structure and interactions. *Prog. Nucl. Magn. Reson. Spectrosc.* **36**, 323–372.

[21] Seavey, B. R., Farr, E. A., Westler, W. M. and Markley, J. L. (1991). A relational database for sequence-specific protein NMR data. *J. Biomol. NMR* **1**, 217–230.

[22] Papadimitriou, C. H. and Steiglitz, K. (1982). *Combinatorial Optimization: Algorithms and Complexity*, Dover Publications, Mineola, NY.

[23] Halgren, T. A., Murphy, R. B., Friesner, R. A., Beard, H. S., Frye, L. L., Pollard, W. T. and Banks, J. L. (2004). Glide: a new approach for rapid, accurate docking and scoring. 2. Enrichment factors in database screening. *J. Med. Chem.* **47**, 1750–1759.

[24] Stahl, M. T., Skillman, A. G. and Sayle, R. (2002). OEChem. *Abstracts of Papers, 224th ACS National Meeting, Boston, MA, 18–22 August 2002*, COMP-175.

[25] Stahl, M. T., Nicholls, A., Sayle, R. A. and Grant, J. A. (1999). Rapid conformation search applied to ligand discovery. *Book of Abstracts, 217th ACS National Meeting, Anaheim, CA, 21–25 March 1999*, COMP-026.

[26] Gray, N. S., Wodicka, L., Thunnissen, A.-M. W. H., Norman, T. C., Kwon, S., Espinoza, F. H., Morgan, D. O., Barnes, G., LeClerc, S., Meijer, L., Kim, S.-H., Lockhart, D. J. and Schultz, P. G. (1998). Exploiting chemical libraries, structure and genomics in the search for kinase inhibitors. *Science* **281**, 533–538.

[27] Nilges, M., Gronenborn, A. M., Brunger, A. T. and Clore, G. M. (1988). Determination of three-dimensional structures of proteins by simulated annealing with interproton distance restraints: Application to crambin, potato carboxypeptidase inhibitor and barley serine proteinase inhibitor 2. *Protein Eng.* **2**, 27–38.

[28] Neal, S., Nip, A. M., Zhang, H. and Wishart, D. S. (2003). Rapid and accurate calculation of protein ^{1}H, ^{13}C and ^{15}N chemical shifts. *J. Biomol. NMR* **26**, 215–240.

[29] Zhang, H., Neal, S. and Wishart, D. S. (2003). RefDB: a database of uniformly referenced protein chemical shifts. *J. Biomol. NMR* **25**, 173–195.

[30] Chothia, C. (1976). The nature of the accessible and buried surfaces in proteins. *J. Mol. Biol.* **105**, 1–12.

[31] McCoy, M. A. and Wyss, D. F. (2000). Alignment of weakly interacting molecules to protein surfaces using simulations of chemical shift perturbations. *J. Biomol. NMR* **18**, 189–198.

[32] Wang, B., Raha, K. and Merz, K. M., Jr. (2004). Pose scoring by NMR. *J. Am. Chem. Soc.* **126**, 11430–11431.

[33] Muchmore, S. W., Sattler, M., Liang, H., Meadows, R. P., Harlan, H. E., Yoon, H. S., Nettesheim, D., Chang, B. S., Thompson, C. B., Wong, S.-L., Ng, S.-C. and Fesik, S. W. (1996). X-ray and NMR structure of human Bcl-x_L, an inhibitor of programmed cell death. *Nature* **381**, 335–341.

[34] Sattler, M., Liang, H., Nettesheim, D., Meadows, R. P., Harlan, J. E., Eberstadt, M., Yoon, H. S., Shuker, S. B., Chang, B. S., Minn, A. J., Thompson, C. B. and Fesik, S. W. (1997). Structure of Bcl-x_L–Bak peptide complex: recognition between regulators of apoptosis. *Science* **275**, 983–986.

[35] Metzler, W. J., Wittekind, M., Goldfarb, V., Mueller, L. and Farmer, B. T., II (1996). Incorporation of $^{1}H/^{13}C/^{15}N$-{Ile,Leu,Val} into a perdeuterated ^{15}N-labeled protein: potential in structure determination of large proteins by NMR. *J. Am. Chem. Soc.* **118**, 6800–6801.

[36] Rosen, M. K., Gardner, K. H., Willis, R. C., Parris, W. E., Pawson, T. and Kay, L. E. (1996). Selective methyl group protonation of perdeuterated proteins. *J. Mol. Biol.* **263**, 627–636.

[37] Gardner, K. H., Rosen, M. K. and Kay, L. E. (1997). Global folds of highly deuterated, methyl-protonated proteins by multidimensional NMR. *Biochemistry* **36**, 1389–1401.

[38] Goto, N. K., Gardner, K. H., Mueller, G. A., Willis, R. C. and Kay, L. E. (1999). A robust and cost-effective method for the production of Val, Leu, Ile (δ1) methyl-protonated ^{15}N-, ^{13}C-, ^{2}H-labeled proteins. *J. Biomol. NMR*, **13**, 369–374.

[39] Constantine, K. L., Mueller, L., Goldfarb, V., Wittekind, M., Metzler, W. J., Yanchunas, J., Jr., Robertson, J. G., Malley, M. F., Friedrichs, M. S. and Farmer, B. T., II. (1997). Characterization of $NADP^+$ binding to perdeuterated MurB: backbone atom NMR assignments and chemical-shift changes. *J. Mol. Biol.* **267**, 1223–1246.

[40] Pellecchia, M., Meininger, D., Dong, Q., Chang, E., Jack, R. and Sem, D. S. (2002). NMR-based structural characterization of large protein–ligand interactions. *J. Biomol. NMR* **22**, 165–173.

[41] Hajduk, P. J., Mack, J. C., Olejniczak, E. T., Park, C., Dandliker, P. J. and Beutel, B. A. (2004). SOS-NMR: a saturation transfer NMR-based method for determining the structures of protein–ligand complexes. *J. Am. Chem. Soc.* **126**, 2390–2398.

[42] Mueller, L. and Kumar, N. V. (1996). Multidimensional NMR of macromolecules. In *NMR Spectroscopy and Its Application to Biomedical Research*, ed. S. S. Sarkar, Elsevier, Amsterdam, pp. 85–157.

[43] Spector, S., Wang, M., Carp, S. A., Robblee, J., Hendsch, Z. S., Fairman, R., Tidor, B. and Raleigh, D. P. (2000). Rational modification of protein stability by the mutation of charged surface residues. *Biochemistry* **39**, 872–879.

[44] Eijsink, V. G. H., Bjork, A., Gaseidnes, S., Sirevag, R., Synstad, B., van den Burg, B. and Vriend, G. (2004). Rational engineering of enzyme stability. *J. Biotechnol.* **113**, 105–120.

[45] Bommarius, A. S., Broering, J. M., Chaparro-Riggers, J. F. and Polizzi, K. M. (2006). High-throughput screening for enhanced protein stability. *Curr. Opin. Biotechnol.* **17**, 606–610.

[46] Thompson, M. J. and Eisenberg, D. (1999). Transproteomic evidence of a loop-deletion mechanism for enhancing protein thermostability. *J. Mol. Biol.* **290**, 595–604.

[47] Matthews, S. J. and Leatherbarrow, R. J. (1993). The use of osmolytes to facilitate protein NMR spectroscopy. *J. Biomol. NMR* **3**, 597–600.

[48] Street, T. O., Bolen, D. W. and Rose, G. D. (2006). A molecular mechanism for osmolyte-induced protein stability. *Proc. Natl Acad. Sci. USA* **103**, 13997–14002.

[49] Wand, A. J., Ehrhardt, M. R. and Flynn, P. F. (1998). High-resolution NMR of encapsulated proteins dissolved in low-viscosity fluids. *Proc. Natl. Acad. Sci. USA* **95**, 15299–15302.

[50] Peterson, R. W., Lefebvre, B. G. and Wand, A. J. (2005). High-resolution NMR studies of encapsulated proteins in liquid ethane. *J. Am. Chem. Soc.* **127**, 10176–10177.

[51] Huse, M. and Kuriyan, J. (2002). The conformational plasticity of protein kinases. *Cell* **109**, 275–282.

[52] Dubinina, G. G., Chupryna, O. O., Platonov, M. O., Borisko, P. O., Ostrovska, G. V., Tolmachov, A. O. and Shtil, A. A. (2007). *In silico* design of protein kinase inhibitors: successes and failures. *Anti-Cancer Agents Med. Chem.* **7**, 171–188.

[53] Gronwald, W. and Kalbitzer, H. R. (2004). Automated structure determination of proteins by NMR spectroscopy. *Prog. Nucl. Magn. Reson. Spectrosc.* **44**, 33–96.

[54] Grishaev, A. and Llinas, M. (2002). CLOUDS, a protocol for deriving a molecular proton density via NMR. *Proc. Natl. Acad. Sci. USA* **99**, 10941.

[55] Grishaev, A. and Llinas, M. (2002). Protein structure elucidation from NMR proton densities. *Proc. Natl. Acad. Sci. USA* **99**, 6713–6718.

[56] Grishaev, A. and Llinas, M. (2005). Protein structure elucidation from minimal NMR data: The CLOUDS approach. *Methods Enzymol.* **394**, 261–295.

[57] Ikeya, T., Terauchi, T., Guntert, P. and Kainosho, M. (2006). Evaluation of stereo-array isotope labeling (SAIL) patterns for automated structural analysis of proteins with CYANA. *Magn. Reson. Chem.* **44**, S152–S157.

[58] Kainosho, M., Torizawa, T., Iwashita, Y., Terauchi, T., Mei Ono, A. and Guentert, P. (2006). Optimal isotope labelling for NMR protein structure determinations. *Nature* **440**, 52–57.

[59] Lyne, P. D., Lamb, M. L. and Saeh, J. C. (2006). Accurate prediction of the relative potencies of members of a series of kinase inhibitors using molecular docking and MM-GBSA scoring. *J. Med. Chem.* **49**, 4805–4808.

[60] Kuhn, B., Gerber, P., Schulz-Gasch, T. and Stahl, M. (2005). Validation and use of the MM-PBSA approach for drug discovery. *J. Med. Chem.* **48**, 4040–4048.

[61] Teramoto, R. and Fukunishi, H. (2007). Supervised consensus scoring for docking and virtual screening. *J. Chem. Inf. Model.* **47**, 526–534.

[62] Marin, A., Malliavin, T. E., Nicolas, P. and Delsuc, M.-A. (2004). From NMR chemical shifts to amino acid types: investigation of the predictive power carried by nuclei. *J. Biomol. NMR* **30**, 47–60.

6

Target-immobilized NMR Screening: Validation and Extension to Membrane Proteins

Virginie Früh, Robert J. Heetebrij and Gregg Siegal

6.1 Introduction

Fragment-based drug discovery (FBDD) methods have been widely embraced in the last few years. Nearly all of the major pharmaceutical firms have developed fragment screening and evolution programs and a number of biotech firms have sprung up that make exclusive use of the approach to develop small-molecule therapeutics. Among the variety of fragment screening and evolution methods that have been implemented, there are two common themes. First, the collection of compounds to be screened consists of small (typically less than 300 Da), highly soluble molecules. As such, they typically interact with the target weakly, with binding constants in the range 2–5000 μM. Second, the low-affinity hits discovered by screening such a collection must be developed into high-affinity, high-specificity ligands. This process is much more successful when 3D structures of target–compound complexes are available.[2]

The promise of FBDD, that is, compounds that through obeying Lipinski's rules[3] are more likely to make orally bioavailable, safe drugs, is starting to be put to the test as compounds begin to move into clinical trials. The number of such compounds is rising rapidly due to the successes of Plexxikon, Astex, Sunesis, SGX Pharma and a host of other biotech companies that place FBDD at the core of their activities. However, a third common theme that applies to all FBDD to date is that is has been strictly applied to

Fragment-Based Drug Discovery: A Practical Approach Edited by Edward R. Zartler and Michael J. Shapiro

soluble targets. On the other hand, the attraction of membrane proteins as pharmaceutical targets has been well documented,[4] with approximately 60% of all current targets being membrane proteins. Hence it would be a significant advantage to be able to apply FBDD to the class of targets that includes integral and membrane-associated proteins.

We have developed a technology called target-immobilized NMR screening (TINS)[5, 6] that in principle can be applied to screening of membrane proteins. In TINS, the target to be screened is immobilized on a commercially available chromatography resin in a simple and efficient process. The immobilized target, along with a second, reference sample, is placed in a flow-injection, dual-cell sample holder in the magnet and the compounds to be screened are injected in mixes of about five compounds each.[6] Spatially selective spectroscopy[1] is then used to acquire independently a 1D ^{1}H spectrum of the compounds in the presence of the target or the reference. Comparison of the two spectra directly yields the identity of any compound that binds the target due to the simple reduction in peak amplitude of all resonances from the ligand. This configuration yields a number of advantages for ligand screening. The combination of effective T_2 relaxation and chemical exchange endows the method with great sensitivity with specific binding as weak as 5–10 mM (K_D) being readily detected. On the other hand, the presence of a reference sample in routine use cancels the weak, nonspecific interactions typically observed between many of the compounds to be screened and the target. Thus the presence of artifacts in TINS screens is greatly reduced, as is the false positive rate. The sensitivity can also be used to reduce the concentration of immobilized target to as low as 5 μM solution equivalent, which combined with the fact that the entire compound collection is routinely screened with a single sample, means the screening can be carried out with as little as 5 nmol of the target.

TINS has been applied to a variety of soluble proteins and in this chapter we will present some of these results. In principle, immobilization should allow an extension of the range of targets to which TINS can be applied to include insoluble membrane proteins. This idea is not new and others have attempted to apply biophysical methods for detecting ligand binding to immobilized membrane proteins.[7] In particular, surface plasmon resonance (SPR) has been used for this application. Membrane proteins represent difficult targets for *in vitro* ligand screening studies, however, since they are insoluble, often require the presence of specific lipids for proper function, are highly challenging to purify and rarely amenable to high-resolution structural analysis. Furthermore, a general limitation that has always been encountered is the difficulty of functionally immobilizing membrane proteins in a form appropriate for the assay. SPR for instance requires a flat surface with an underlying metal layer (to provide the material with dielectric constant opposite that of water). Although a few cases of successful immobilization of membrane proteins have been reported under these conditions, a widely applicable method is still lacking. Here we will report on our initial efforts in two areas, the ultimate goal of which is to allow routine *in vitro* fragment screening of a wide variety of membrane proteins.

6.2 General Considerations for Fragment Screening

6.2.1 Fragments

Since an entire chapter of this book is devoted to fragment library design, it is not our intention to recapitulate this information here. Instead we will focus on the principles and benefits of the TINS fragment library designed and tested as collaborative effort between ZoBio (www.zobio.com) and Pyxis Discovery (www.pyxis-discovery.com) of Delft, The Netherlands.[8]

It is now a well-accepted principle that the 'rule of three'[9] forms an approximate limit guiding the chemical nature of compounds that should be considered as a fragment for inclusion in a collection for ligand screening. At the other end of the spectrum, recent work from the Shoichet[10] laboratory suggests that including very simple fragments of less than approximately 150 Da could cause difficulties downstream during the lead evolution process. Clearly, a number of *in silico* filters must also be employed to remove undesirable compounds such as known toxicophores or reactive groups. In our efforts, we also placed great emphasis on water solubility of the compounds. In one of the first publications concerning fragment library design, only about 50% of the selected fragments possessed sufficient solubility (1 mM) to be screened.[11] In more recent publications, better results for the water solubility of fragment libraries have been reported.[12, 13] The prediction of water solubility, however, remains a challenge because one has to take into consideration both the crystal and solution states of the compound. Moreover, in our own analysis, we have not been able to find a simple correlation between the number of hydrogen bond donors/acceptors and water solubility. Since computational methods for better prediction of water solubility are still under development, one must determine experimentally the solubility of a given fragment. However, by applying cut-off values based on experience, for properties that can be better predicted, such as Clog*P* and the number of hydrogen bond donors and acceptors, which have a profound influence on water solubility, the fraction of water-soluble fragments can be increased considerably. In our own efforts, about 90% of compounds that were selected were soluble as singletons at 500 μM in phosphate-buffered saline and 5% DMSO. Evotec has recently mentioned an in-house QSAR model to predict solubility which is claimed to be useful, but no data are currently available.[14] While originally our emphasis on water solubility stemmed from practical aspects of making mixes of compounds at 500 μM each in aqueous buffer, this effort has been well served when screening membrane proteins, since we feel that it is one of the important reasons that we have so far experienced a very low false positive rate.

Our library, which is intended to serve as a source of chemical diversity, is composed of compounds selected from four themes: (1) diversity using the scaffold-based classification approach (SCA),[15] (2) amino acid derivatives, (3) scaffolds found in natural products and (4) shape diversity. All compounds were selected from a carefully prepared database representing 70 000 compounds that would make desirable starting points for drug discovery, including rule of three compliance, and were commercially available from reliable suppliers. One of our explicit intentions in forming the library upon these design principles is to evaluate the performance of the various classes of compounds against different targets, both soluble and membrane bound. Although it remains too early to draw sensible conclusions from the roughly 10 targets that have been screened to date, in many cases there are up to twofold differences in hit rates between the different themes for a given target.

6.2.2 Immobilization and Reference Protein

The strength of TINS lies in the fact that it is a referential system. That is, the signal acquired in the presence of the target protein is compared with the signal acquired in the presence of a reference sample consisting of a known protein immobilized at approximately the same density as the target. The requirement for a reference protein comes from the fact that TINS is highly sensitive to even very weak interactions between the compounds and the immobilized target. Therefore, the choice of reference protein is important. Ideally, one would like to have a reference protein which is convenient to produce in large quantities, can be readily immobilized, has the roughly 'typical' amounts of exposed surface charge and hydrophobicity and has essentially no small-molecule binding capacity. The pH domain of the cellular kinase AKT is a nearly ideal candidate which we use for screening of all soluble targets. Hajduk *et al.* showed that this protein was essentially refractory to small-molecule binding using their well-known SAR by NMR assay.[16] Although we initially had concerns that this small protein would be unrepresentative of larger, potentially multi-domain targets or that proper cancellation of nonspecific binding would require accurate matching of total surface area, this turns out not to be the case, as shown in Figure 6.1.

Immobilization is a constant source of questions with regard to TINS screening. In principle, one is free to choose any immobilization approach which is compatible with

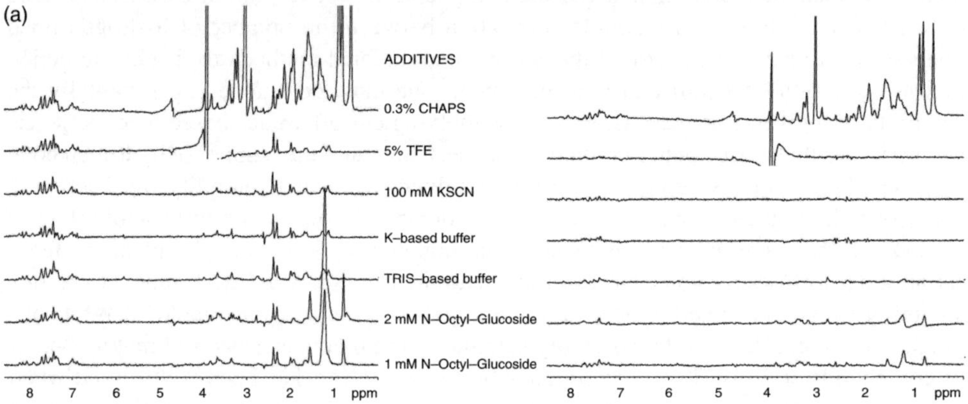

Figure 6.1 *(A) Cancellation of nonspecific binding by the reference sample in TINS screening. The left-hand panel shows difference ^{1}H NMR spectra of a mixture of nonbinding compounds acquired in the presence of Sepharose resin to which 6 mg mL^{-1} of an SH2 domain (111 amino acid residues) had been immobilized or just the resin itself. The indicated additive was included with each of the compound mixtures. The right-hand panel shows the same difference spectra, but the second spectrum was acquired in the presence of a resin to which 6 mg mL^{-1} of FKBP had been immobilized. The improvement in cancellation when an immobilized protein is used as a reference is clear. (B) In this example, taken from a screen of a soluble target, both the target and the reference protein (the pH domain of the kinase AKT) were immobilized on Actigel ALD (Sterogene Bioseparations, Carlsbad, CA, USA) at a solution equivalent of 100 μM. A mix consisting of three different compounds (upper three 1D ^{1}H NMR spectra are of each compound in the mix separately) was applied simultaneously to the sample of immobilized target and reference protein in the dual-cell sample holder. Spatially selective Hadamard spectroscopy[1] was used to acquire simultaneously separate spectra of the compound mix in the presence of the immobilized target and reference. These spectra are overlaid at the bottom of the figure. The similarity of the two spectra indicates that none of the compounds specifically binds the target. The weak interactions with any immobilized protein that are observed for most compounds in the library are approximately the same for both the reference and target.*

(b)

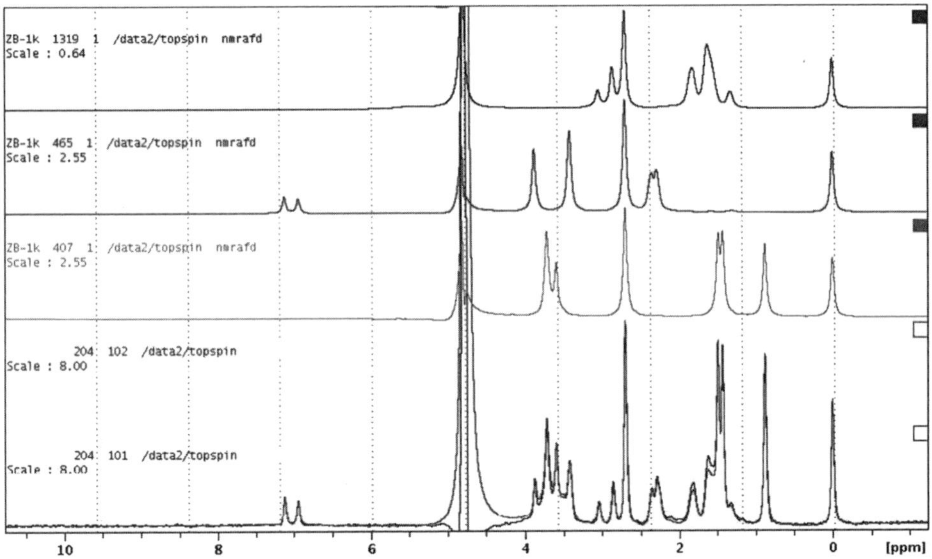

Figure 6.1 *(Continued).*

(a) the biochemical function of the protein and (b) the constraints of NMR. Specifically, the major concern related to NMR is susceptibility mismatch between the solid support and the surrounding aqueous environment. Meyer's group had originally demonstrated ligand binding to targets immobilized on glass beads.[17] However, the susceptibility mismatch was so severe in this case that magic angle spinning NMR was necessary to average out the inhomogeneity. Clearly, this arrangement would not be compatible with flow-injection NMR, so we sought a solid support which would not bind the compounds, would provide high capacity to immobilize proteins and would minimize susceptibility differences. Sepharose-based affinity resins turned out to be very useful in that they are very good matches for this list of requirements. In contrast to glass beads, Sepharose beads can be more readily described as a three-dimensional, biocompatible mesh which is highly hydrated, yet sufficiently rigid to maintain good flow characteristics even after 300 applications of compound mixes. The susceptibility mismatch is minimal such that under our current screening setup, using the dual-cell sample holder made from KelF, we routinely obtain a linewidth of about 12 Hz. However, the nature of the immobilization chemistry of the Sepharose bead also appears to play a role in the linewidth observed for the compounds, as can be seen in Figure 6.2.

A wide range of immobilization chemistries are commercially available in conjunction with Sepharose beads. We have investigated a limited subset of these possibilities which include: direct, nonoriented immobilization via Schiff's base chemistry, oriented noncovalent immobilization via immobilized metal affinity chromatography resins and oriented noncovalent immobilization via biotin–streptavidin binding. At present we favor direct, covalent attachment of proteins via primary amines since it is highly efficient (typically better than 85% yield), minimizes leaching and provides the best NMR results (Figure 6.2).

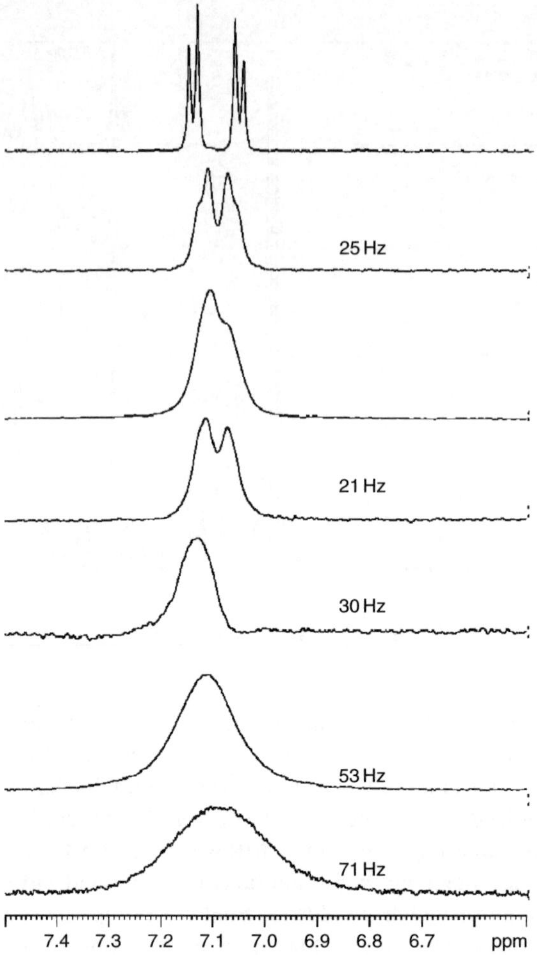

Figure 6.2 *Effect of immobilization chemistry on the linewidth of compound s in solution. 1D* ^{1}*H spectra of the aromatic protons of phosphotyrosine (pY) are shown with the fitted linewidth. From top to bottom, pY in solution, in the presence of Actigel ALD, streptavidin Sepharose, Zn-IDAA Sepharose, Zn-NTA Sepharose, Zn-NTA silica and controlled-pore glass beads (for comparison).*

At the pH at which we typically carry out immobilization (7.4), this reaction is fairly specific for the amino terminus. In principle, one could imagine that immobilization might interfere with the functionality of certain proteins, such as kinases that contain a lysine at an active site. Thus far we have not encountered this issue, but it is always possible to block access to this lysine by immobilizing in the presence of high levels of an ATP mimic such as AMPPNP. Kinases have been successfully immobilized for Biacore studies using related chemistry.[18] We have investigated the use of IMAC resins to immobilize proteins via a 6-His tag. Although this method is convenient, it is not possible to use Ni^{2+} as the ion for chelating the tagged protein due to the potent paramagnetic relaxation. It is possible to immobilize His-tagged protein using Zn^{2+} instead and leaching does not pose a problem. However, despite the fact that a Sepharose resin is used in conjunction with a diamagnetic ion, there appears to be additional line broadening effects (Figure 6.2). These may result from nonspecific interactions with available NTA sites on the resin which turn out to be difficult to block. We have also used streptavidin Sepharose to immobilize biotinylated ribonucleotides for ligand binding studies. This system is convenient and yields high quality NMR spectra. By blocking unoccupied binding sites with free biotin (and naturally using streptavidin Sepharose as the reference sample) one should be able to limit small-molecule binding to sites that are not on the target; however, we have not carried out a full screen on such a system so it not possible to make a definitive statement at this time. Other affinity tags can also form the basis of successful, NMR-compatible immobilization. For example, Haselhorst *et al.* have recently reported the use of Strep-tactin Sepharose, a variant of streptavidin Sepharose, to perform saturation transfer difference (STD) studies.[19]

6.2.3 Ligand Screening

We decided to carry out our ligand screening studies using mixes of compounds at a very early stage in the process of developing TINS. This decision was made on the basis of throughput and robustness. Since our mixes consist of on average five compounds, obviously throughput is increased by a factor of five with respect to screening singletons. Also, since it is expected that only one compound (and occasionally two) per mix bind to the target, most peaks in the reference and target spectra should be of the same amplitude. If this is not the case, it may be a sign that there is a problem with the screening sample. The use of mixes requires a strategy to design them properly. Given the constraint of increased linewidth generated by the heterogeneous TINS system, the primary factor governing the selection of compounds for a mix is the number of well-resolved peaks for each. We have therefore recorded a reference 1D ^{1}H spectrum of every compound in the ZoBio/Pyxis fragment collection at 500 μM in phosphate-buffered saline (PBS) in the presence of a fixed amount of TSP. The reference spectra also serve the dual role of quality control. The reference spectra are automatically peak picked and the peak positions stored in our database. We have developed an in-house algorithm to select compounds randomly from the collection and test them rapidly for TINS compatibility, that is, at least three well-resolved peaks for each compound (when available). This allows us to read out the ligand from the mix directly without further deconvolution (see below). The algorithm also places explicit limits on the number of aromatic compounds per mix and avoids mixing compounds with pK_a extrema. Once designed, the mixes are then made at 500 μM for each compound in PBS. The mixes are stored at room temperature and subsequently inspected visually for signs of precipitation. About one-third of mixes are rejected at this point. Mixes that do not precipitate are subjected to ^{1}H NMR analysis, where we expect to see that the NMR spectrum of the mix is a simple sum of the NMR spectra of the individual compounds using TSP as a reference. Changes to the NMR spectrum of the mix, which we rarely observe, are indicative of possible aggregation behavior of the compounds.

In order to carry out a ligand screen, the resin bearing the target and reference proteins, which have been immobilized at a solution equivalent of about 100 μM, must be packed into the dual-cell sample holder. A home-made packing reservoir has been built to fit on top of the dual-cell sample holder and double the volume of each cell. The resin (as a 50 %

slurry) is pipetted into each cell one at a time, allowed to settle by gravity and packed at a pressure of 0.5 bar. Once packed, the cell can be connected to the sample delivery system via PEEK capillary tubes and inserted into the magnet using an aluminum arm. By attaching the cell to the aluminum arm, we can readily orient it such that the plane that bisects each of the two cylindrical cells is parallel to one of the transverse gradients in our triple-gradient flow-injection probe.[6] In this way, optimization of the NMR experiment for each screen is minimized. All that is necessary is to perform routine tuning and matching and shim, which we do using the FID of water. When known ligands are available, initial tests are performed to insure the integrity of the immobilized sample. This same experiment is repeated 4–5 times throughout and after the screen to detect possible target degradation (Figure 6.3A.) Once prepared, the mixes are placed in the Gilson autosampler in deep 96-well plates and the Bruker HyStar software is programmed for each. We also use standard ICON NMR in Topspin to acquire the TINS data. A complete screen of about 1500 unique compounds (including some replicates for quality assurance) requires about 7 days and runs without human intervention. Having evaluated a variety of different spatially selective NMR experiments, we have settled on the Hadamard sampling approach. The quality of the data using this experiment with carefully designed mixes is fairly high, as can be seen in Figure 6.3B.

We have now screened a number of different targets, both soluble and membrane bound, using TINS. The hit rate for targets has varied from a low of 3 % to a high of about 10 %, where we define a hit as having at least a 30 % difference in amplitude between the reference

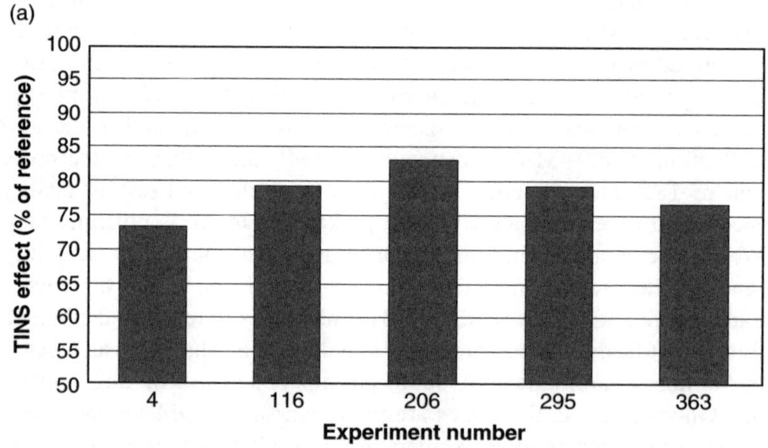

Figure 6.3 (A) Determination of target integrity during a TINS ligand screen. A known ligand was applied to both the target and reference cells and the reduction in peak amplitude was measured ('TINS effect'). This experiment was carried out serially after the indicated number of mixes had been applied to the immobilized target. (B) Direct determination of ligand identity using TINS. A mix of five compounds was applied to the dual sample holder containing immobilized target and the pH domain of AKT, both at 100 μM solution equivalent. The individual spectra of each cell, acquired with a 30 min measuring time, are overlaid at the bottom of the figure. The ¹H spectra of four of the five compounds are shown above for reference. The identity of the ligand (fourth spectrum identifier 1059) is readily obtained by simple inspection.

(b)

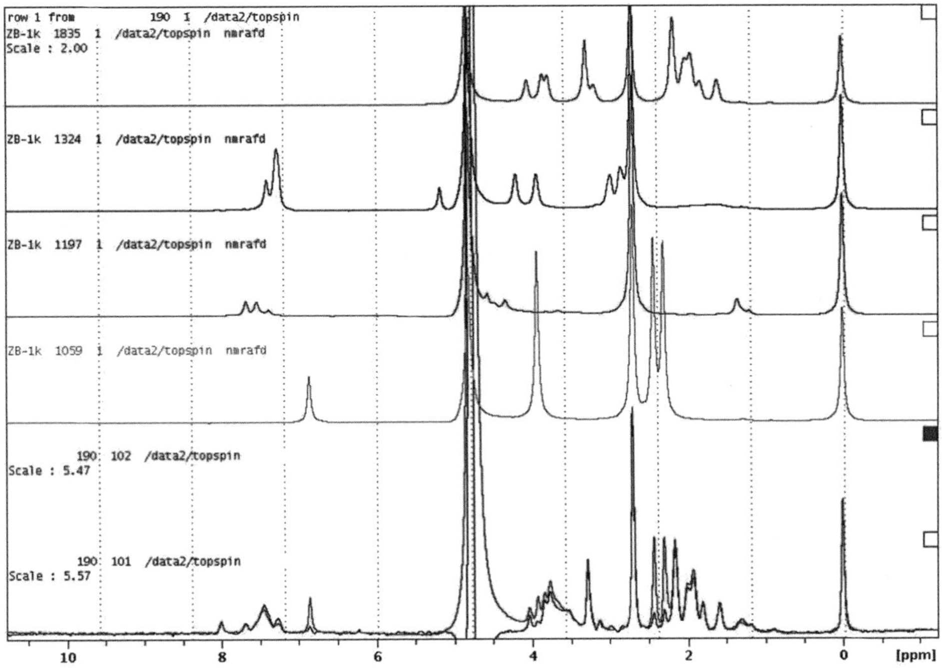

Figure 6.3 *(Continued).*

and target spectra for all well-resolved peaks. This cut-off was chosen for practical reasons based on the fact that the difference was sufficiently large to overcome artifacts related to spectral noise, minor lineshape differences between the two samples and spectral crowding, and therefore allowed reliable detection of a hit. This last fact is particularly important since we wish to automate the data analysis process. Since screening on these targets has only been carried out using TINS, it is not possible to compare directly the observed hit rates with other methods, including high-concentration screening (i.e. screens based on inhibiting and enzymatic activity). Where Hajduk *et al.* reported essentially a 0 % hit rate for the pH domain of AKT.[16] we in fact do detect some compounds binding, but our 'hit rate' is about 0.2 %, some 10-fold lower than the lowest rate obtained for a target that is expected to be 'druggable'. In their work, Hajduk *et al.* reported hit rates of up to 1 % for SAR by NMR. Interestingly, the 3 % hit rate for TINS was found when screening a soluble 'NTPase' in the NDP-bound form. The hit rate for the apo-protein was about 9 %. The low hit rate found when the nucleotide binding pocket is occupied is expected and suggests that the high hit rates that we observe are not due to artifacts, but rather to reliable sensitivity to binding events. This idea is further supported by follow-up biochemical studies that we have now performed for two targets with enzymatic activity. Considering a soluble enzymatic target for which we found a hit rate of 9.5 %, approximately 50 % of the TINS hits showed significant inhibitory activity at 500 μM, and we would expect this number to increase even further if tested at the 1–2 mM typically used in high-concentration screening. A similar pattern has been observed for membrane proteins (see below).

6.3 Membrane Protein Considerations

6.3.1 Quantity Limitations

Although TINS removes limitations such as size and solubility of the target protein to be applied, there still remain quantity limitations with regard to membrane proteins. At present, the practical lower limit for screening is roughly 25 μM solution equivalent (e.g. nmol mL^{-1} settled bed volume). Since we typically prepare 500 μl of immobilized resin to fill one cell of the sample holder, we require about 15 nmol of target. For a 50 kDa protein, this works out to slightly under 1 mg and therefore it is safe to use 1 mg as a lower limit. For soluble proteins in which structure-guided hit optimization is the primary means for evolving fragments, this limit does not generally present a problem. However, for many membrane proteins, formidable efforts are required to produce even this quantity. Accordingly, efforts are under way in our laboratory to enhance the sensitivity of TINS towards an eventual goal of being able to screen recombinantly expressed proteins in their native membrane environment, that is, without purification. Below we present data demonstrating the feasibility of immobilizing such native membrane fragments. Since this approach is beyond the present sensitivity limits of our TINS ligand screening station, however, current efforts utilize highly expressed, purified and functionally solubilized membrane proteins.

Given the current requirement for about 1 mg of functional protein to carry out ligand screening, it is clear that an appropriate system must be available to produce large quantities. Due to the interest in pharmacology and structure of membrane proteins, tremendous efforts have been made in recent years in developing new means to express, purify and solubilize them. It is not our intention to catalogue these approaches here, merely to mention some which show promise with respect to producing sufficient quantities for ligand screening and subsequent structural studies. Conceptually the simplest method for membrane protein production is via cell-free expression. Recently, six different GPCRs have been produced in milligram quantities using an *E. coli*-based expression system that included Brij78 as a solubilizing detergent.[20] Studies were performed to show that at least one of the *in vitro* expressed GPCRs was functional. Interestingly, all appeared to be dimeric. Bacterial expression of membrane proteins typically results in the protein being unfolded and located in inclusion bodies. Although purification of proteins from inclusion bodies is easy, the requirement for refolding can represent a considerable hurdle. Nonetheless, companies such as M-fold have successfully produced isotope-labeled GPCR using this approach and showed that the protein was amenable to NMR studies.[21]

Beyond bacterial expression systems, a number of eukaryotic expression systems have also been developed. One simple method of producing functional membrane proteins is to generate recombinant transient or stable cell lines based on CHO or HeLa cells. Such cell lines have the benefit of providing appropriate post-translational modifications such as glycosylation which are not available in prokaryotic expression systems.[22] Often these modifications are required for protein function, as shown for rhodopsin, where folding is inefficient when the glycosylation site at its N-terminus is suppressed.[23] Unfortunately the yield of proteins from stable cell lines is more often than not insufficient for ligand screening studies. Transient expression of membrane proteins can increase the yield by as much as a factor of 10, but results in other inconveniences such as repeatability issues. Alternatives that have seen increasing success include recombinant expression in insect SF9 cells,[24] use of Semliki Forest virus-infected cells [25] and expression in the yeast *Pichia pastoris*.[26, 27] All of these systems are capable of yielding sufficient quantities of folded, functional membrane proteins for ligand screening and structural studies. Unfortunately none is perfectly general and the rate-limiting step remains finding the best system for a particular target of interest.

6.3.2 The Membrane Environment

Membranes are structured as stable phospholipids bilayers which delimit the boundaries of the organelle or the cell. The membrane provides an environment where chemical signals can be emitted and detected, where energy can be converted into inter- and intracellular functions and through which materials can be transported. For all these activities, there are complex networks of interactions between the membrane-associated proteins, such as receptors, ion channels and enzymes, and the ligands which stimulate or inactivate them. The membrane itself plays more than a passive role in these processes. Current understanding suggests that interaction between the membrane and embedded proteins is at least required for and may regulate protein function. Therefore, the ultimate goal of research in our group is to be able to perform NMR-based ligand screening studies on membrane proteins in their native environment. However, in the light of the discussion in the preceding section, it is clear that this is not yet possible and therefore membrane proteins must be recombinantly expressed and purified. Given the intimate interaction between protein and membrane, functional solubilization represents a major hurdle.

In order to retain functionality of a membrane protein, it is imperative to refold it or reconstitute it into a synthetic lipid environment which mimics the properties of its natural membrane as closely as possible.[28] Integral membrane proteins must be solubilized before being purified and this often calls for addition of detergents after the initial centrifugation steps. For example, the potassium channel KcsA was extracted from the cell membrane by addition of Foscholine-12 prior to purification using IMAC and gel filtration chromatography.[29] Transmembrane proteins have large hydrophobic domains which can cause aggregation during purification. This can be avoided by using high concentrations of urea to prevent random folding before reconstitution in lipids.[30] These solubilization and purification steps are important because the success of lipid reconstitution depends on the state of the protein at this point. Organic solvents are the simplest approach to mimicking a membranous environment, but it has only been possible to use them with proteins with stable native folds such as ATP synthase[31] or colicin E1 immunity proteins.[32] The simplest true mimic of a membrane occurs when ionic or nonionic surfactants in organic solvents or water create micellar vesicles.[33] Micelles, which are 10–100 kDa in size when there is low ionic concentration, are very convenient since they are readily formed and can be used to solubilize membrane proteins in a monomeric form amenable to high-resolution structural studies. To date, all TINS screening has been applied to micelle-solubilized membrane proteins. However, due to, at least in part, the monolayer and the extreme curvature of micelles, they are only rarely compatible with native functioning of membrane proteins. Surfactants used for such preparations include, but are certainly not limited to, sodium dodecyl sulfate (SDS), cetyltrimethylammonium chloride and bromide (CTAC and CTAB), lysophosphatidylcholine (LPC), Triton X-100 and dodecylphosphocholine (DPC).[34] For NMR studies, deuterated surfactants are at least convenient and many times may be required. At present, only DPC and SDS are commercially available in this form, although the latter tends to denature some proteins.[35] Micelles are formed when the surfactant is in a higher concentration than its critical micellar concentration (CMC), which can vary from 0.01 mM for nonionic surfactants to 10 nM for short-chain ionic surfactants, such as SDS.[36] The equilibrium shifts from micellar to monomeric forms of the surfactant when diluted with buffers that do not contain the detergent and therefore buffers must always contain a concentration of surfactant above the CMC to prevent micelle disruption and loss of protein conformation. In our hands, there is rapid exchange of surfactant molecules from the micellar to the monodispersed form, resulting in rapid breakdown of micelle-bound proteins when the surfactant is not included (see below). Bicelles are micelles which are composed of phospholipids rather than detergents and are slightly more complex than micelles. Usually bicelles are composed of long-chain phospholipids such as dimyristoylphosphatidylcholine (DMPC), forming bilayers, and one shorter chain phospholipid such as dihexanoylphosphatidylcholine (DHPC), which lines the hydrophobic edges of the bilayer.[37] Bicelles, being mostly planar, represent a better membrane mimic than micelles and should be more compatible with protein function. The utility of bicelles for functionally solubilizing membrane proteins has recently been demonstrated by their use in crystallization of the GPCR, β_2-adrenergic receptor.[38] However, we have not yet tested bicelles for compatibility with TINS. In addition, there are more complex stable bilayer or multilayer vesicles of synthetic phospholipids which can be used to immobilize and orient membrane proteins on glass slides in solid-state NMR,[39] but their usefulness for membrane protein immobilization on supports that are compatible with static NMR studies is not yet known.

6.3.3 Immobilization

The TINS methodology, by definition, requires immobilized protein to allow flow-through screening of ligands. Clearly, the choice of the surface upon which the protein will be immobilized and the choice of the immobilization chemistry have to be made within the limitations of the TINS equipment. The general requirements for immobilization compatible with high-resolution NMR have been discussed, so here we focus on issues specifically related to membrane proteins.

We have taken a pragmatic approach when attempting to apply the TINS methodology to membrane proteins by beginning with what has worked for soluble proteins. To date we have immobilized three purified, micelle-solubilized membrane proteins, KcsA, OmpA and DsbB, all of which are from bacterial sources. All three membrane proteins were solubilized in dodecylphosphocholine micelles (DPC).[40] In all three cases we have simply utilized the same immobilization scheme that has been successfully applied to soluble proteins, i.e. Schiff's base chemistry, to primary amines. We have found that the yield of immobilized micelle-solubilized protein is nearly identical with that of soluble proteins. Further, immobilization has not had any detectable effect on the functionality of the immobilized, micelle-solubilized proteins. This has been checked in two ways. For KcsA a panel of known ligands was available and we simply assayed for binding using TINS. Since DsbB has an enzymatic activity, we adapted a spectrophotometric assay[41] for use with beads containing immobilized protein. Enzyme inhibition studies were carried out by adding a reduced partner enzyme and ubiquinone, the reduction of which can be monitored by measuring the absorption decrease at 275 nm over time. In order to reduce nonspecific interactions to the resin and thus to compare enzymatic activity of the target prior to and after immobilization, there was an equivalent presence of resin in both cases. Results showed an efficient enzymatic activity post-immobilization. Considering the imprecision in determining the amount of immobilized enzyme, the rate of the reaction of immobilized enzyme (3 M Ubiquinone-5/M DsbB s^{-1}) was close to that of the enzyme in the presence of, but not immobilized on, resin (4 M Ubiquinone-5/M DsbB s^{-1}) (Figure 6.4).

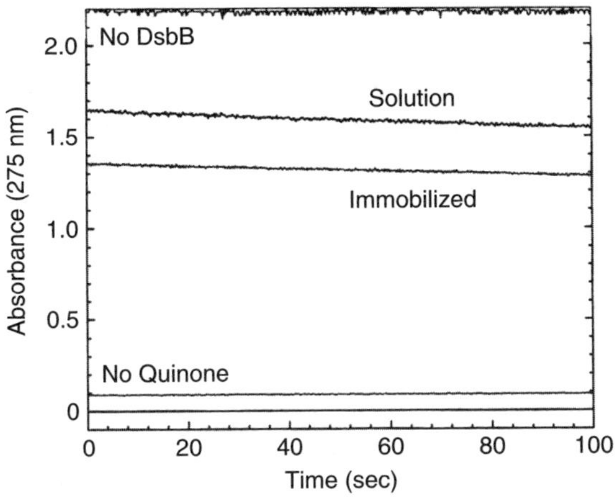

Figure 6.4 *The target immobilized to the resin shows similar enzymatic activity of to the target in the presence of, but not immobilized to, the resin.*

Naturally, more complex strategies can be envisioned and may prove necessary for membrane proteins that are less robust than those used so far. One interesting strategy immobilizes protein first, followed by subsequent reconstitution into a synthetic lipid environment.[42] As with soluble proteins, active site blockers may be necessary in cases where illicit immobilization of lysine side-chains in close proximity to the binding site may occur and thereby inhibit protein function. Various native or synthetic lipid assemblies have been extended to encompass the use of high-affinity immobilization reagents such as biotin and streptavidin,[43–46] antibodies[47–49] or metal affinity[50, 51] in order to immobilize the protein in more oriented manners. Therefore, as with soluble proteins, these approaches should also be compatible with TINS.

As a first step along the road to enabling TINS ligand screening for a truly broad range of membrane targets, we have begun to immobilize GPCRs in native membrane fragments (Früh *et al.*, in preparation). In this experiment, the idea was to use standard, stable animal cell expression systems such as CHO or HeLa cells as a source of material. In this way, all membrane proteins that can be recombinantly expressed in these simple systems could potentially be used in fragment screening campaigns. Thus far we have succeeded in immobilizing membrane fragments produced by pottering (gentle disruption of animal cells) of post-centrifugation membrane preparations. We have applied the procedure to both histamine receptors and adenosine receptors and, in both cases, the pharmacology of immobilized receptors was similar to that of nonimmobilized receptors. The efficiency of immobilization is reasonable, with approximately 20 % of total receptors functionally immobilized and, in comparison with nonimmobilized receptors, the immobilized receptors appear significantly more stable. At present the density of receptors is insufficient to perform NMR ligand screening, but work is in progress to address this issue.

6.3.4 Screening

We have developed a diversity library for use in TINS and it is our intention to screen it against all targets. The design requirement for high solubility (to maximize oral bioavailability) pays dividends when used in membrane protein ligand screening, since partitioning to the lipid phase is minimized. Nonetheless, as with soluble proteins, it remains important to use an appropriate reference system to cancel out nonspecific binding events. We have used the *E. coli* protein OmpA as a successful reference protein in one partial screen of about 200 compounds and one complete screen of about 1300 compounds. Its advantages include easy expression and purification, solubility in DPC and low small-molecule binding. One potential way to avoid the use of a reference protein would be to screen using a known, competitive ligand. We are currently adapting the hardware of the TINS ligand screening station to permit competition ligand screening studies. In this arrangement, the target is immobilized in both cells of the sample holder and the same mix is applied to both cells whereas the competitor is added to only one of the cells. Competition ligand screening will eliminate the need for a separate reference protein but has the drawback that one can only find ligands to known binding pockets. When it becomes possible to screen proteins in native membrane vesicles, then a preparation of membrane vesicles of parental cell lines not expressing the target should serve as an ideal reference.

In order to improve the robustness of TINS further, we include a reference compound in all mixtures that can be used to scale the two spectra post-acquisition. With membrane proteins, even more so than with soluble proteins, it is important to ascertain whether the reference compound interacts with the target or the surfactant used to solubilize it. The ideal reference compound has only one peak outside the spectral range of all compounds and, naturally, does not interact with the reference, target or surfactant. TSP fulfils most of these requirements but does bind to some targets. Alternatives that we have used include glycine and tetramethylammonium chloride (TMA). A crude scaling factor for the two cells can be determined experimentally by integrating the water signal from each cell using a standard 1D imaging experiment with a single scan. Binding of potential reference compounds can readily be established by simply conducting TINS experiments on all, applying the scaling factor and analyzing the spectra for equal peak intensity in both cells. So far we have not encountered a case where more than one of the three potential reference compounds bound to the target.

As noted previously, individual detergent molecules rapidly exchange between the micellar and monomeric forms. Thus, washing of immobilized micelles in buffer without detergent leads to rapid loss of protein functionality, as shown in Figure 6.5. At least for the case of KcsA, which consists of a single polypeptide, the loss of functionality (as measured by binding of a known ligand) appears to be perfectly reversible. Nonetheless, it is clear that DPC must be applied throughout the screening procedure. Since DPC is available in deuterated form, its presence does not interfere with the acquisition of the NMR spectra of the compounds. For convenience we chose to include DPC only in the buffer used to wash the compounds out of the cells of the sample holder and not in the mixes themselves. Since this approach has led to two successful screens of membrane proteins, we are optimistic that it will be general. In this way, it may prove possible to acquire NMR spectra even in the presence of nondeuterated detergents, since the concentration of the monomer is reduced by application of the compound mix in the absence of detergent. However, we have yet to test this hypothesis. Once the immobilized protein functionality has been verified, it is also important to create checkpoints at different time points of the screen with mixes containing a known binder as a positive control to check that protein functionality and thus conformation is maintained through the screen.

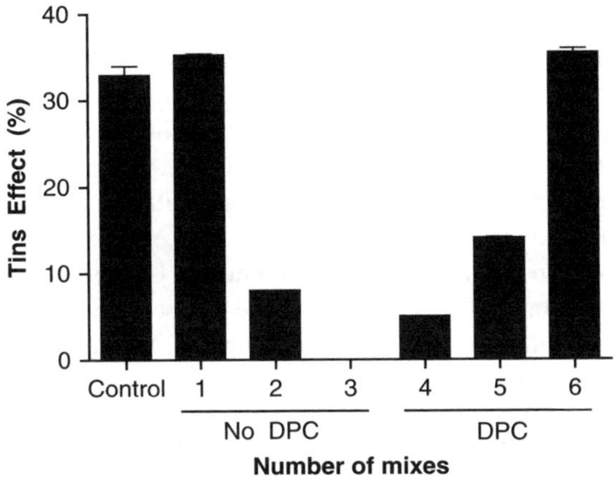

Figure 6.5 *Requirement for the presence of detergent while screening micelle-solubilized membrane proteins. In this series of experiments both the target (KcsA) and the reference (OmpA) were immobilized at a solution equivalent of 150 μM. The histogram represents the fractional difference in peak amplitude of a known ligand of KcsA in the presence of KcsA and OmpA. The bar labeled control represents the first application of the ligand. Subsequently three injections of the ligand were performed using buffers that contained no detergent. A further three injections were performed where the buffer used to wash the immobilized samples contained deuterated DPC.*

One final issue deserves special attention when considering carrying out ligand screening studies on a membrane protein, namely the kinetics of ligand binding. Although low-affinity ligands for soluble proteins nearly always exhibit rapid exchange kinetics on the NMR time-scale, this may not be the case for membrane proteins. For example, histamine binds the human H1 receptor with a K_d of 20 μM.[52] Such a small molecule (histamine fits well within the definition of a 'fragment'), binding with moderate affinity would normally imply a fast on-rate. However, in this solid-state NMR study, the on-rate was found to be of the order of minutes! Likely mechanisms for such slow binding include access to the active site of the protein via the membrane or slow conformational exchange of the protein due to interaction with the membrane (or membrane mimetic). Since the dynamic behavior of detergents and phospholipids is strongly temperature dependent, it may be necessary to carry out screening at near physiological temperature, where the long-term stability of the target may be less than optimal. In such situations, it may be necessary to prepare multiple samples in order to carry out successfully a screen of a complete fragment library.

6.4 Application of TINS to Ligand Discovery

6.4.1 Soluble Targets

To date TINS has been applied to more than ten different soluble targets. We have immobilized the target at a range of concentrations for the various screens, from as high as 500 μM to as low as 100 μM solution equivalent. We now typically screen at 100–150 μM, which represents an optimal balance between sensitivity, artifact suppression and protein consumption. In all cases we have used the pH domain of AKT as the reference. Typically we immobilize the target and reference on the activated Sepharose, Actigel ALD. The efficiency of immobilization is monitored by UV absorption of the supernatant and visual inspection to insure that no precipitation has occurred. If an enzymatic assay of the target is available, we use it at this stage to confirm that the immobilized protein remains functional. The derivatized supports are subsequently packed into the dual-cell sample holder under pressure (0.5 bar per cell), connected to the solvent delivery lines from the sample handling system and then placed in the magnet. In most cases a small number of known weak ligands (up to six) are available to test whether the target has been functionally immobilized and to demonstrate that we can indeed detect ligand binding. One of the known ligands is then selected for use in monitoring the condition of the target during screening. We routinely monitor the condition of the immobilized target through repeated injection of the known ligand throughout the screen.

Once the immobilized target has been deemed functional, we carry out the actual screen. The mixes are delivered in 1 mL volumes in deep 96-well plates to the Gilson autosampler. Sample handling is controlled by Bruker HyStar software, which communicates with Bruker TopSpin to acquire the NMR data. Using the Hadamard sampling experiment described earlier, we currently acquire data for 30 min with an additional 5 minutes for sample handling resulting in a cycle time of about 35 min. In a recent screen, 324 experiments were run in total to assess the binding of 1393 compounds from our fragment collection. This number includes repeated assaying of the positive control to assess target condition and some overlap of compounds (e.g. compounds appear in two different mixtures). This design allows us to assess the repeatability of the screening data. Such a screen was carried out without human intervention in under 8 days. Finally, since the target and reference are immobilized, it is possible to change the buffer conditions to match closely the crystallography conditions without regard to protein stability. We routinely screen under solution conditions in which the reference protein would precipitate if not immobilized. Nonetheless, its ligand binding characteristics vary only very moderately from one set of solution conditions to the next.

6.4.2 TINS Proof of Principle Application to a Bacterial Membrane Protein

TINS is a comparative method, where detection of ligand binding to the immobilized target is quantitated by comparison with an immobilized reference. With membrane proteins, partitioning of ligands can occur on the native or synthetic lipids surrounding the target present on the resin. An appropriate reference system had to be developed to control for nonspecific binding of hydrophobic compounds to lipids or detergents used to solubilize the membrane proteins. An appropriate choice for such a reference protein would be one with few known binders, in order to minimize the chances of nonspecific binding. The *E. coli* Outer membrane protein A (OmpA) was chosen for such qualities. This reference protein was of similar size to our intended target and also refolded in DPC micelles. To get an initial feel for whether we could detect specific binding to a membrane protein using TINS, we conducted a proof of principle study with a screen of a small subset (about 100 compounds) of our compound library using KcsA from *Streptomyces* as the target and OmpA as the reference.

Prior to screening, it was necessary to establish an appropriate (1) level of DPC to include in the wash buffer to maintain the integrity of the immobilized, micelle solubilized target and (2) internal reference compound. If the DPC concentration in the environment of the target decreased to below its CMC, the micelles formed by DPC would start to dissociate slowly into monomers and be flushed away. Simple calculation suggested that it was necessary to use DPC at 5 mM in the wash buffer to in order to maintain the concentration above the CMC (1 mM) upon dilution with the compound mix with DPC absent. We tested both TSP and TMA as a possible internal standard by including both in a mixture with (4-fluorophenyl)methylsulfanylmethanimidamide (FPMSMA) (Figure 6.6A), a known

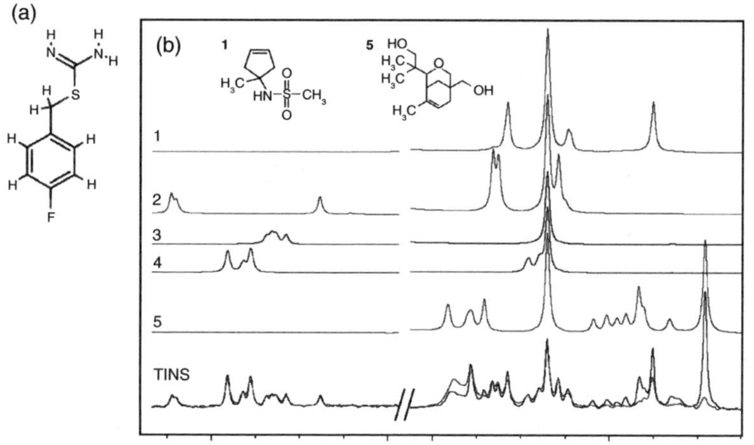

Figure 6.6 *Proof of principle ligand screen against a bacterial membrane protein. (A) Structures of the known ligand (4-fluorophenyl)methylsulfanylmethanimidamide used to determine the integrity of the immobilized KcsA. (B) Detection of ligand binding in one mix during the screen. A mix containing five different compounds was applied simultaneously to the cell containing immobilized KcsA and to the cell containing OmpA. The individual ^{1}H NMR spectra of each cell are overlaid (labeled TINS). The ^{1}H NMR spectrum of each individual compound, which has been intentionally line broadened to approximately match the linewidth of the TINS spectra, is shown above (numbered). All peaks from compounds **1** and **5** were reduced in amplitude in the presence of the immobilized KcsA with respect to OmpA, indicating that these compounds bind to KcsA. The structures of compounds **1** and **5** are shown.*

ligand for KcsA. These tests indicated that both TSP and the known ligand FPMSMA specifically bind KcsA and we therefore chose to use TMA as an internal standard. Repeated application of TMA and FPMSMA, followed by washing with buffer plus 5 mM DPC, demonstrated stability of the immobilized KcsA and so these conditions were used for the limited library screen. During the screen, the immobilized target showed insignificant loss of binding capacity for the control compound and only 12 % loss after 3 months of storage. Of the 95 fragments that were screened, 7 % showed substantial changes in the NMR spectrum that were specific to the target and were considered binders after analysis of spectral intensities (Figure 6.6B). This is in line with target hit rates obtained for soluble proteins applied to TINS. Of the potential new hits, two structures had a similar scaffold to the known binder. The other hits had a variety of scaffolds with a variety of shapes and numbers of rings.

6.4.3 Development of a High-affinity Inhibitor of Bacterial Membrane Protein DsbB Using TINS

Very recently we have undertaken a program to develop high-affinity inhibitors to the bacterial inner membrane protein DsbB in collaboration with Bushweller's group at the University of Virginia (USA). DsbB is a redox enzyme involved in the production of toxin in Gram-negative bacteria[53] and as such is a potentially medically interesting target. The crystal structure of DsbB bound to its redox partner, DsbA, has been solved[54] and Bushweller's group has solved the solution structure of a disulfide mutant of DPC solubilized DsbB (in preparation). Since this is very much a research project in progress at the time of writing, we provide only an overview of the current status here (we will provide a full report when completed).

For ligand screening, we immobilized both the functional wild-type DsbB (see above) and OmpA (as a reference) at a solution equivalent of 100 μM. We used the compound Ubiquinone-5 (Figure 6.7) which binds competitively with the native DsbB ligand and is in rapid exchange on the NMR time-scale to report on the condition of immobilized DsbB throughout the screen. Similarly to KcsA, deuterated DPC was included only in the wash buffer. Using this arrangement, 1270 compounds were screened in mixtures that averaged a little over five compounds each. Figure 6.7 demonstrates that the immobilized DsbB remains intact throughout the screen. In the screen we found 93 compounds that specifically bind DsbB for a hit rate of 7.3 %. Follow-up biochemical studies are currently under way. To date, 41 of the 93 hits have been investigated for enzyme inhibition at 250 μM, where we find that nearly half are substantially potent (greater than 20 % inhibition). The best nine of these compounds have IC_{50}s of 150 μM or better, and a representative curve is shown in Figure 6.7. We have carried out both competition binding and competition enzyme inhibition analyses on a limited subset of the hits. Most of the hits are competitive with ubiquinone binding and this seems to represent the major small-molecule binding pocket. However, a subset of hits are not competitive with ubiquinone. Docking studies place the noncompetitive compounds in a secondary pocket which is excitingly, about 7 Å away.

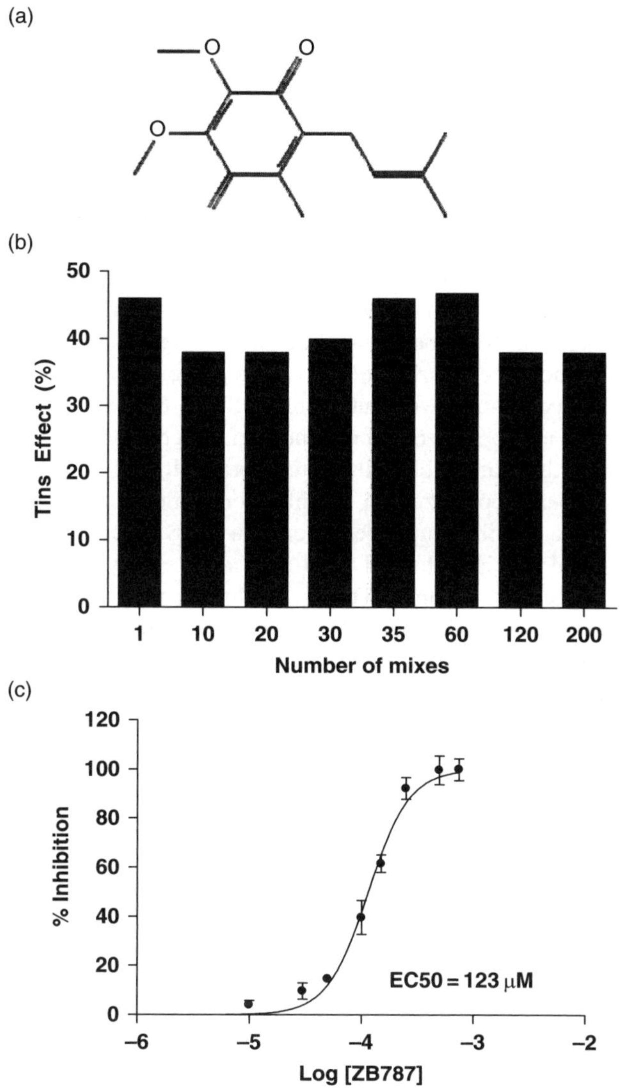

Figure 6.7 *Ligand screening of a bacterial membrane protein. (A) The structure of Ubiquinone-5 used to asses the integrity of immobilized DsbB during the screen. (B) Ubiquinone-5 binding to immobilized DsbB during the screen. Binding is defined as in Figure 6.5. (C) Enzyme inhibition curve of a hit from the screen.*

6.5 Outlook

In the past decade, an impressive repertoire of methods has been developed to permit drug development against soluble targets at the molecular level. In addition to fragment screening methods, structural biology has played a key role in this process. Although at present no drugs are marketed that are the exclusive result of the fragment approach, the principles can clearly be seen in the remarkable specificity and potency of recently marketed kinase inhibitors such as imatinib and gefitinib and, indeed, many fragment-based drugs are in the late stages of clinical trials.[2] Membrane proteins represent a similar pharmacological challenge in that one would like to be able to address specifically individual targets from amongst large numbers of closely related members of a protein family. However, it is currently not possible to use the molecular methods developed for soluble proteins for drug discovery efforts on membrane proteins.

A major goal of the research in our laboratory is to adapt methods developed for soluble targets to membrane proteins or to develop alternative ones. Although we are clearly only at the beginning stages of this process, we have nonetheless made a promising start. We have been able to immobilize a variety of membrane proteins in functional form and have carried out ligand screening on two. Our current efforts are geared towards finding new ways to solubilize and immobilize membrane proteins that can be more widely applied. We are also looking towards a variety of methods to improve the sensitivity of TINS, including experiments that are better optimized for the diffusion-limited nature of the heterogeneous system we employ and possible implementation of a TINS cryoprobe.

Once one finds and validates hits, it is of course necessary to evolve these towards high-affinity, high-specificity ligands. The hit evolution process is greatly aided by the availability of three-dimensional structural information of target–ligand complexes for soluble targets. Since crystallography of membrane proteins is not yet widely applicable, it will be imperative to develop alternative approaches. We envision a number of such approaches that utilize the power of liquid- or solid-state NMR. In recent years, both solid-state NMR[55] and solution-state NMR[56] have made significant progress in elucidating 3D structures of either the membrane protein itself or ligands bound to membrane proteins. Although it is vital that these efforts continue, it is also logical that NMR should be employed to take advantage of its unique ability to rapidly generate local, low-resolution structural information. For this we foresee new applications in chemical shift perturbation-based modeling of protein–ligand complexes,[57] sparse NOE-based methods[58, 59] and paramagnetic NMR.[60] With the foreseeable advancements in ligand screening and structural analysis, the era of molecular drug discovery on membrane protein targets should soon be upon us.

6.6 Abbreviations

AMPPNP	aden-5′-yl imidodiphosphate
ATP	adenosine triphosphate
CLog*P*	logarithm of the partition coefficient between *n*-octanol and water
CMC	critical micellar concentration
CTAB	cetyltrimethylammonium bromide
CTAC	cetyltrimethylammonium chloride
DHPC	dihexanoylphosphatidylcholine
DMPC	dimyristoylphosphatidylcholine
DMSO	dimethyl sulfoxide
DPC	dodecylphosphocholine

FBDD	fragment-based drug discovery
FID	free induction decay
FPMSMA	(4-fluorophenyl)methylsulfanylmethanimidamide
GPCR	G-protein coupled receptor
IMAC	immobilized metal affinity chromatography
LPC	lysophosphatidylcholine
NTA	nitrilotriacetic acid
PBS	phosphate-buffered saline
PEEK	polyether ether ketones
QSAR	quantitative structure–activity relationship
SCA	scaffold-based classification approach
SDS	sodium dodecyl sulfate
SPR	surface plasmon resonance
STD	saturation transfer difference
TINS	target-immobilized NMR screening
TMA	tetramethylammonium chloride
TSP	trimethylsilyl-2,2,3,3-tetradeuteropropionic acid

References

[1] Murali, N., Miller, W. M., John, B. K., Avizonis, D. A. and Smallcombe, S. H. (2006) *J. Magn. Reson.* **179**, 182–189.
[2] Hajduk, P. J. and Greer, J. (2007) *Nat. Rev. Drug Discov.* **6**, 211–219.
[3] Lipinski, C. A., Lombardo, F., Dominy, B. W. and Feeney, P. J. (1997) *Adv. Drug Deliv. Rev.* **23**, 3–25.
[4] Overington, J. P., Al Lazikani, B. and Hopkins, A. L. (2006) *Nat. Rev. Drug Discov.* **5**, 993–996.
[5] Vanwetswinkel, S., Heetebrij, R. J., van Duynhoven, J., Hollander, J. G., Filippov, D. V., Hajduk, P. J. and Siegal, G. (2005) *Chem. Biol.* **12**, 207–216.
[6] Marquardsen, T., Hofmann, M., Hollander, J. G., Loch, C. M. P., Kiihne, S. R., Engelke, F. and Siegal, G. (2006) *J. Magn. Reson.* **182**, 55–65.
[7] Minic, J., Grosclaude, J., Aioun, J., Persuy, M. A., Gorojankina, T., Salesse, R., Pajot-Augy, E., Hou, Y. X., Helali, S., Jaffrezic-Renault, N., Bessueille, F., Errachid, A., Gomila, G., Ruiz, O. and Samitier, J. (2005) *Biochim. Biophys. Acta Gen. Subj.* **1724**, 324–332.
[8] Siegal, G., AB, E. and Schultz, J. (2007) *Drug Discov. Today* **12**, 1032–1039.
[9] Congreve, M., Carr, R., Murray, C. and Jhoti, H. (2003) *Drug Discov. Today* **8**, 876–877.
[10] Babaoglu, K. and Shoichet, B. K. (2006) *Nat. Chem. Biol.* **2**, 720–723.
[11] Lepre, C. A. (2001) *Drug Discov. Today* **6**, 133–140.
[12] Baurin, N., Aboul-Ela, F., Barril, X., Davis, B., Drysdale, M., Dymock, B., Finch, H., Fromont, C., Richardson, C., Simmonite, H. and Hubbard, R. E. (2004) *J. Chem. Inf. Comput. Sci* **44**, 2157–2166.
[13] Jacoby, E., Davies, J. and Blommers, M. J. J. (2003) *Curr. Top. Med. Chem.* **3**, 11–23.
[14] Carney, S. (2007) *Drug Discov. Today* **12**, 789–793.
[15] Xu, J. (2002) *J. Med. Chem.* **45**, 5311–5320.
[16] Hajduk, P. J., Huth, J. R. and Fesik, S. W. (2005) *J. Med. Chem.* **48**, 2518–2525.
[17] Klein, J., Meinecke, R., Mayer, M. and Meyer, B. (1999) *J. Am. Chem. Soc.* **121**, 5336–5337.
[18] Nordin, H., Jungnelius, M., Karlsson, R. and Karlsson, O. P. (2005) *Anal. Biochem.* **340**, 359–368.

[19] Haselhorst, T., Munster-Kuhnel, A. K., Oschlies, M., Tiralongo, J., Gerardy-Schahn, R. and von Itzstein, M. (2007) *Biochem. Biophys. Res. Commun.* **359**, 866–870.
[20] Klammt, C., Schwarz, D., Eifler, N., Engel, A., Piehler, J., Haase, W., Hahn, S., Dotsch, V. and Bernhard, F. (2007) *J. Struct. Biol.* **159**, 194–205.
[21] Park, S. H., Prytulla, S., De Angelis, A. A., Brown, J. M., Kiefer, H. and Opella, S. J. (2006) *J. Am. Chem. Soc.* **128**, 7402–7403.
[22] Parker, E. M., Kameyama, K., Higashijima, T. and Ross, E. M. (1991) *J. Biol. Chem.* **266**, 519–527.
[23] Kaushal, S., Ridge, K. D. and Khorana, H. G. (1994) *Proc. Natl. Acad. Sci. USA* **91**, 4024–4028.
[24] Akermoun, M., Koglin, M., Zvalova-Iooss, D., Folschweiller, N., Dowell, S. J. and Gearing, K. L. (2005) *Protein Express. Purif.* **44**, 65–74.
[25] Lundstrom, K., Wagner, R., Reinhart, C., Desmyter, A., Cherouati, N., Magnin, T., Zeder-Lutz, G., Courtot, M., Prual, C., Andre, N., Hassaine, G., Michel, H., Cambillau, C. and Pattus, F. (2006) *J. Struct. Funct. Genomics* **7**, 77–91.
[26] Fraser, N. J. (2006) *Protein Express. Purif.* **49**, 129–137.
[27] Sarramegna, V., Muller, I., Mousseau, G., Froment, C., Monsarrat, B., Milon, A. and Talmont, F. (2005) *Protein Express. Purif.* **43**, 85–93.
[28] Xu, Y., Yushmanov, V. E. and Tang, P. (2002) *Biosci. Rep.* **22**, 175–196.
[29] Heginbotham, L., Odessey, E. and Miller, C. (1997) *Biochemistry* **36**, 10335–10342.
[30] Kleinschmidt, J. H., Wiener, M. C. and Tamm, L. K. (1999) *Protein Sci.* **8**, 2065–2071.
[31] Girvin, M. E. and Fillingame, R. H. (1993) *Biochemistry* **32**, 12167–12177.
[32] Taylor, R. M., Zakharov, S. D., Heymann, J. B., Girvin, M. E. and Cramer, W. A. (2000) *Biochemistry* **39**, 12131–12139.
[33] Xu, Y., Yushmanov, V. E. and Tang, P. (2002) *Biosci. Rep.* **22**, 175–196.
[34] Xu, Y., Yushmanov, V. E. and Tang, P. (2002) *Biosci. Rep.* **22**, 175–196.
[35] Arora, A. and Tamm, L. K. (2001) *Curr. Opin. Struct. Biol.* **11**, 540–547.
[36] Xu, Y., Yushmanov, V. E. and Tang, P. (2002) *Biosci. Rep.* **22**, 175–196.
[37] Struppe, J. and Vold, R. R. (1998) *J. Magn. Reson.* **135**, 541–546.
[38] Rasmussen, S. G. F., Choi, H. K., Rosenbaum, D. M., Kobilka, T. S., Thian, F. S., Edwards, P. C., Burghammer, M., Ratnala, V. R. P., Sanishvili, R., Fischetti, R. F., Schertler, G. F. X., Weis, W. I. and Kobilka, B. K. (2007) *Nature* **450**, 383–387.
[39] Opella, S. J., Kim, Y. and Mcdonnell, P. (1994) *Nucl. Magn. Reson. C* **239**, 536–560.
[40] Yu, L. P., Sun, C. H., Song, D. Y., Shen, J. W., Xu, N., Gunasekera, A., Hajduk, P. J. and Olejniczak, E. T. (2005) *Biochemistry* **44**, 15834–15841.
[41] Bader, M. W., Xie, T., Yu, C. A. and Bardwell, J. C. A. (2000) *J. Biol. Chem.* **275**, 26082–26088.
[42] Karlsson, O. P. and Lofas, S. (2002) *Anal. Biochem.* **300**, 132–138.
[43] Martinez, K. L., Meyer, B. H., Hovius, R., Lundstrom, K. and Vogel, H. (2003) *Langmuir* **19**, 10925–10929.
[44] Bieri, C., Ernst, O. P., Heyse, S., Hofmann, K. P. and Vogel, H. (1999) *Nat. Biotechnol.* **17**, 1105–1108.
[45] Neumann, L., Wohland, T., Whelan, R. J., Zare, R. N. and Kobilka, B. K. (2002) *ChemBiochem* **3**, 993–998.
[46] Kada, G., Riener, C. K., Hinterdorfer, P., Kienberger, F., Stoh, C. M. and Gruber, H. J. (2002) *Single Mol.* **3**, 119–125.
[47] Nollert, P., Kiefer, H. and Jahnig, F. (1995) *Biophys. J.* **69**, 1447–1455.
[48] Neumann, L., Wohland, T., Whelan, R. J., Zare, R. N. and Kobilka, B. K. (2002) *ChemBiochem* **3**, 993–998.
[49] Stenlund, P., Babcock, G. J., Sodroski, J. and Myszka, D. G. (2003) *Anal. Biochem.* **316**, 243–250.
[50] Friedrich, M. G., Giess, F., Naumann, R., Knoll, W., Ataka, K., Heberle, J., Hrabakova, J., Murgida, D. H. and Hildebrandt, P. (2004) *Chem. Commun.* 2376–2377.

[51] Giess, F., Friedrich, M. G., Heberle, J., Naumann, R. L. and Knoll, W. (2004) *Biophys. J.* **87**, 3213–3220.
[52] Ratnala, V. R. P., Kiihne, S. R., Buda, F., Leurs, R., de Groot, H. J. M. and Degrip, W. J. (2007) *J. Am. Chem. Soc.* **129**, 867–872.
[53] Stenson, T. H. and Weiss, A. A. (2002) *Infect. Immun.* **70**, 2297–2303.
[54] Inaba, K., Murakami, S., Suzuki, M., Nakagawa, A., Yamashita, E., Okada, K. and Ito, K. (2006) *Cell* **127**, 789–801.
[55] Baldus, M. (2006) *Curr. Opin. Struct. Biol.* **16**, 618–623.
[56] Tamm, L. K. and Liang, B. Y. (2006) *Prog. Nucl. Magn. Reson. Spectrosc.* **48**, 201–210.
[57] Wang, B., Westerhoff, L. M. and Merz, K. M. Jr (2007) *J. Med. Chem.* **50**, 5128–5134.
[58] Medek, A., Olejniczak, E. T., Meadows, R. P. and Fesik, S. W. (2000) *J. Biomol. NMR* **18**, 229–238.
[59] Pellecchia, M., Meininger, D., Dong, Q., Chang, E., Jack, R. and Sem, D. S. (2002) *J. Biomol. NMR* **22**, 165–173.
[60] Pintacuda, G., John, M., Su, X. C. and Otting, G. (2007) *Acc. Chem. Res.* **40**, 206–212.

7

In Situ Fragment-based Medicinal Chemistry: Screening by Mass Spectrometry

Sally-Ann Poulsen and Gary H. Kruppa

7.1 Introduction

The task of discovering and then optimizing small molecules that interact with and appropriately modify the activity of large biomolecules such as enzymes is central to the forward progression of the drug discovery pipeline. To transform a small molecule lead into a safe medicine that can be used in people requires conquering a broad spectrum of challenges and it is unsettling that massive investment by the pharmaceutical industry in research efficiencies and platform technologies has not equated with improving the speed of drug discovery.[1] While the reasons behind this limited success are complex and varied (and outside the scope of this chapter), the situation does provide grounds for the industry to address urgently the effectiveness of current medicinal chemistry programmes and to consider exploring alternative avenues for improving the quality of lead discovery outcomes.

Identification of new drug leads by fragment-based screening now has its foundations firmly established as a valuable tool to facilitate drug discovery, and this is evidenced by the content of this book. Target-templated synthetic approaches that covalently link fragments within the confines of a target's binding site have developed alongside the modern fragment-based screening and fragment-based drug discovery concepts. These synthetic approaches include *in situ* dynamic combinatorial chemistry (DCC) and *in situ* Click chemistry. *In situ*

Fragment-Based Drug Discovery: A Practical Approach Edited by Edward R. Zartler and Michael J. Shapiro

medicinal chemistry represents a first-generation extension of fragment-based drug discovery towards the direction of target-guided fragment optimization. The *in situ* linking of fragments was articulated in a proof-of-concept format for DCC in 1997 by Huc and Lehn[2] and for Click chemistry in 2002 by Sharpless, Finn and co-workers.[3] These novel synthetic approaches have since flourished as elegant and vibrant research disciplines in their own right, each with broadly scoped potential applications of which drug discovery is just one.

Performing synthetic chemistry to link fragments covalently within the reaction environment dictated by a target biomolecule in its native state is demanding, but this is now reasonably well established for both DCC and Click chemistry. The introductory material of this chapter will provide an overview into the principles of both *in situ* DCC and Click chemistry approaches as they apply to fragment-based drug discovery. A discussion on the practical considerations that are key to translating these synthetic approaches from proof-of-concept investigations to outcomes-focused applications that may then be considered as useful tools by the pharmaceutical industry for the purpose of drug discovery will follow. Critical to this translational capacity is the ability to identify rapidly and accurately the linked fragments of interest. An account of how mass spectrometry is emerging as the key analytical methodology to fulfil the analytical demands associated with *in situ* medicinal chemistry will be presented. Griffey and Swayze[4] have previously described the principles of electrospray ionization mass spectrometry (ESI-MS) and demonstrated its immense value as a primary screen for fragment-based drug discovery. Here we will focus our discussion on 'practical aspects', including the optimization of experimental conditions that allow mass spectrometry to detect specific target–ligand binding interactions. By way of published examples, we will demonstrate that mass spectrometry is an invaluable tool as a primary screen particularly in the case of *in situ* medicinal chemistry applications where synthesis and screening are integrated into a single step.

7.2 Target-guided *In Situ* Medicinal Chemistry – the Principles

Biomolecular targets exist as an ensemble of equilibrating conformers, and these conformational changes are associated with dynamic disorder, i.e. distributions and fluctuations over time-scales of 1 ms to 100 s.[5] Structural variations can range from subtle to extreme and invariably compromise the outcome of structure-based drug design efforts that are typically guided by a static and/or ensemble-averaged conformation. Even so, applying the principles of molecular recognition may sometimes effectively guide the design of small molecules and the level of success with structure-based drug design is in many respects very good. This approach is plagued, however, by the difficulty of appropriately interrogating the subtleties and complexities of molecular recognition between a dynamic target and a small molecule within the biological context. Structure-based drug design is imperfect and necessarily a resource-intensive approach to drug discovery, requiring many iterative cycles between small-molecule synthesis and biological assay results, intervened by interpretation of structure–activity relationships and refinement of the 'rational' structure-based drug design model.

Target-guided *in situ* medicinal chemistry is a novel means of synthesizing small-molecule ligands for medically relevant biomolecular targets, whereby the target biomolecule directs the outcome of the synthetic efforts. Target-guided synthesis (TGS) has added a new dimension to the synthetic capability of medicinal chemistry as the target assists the medicinal chemist to select and synthesize those reaction products that have a high affinity for the target from all potential reaction products that are accessible through the chemistry employed. Target-guided synthesis is readily conceptualized by the familiar 'lock and key' host–guest descriptors of Emil Fischer (Figure 7.1), wherein the biomolecular target acts as the host to facilitate the 'correct' assembly of fragment components leading to the synthesis of a guest small molecule. This description equally encapsulates the principle of *in situ* chemical linking of fragments for application to drug discovery. The latter can be subdivided into two complementary approaches: *in situ* DCC (linking fragments under thermodynamic control) and *in situ* Click chemistry (linking fragments under kinetic control).

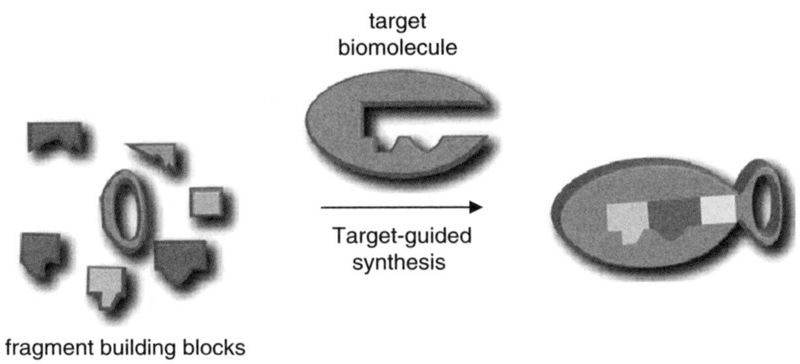

Figure 7.1 *Schematic representation of target-guided synthesis to generate a ligand (guest) templated by a target biomolecule (host) using the 'lock and key' descriptors of Emil Fischer.*

7.3 Dynamic Combinatorial Chemistry – an Overview

Conventional combinatorial chemistry permits the rapid synthesis of small-molecule libraries and the impact of this development has been to revolutionize synthetic chemistry. When applied in a drug discovery setting, the sheer number of compounds generated, when coupled with high-throughput screening, in principle could facilitate a speedier route to drug lead discovery compared with the traditional single compound/single assay strategy. Vast numbers of compounds have not had the level of impact that one might have expected, however, mostly owing to inappropriate compound selection, and this realization has encouraged drug lead discovery programmes to challenge synthetic efforts to capture biological relevance better.[1, 6]

Dynamic combinatorial chemistry (DCC) offers a conceptually different approach to the synthesis and screening of libraries of small molecules for drug discovery compared with conventional combinatorial chemistry.[7–10] These libraries are called dynamic combinatorial libraries (DCLs) and are generated from the reversible reactions between a set of building blocks or fragments to give a library of covalently, but reversibly, linked fragments. In a DCL all linked fragments, i.e. constituents, are thus in equilibrium, with interconversion of constituents possible through reversible chemical reactions between component fragments. The composition of a DCL is governed by the thermodynamic stability of the possible constituents under the conditions of the experiment, meaning that a change in the library environment can instruct and inform changes in the library composition. Libraries are therefore not static (as in conventional combinatorial chemistry) but can alter in composition through the re-equilibration of their reversibly linked fragments, for example in response to the presence of a selection pressure in the immediate environment such as the addition of a target biomolecule. The shift in equilibrium that occurs with this adaptive chemistry leads to an increase in the concentration of the library molecule(s) that best recognizes (through molecular recognition) the target (Figure 7.2). The dynamic chemistry may occur both untemplated, in the bulk solution, or templated, within the target's active site. Either way, the target-selected linked fragments are withdrawn from the solution equilibrium and the DCL then re-equilibrates to produce more of these selected products at the expense of other possible products. Hence the most active DCL constituents are selected and amplified in the presence of the target. The diversity of a DCL is on two distinct levels: the first is fixed and dependent solely upon the number and molecular architecture of the initial fragments, whereas the second is the diversity accessible upon linking these fragments to generate the library constituents and hence can alter under different library conditions (i.e. in the presence of different targets). The composition and diversity of DCLs are *informed* and are driven by the target rather than governed by sheer numbers alone. Provided that a molecule is accessible with the reversible chemistry and available fragments, it may be selected and amplified in the presence of a target, circumventing the need for the synthesis and even representation of every library member. One appealing advantage of DCC is that it readily lends itself to the optimization of fragments identified as privileged for a particular therapeutic target.

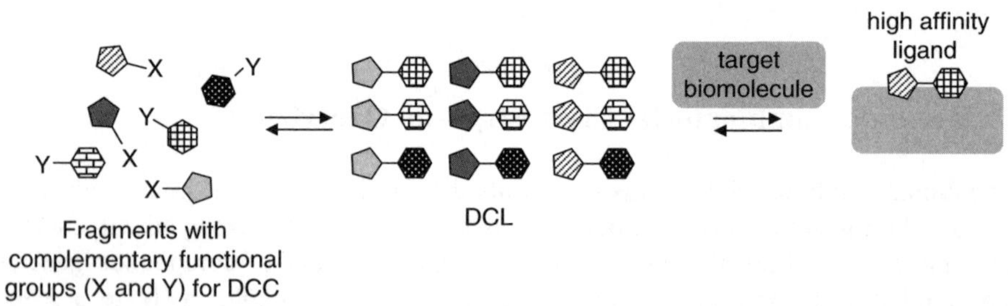

Figure 7.2 *Schematic representation of target-templated* in situ *dynamic combinatorial chemistry.*

7.4 Dynamic Combinatorial Chemistry – Reversible Chemistry

The key feature of DCC is the reversible chemical reaction that mediates exchange of the building block fragments.[7–9] For *in situ* drug discovery applications of DCC, the selection of linked fragments occurs in the same environment as the equilibration reaction and this demands that the reaction fulfils several requirements in addition to reversibility: (i) the exchange reaction should occur at a rate that allows equilibrium to be reached within an acceptable time-scale; (ii) it should be bioorthogonal with the template biomolecule, i.e. with reactivity inert to the functional groups of the biomolecule; (iii) it must operate under conditions that preserve native state (or other biologically relevant conformations) of the biomolecule – typically aqueous buffer at physiological temperature and pH; (iv) the reaction conditions should not inhibit formation of noncovalent interactions involved in molecular recognition between target biomolecule and fragments; and (v) ideally the reactivity of all fragments should be similar to allow access to unbiased DCL compositions. The reversible reactions that meet these requirements have been detailed in a recent review;[7] here we will present only a specific example, the hydrazone exchange reaction (Scheme 7.1). This C=N exchange reaction has been widely used in DCC with multiple examples of drug discovery applications.[11–15] Hydrazone exchange is the reversible reaction between a hydrazine or hydrazide [$R^1R^2N–NH_2$ or $R^1(C=O)NH–NH_2$] and an aldehyde [$R^3(C=O)H$]. Other C=N exchange reactions include imine and oxime exchange; however, to ensure a balanced product distribution, libraries relying on C=N are ideally prepared from closely related $X–NH_2$ fragments.[7]

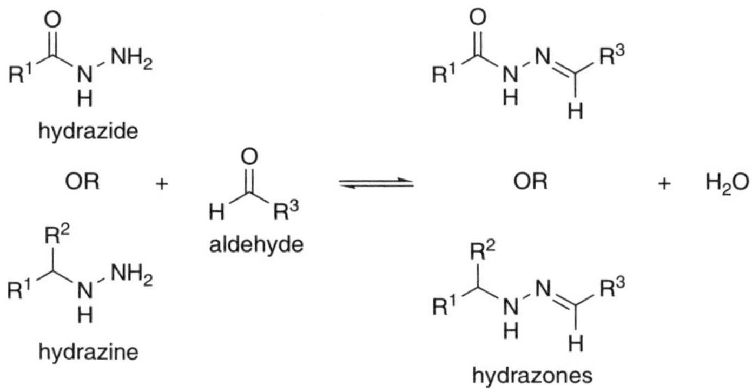

Scheme 7.1 *Reversible hydrazone formation and exchange.*

Hydrazone formation is rapid under acidic conditions (typically fastest at pH $\approx$ 4), but importantly for *in situ* DCC applications it can also occur under mild conditions, such as in aqueous environments at neutral pH.[16–18] The amino functional groups on proteins are predominantly protonated in this pH range so that this reaction is essentially bioorthogonal and potential imine products, from the reaction of aldehyde fragments with amino groups on the protein target, are not detected at neutral or acidic pH. If the aldehyde is present as a hydrate (reversibly formed from the aldehyde), the aldehyde along with the hydrate is consumed within minutes of initiating the exchange reaction.[19] The equilibration kinetics of hydrazone exchange are fairly slow and it may take days to reach equilibrium, whereas this is not necessarily a barrier for drug discovery applications: an increase in the kinetics of reaction while preserving the bioorthogonal reaction attributes is indeed desirable. A recent investigation by Dawson and colleagues demonstrated that the equilibration kinetics of hydrazone formation from α-carbonyl aldehydes and subsequent transimination could be significantly accelerated (reaching equilibrium in hours) by using aniline as a nucleophilic catalyst.[20] It remains to be verified if this catalyst is compatible with *in situ* DCC; however, we can expect that future refinements of this finding will facilitate a greater use of the hydrazone exchange reaction in DCC drug discovery applications.

7.5 Click Chemistry – an Overview

In the past few years, there has been a flurry of activity in the literature concerning the 1,3-dipolar cycloaddition reaction (1,3-DCR) of organic azides with terminal acetylenes yielding 1,2,3-triazoles, i.e. the Huisgen reaction.[21] This renewed interest stems largely from the optimization of this 1,3-DCR, independently by the groups of Meldal[22] and Sharpless,[23] with respect to ease and efficiency of catalysis and regioselectivity to form exclusively the 1,4-disubstituted 1,2,3-triazole (or *anti*-triazole) product (Scheme 7.2). The reaction involves a stepwise Cu(I)-catalysed dipolar cycloaddition of a terminal acetylene to an organic azide. The highly exothermic and kinetically controlled reaction is conducted favourably in water at a physiologically relevant temperature and the reactants are bioorthogonal to biological systems. For these reasons, the reaction is now the premier transformation of *in situ* Click chemistry reactions, wherein complementary fragments bearing either azide or acetylene moieties are combined in the presence of a target biomolecule.[24]

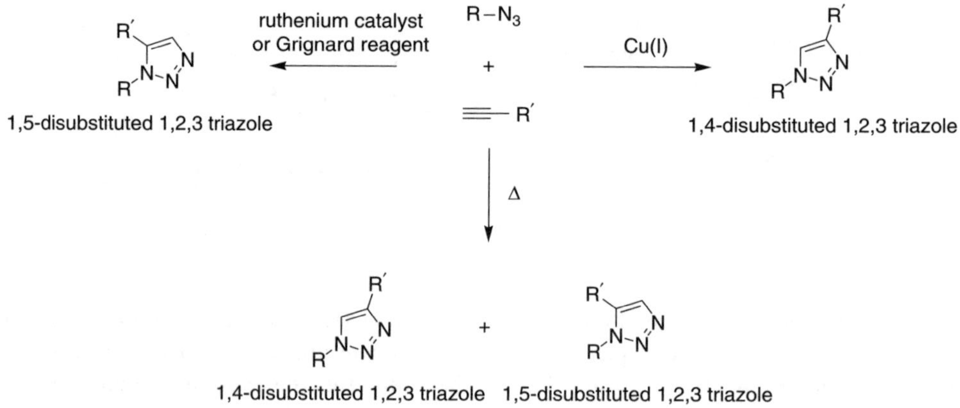

Scheme 7.2 *Synthesis of 1,2,3-triazoles by the 1,3-dipolar cycloaddition reaction of organic azides with terminal acetylenes.*

The 1,5-disubstituted 1,2,3-triazole (*syn*-triazole) regioisomer may be regioselectively synthesized by using magnesium acetylides[25] or the more recently discovered catalysis by ruthenium complexes.[26] Almost equimolar *syn*- and *anti*-triazole mixtures are obtained by heating neat mixtures of the corresponding azides and alkynes at elevated temperatures (Scheme 7.2).[27]

Like *in situ* DCC, *in situ* Click chemistry permits synthesis and screening to be combined into a single step with the target biomolecule guiding the assembly of fragments. A key differentiating attribute of these complementary concepts is that Click chemistry utilizes an irreversible reaction to lock the fragments together whereas DCC utilizes a reversible reaction to link fragments (Figure 7.3). Another important aspect of *in situ* Click chemistry is that the reaction avoids the combination of strong nucleophilic and electrophilic functional groups typical in DCC. The reactive partner fragments for Click chemistry are thus bioorthogonal under any reaction conditions.[24] The Click reaction occurs almost exclusively within the target's binding site, with minimal background reaction in the bulk solution, meaning that the formation of a triazole from Click chemistry fragments *in situ* virtually guarantees that the resulting triazole will be a potent lead compound for drug discovery;[24] the examples described later in this chapter exemplify this. A potential disadvantage of *in situ* Click chemistry for drug lead compound discovery is the possibility that effective inhibitors are not assembled in the presence of the target and are thus 'missed opportunities' in the screening campaign, i.e. false negatives.

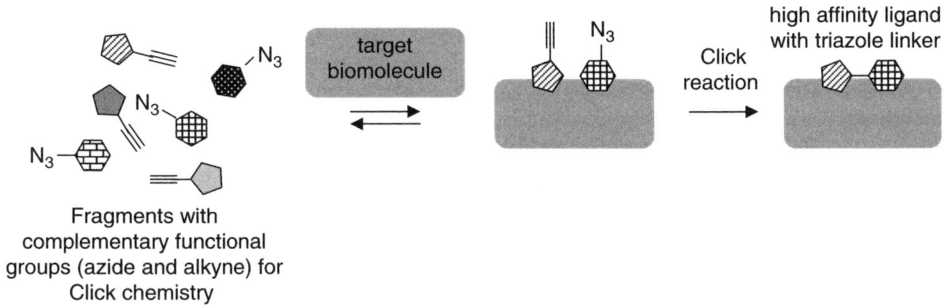

Figure 7.3 *Schematic representation of target-templated* in situ *Click chemistry.*

7.6 *In Situ* Medicinal Chemistry: Current State of Play

Both of the *in situ* synthesis approaches described here have had considerable success with respect to the discovery of potent hit compounds for the target interrogated. DCC, benefiting from about a 5 year head start on Click chemistry, has so far been more broadly applied with respect to target diversity and number of active research groups. The entries in Table 7.1 list the most recent published examples of *in situ* DCC (2006–mid-2007) and *in situ* Click chemistry since its inception (2002 onwards). These publications exemplify the growing academic interest in the application of these informed chemistries for drug discovery; however, it is not readily apparent what may be occurring in the pharmaceutical industry, nor is it evident what progress either of these approaches has made towards advancing compounds into the drug discovery pipeline. It is, however, arguably an area that the pharmaceutical industry is unlikely to ignore given its tremendous potential as outlined here.

Table 7.1 *Published examples of* in situ *medicinal chemistry:* in situ *DCC (2006–mid-2007) and* in situ *Click chemistry (since its inception in 2002–mid-2007).*

Biomolecular target	Reversible chemistry	Maximum fragment size (Da)[a]	Analytical method for hit identification[a]
In situ DCC			
Carbonic anhydrase[15]	Acyl hydrazone exchange	<300	ESI-FTICR-MS
Metallo-β-lactamase[28]	Disulfide exchange	<200	ESI-MS
Transactivation-response element of HIV-1[29]	Imine exchange	<230	SELEX, HPLC and MALDI-TOF MS
α-1,3-Galactosyltransferase; β-1,4-galactosyltransferase[30]	Imine exchange	<250	HPLC
α-1,3-Galactosyltransferase[31]	Imine exchange	<290	HPLC
Concanavalin A[32]	Disulfide exchange	~220	Quartz crystal microbalance
Galectin-3; *Viscum album* agglutinin; *Ulex europaeus* agglutinin[33]	Disulfide exchange	~220	Bioassay
Schistosoma japonica glutathione-*S*-transferase[34]	Conjugate addition of thiols to enones	~300	HPLC
Calmodulin[35]	Disulfide exchange	~300	HPLC
Subtilisin; albumin[36]	Disulfide exchange	~500	ESI-MS, HPLC
In situ Click chemistry			
Acetylcholinesterase[37–39]	—	~230–440	DIOS-MS, LC-MS-SIM
Carbonic anhydrase[40, 41, 44]	—	~180–350	LC-MS-SIM
HIV-1 protease[42]	—	~190–430	LC-MS-SIM
Plasmodium falciparum tryptophanyl-tRNA synthetase[43]	—	NA	NA
Cyclooxygenase-2[44]	—	NA	NA

[a] NA, not available (targets described in published conference abstracts[43, 44]).

7.7 Practical Considerations for Drug Discovery Applications

In principle, the *in situ* medicinal chemistry approaches described here have the potential to circumvent the need to synthesize, characterize and screen each possible library constituent individually for the purpose of drug discovery. The chemistry, together with the target acting as a template, permits self-screening owing to selection and/or amplification of the 'best binders' by molecular recognition events between fragments and the target. This qualitative description is adequate to convey the concepts of target-guided synthesis; however, in practice it is essential to know more about the level of product formation to expect as this is critical to the successful application of analytical methods to detect and identify these best binder(s).

The synthetic advantages and target-informed product formation with *in situ* DCC and Click chemistry are of little consequence unless we have the ability to identify and characterize readily the linked fragments of interest. Accomplishing the rapid detection of these enriched product(s) is clearly a make-or-break factor to *in situ* DCC and Click chemistry taking a prominent place in fragment-based optimization for drug discovery.[45] The standard DCL screening approach is indirect and must carry out the screening of identical DCLs twice, with and without the target. The equilibrium concentration profiles of all library ligands are then compared, typically following disruption of the ligand–target noncovalent complexes of the target protein. The detection of ligand enrichment in the targeted DCL is the basis of identifying the 'best binders'. The complexity of screening with this indirect method increases with the size of the DCL owing to chromatographic (e.g. HPLC) and/or spectral (e.g. NMR) overlap so that either the need for the synthesis of individual library components to validate assignments and/or the preparation and screening of DCL sublibraries becomes necessary for deconvolution of larger DCLs; this is both labour and time intensive and has the effect of undermining the promoted advantages of DCC. For DCC to have a significant impact on drug discovery necessitates the development of a direct screening protocol(s) for rapid identification and characterization of those ligands with affinity for the target protein, against the background of inactive DCL constituents. It would be a tremendous advantage to drug discovery applications of DCC if DCL screening methodologies could proceed without the need for chromatography, conversion to a static library, preparation of sublibraries, prior disruption of the protein–ligand complexes, synthesis of individual library ligands to validate spectral assignments or for the preparation and duplicate screening of identical DCLs (with and without the target). In pursuing this, we must relinquish our 'need to know everything' philosophy of medicinal chemistry and instead settle for what we 'really need to know', that being the molecular structure of the linked fragments that are selected and amplified in the system under investigation.

The strong amplification of selected DCL constituents has been observed in many examples, particularly from smaller DCLs. For drug discovery we may sometimes also want to work with larger libraries, so what can we expect of amplification levels when we have an increased number of initial library fragments? Fortunately, the question of what the limit to the size of effective DCLs is has already received serious consideration, with various groups having modelled DCL product distributions under a variety of library scenarios. This material has been reviewed recently and a complete discussion is beyond the scope of this chapter.[46, 47] The results from one of the most relevant models to drug discovery applications are summarized in Figure 7.4. This model confirms the intuitive reduction in the observed concentration of the best binders that occurs with increasing number of DCL fragments (and possible products) that is a consequence of spreading the fragments out over an increasing number of possible library constituents.[46] However, this decrease in concentration is not as sharp as expected; for example, in a 1000 compound library the best binder will become 13 % (on average) of the library composition and in a 10 000 compound library the level is 8 % (on average) (Figure 7.5). Several good binders are a probable outcome from *in situ* DCC; with this scenario, amplification levels for the good binders are likely to remain within the detection capability of modern analytical equipment.

Like DCC, applying *in situ* Click chemistry against drug discovery targets necessitates the identification of the best binding ligands. The experimental reaction mixture for *in situ* Click chemistry shares many component attributes in common with DCC, namely

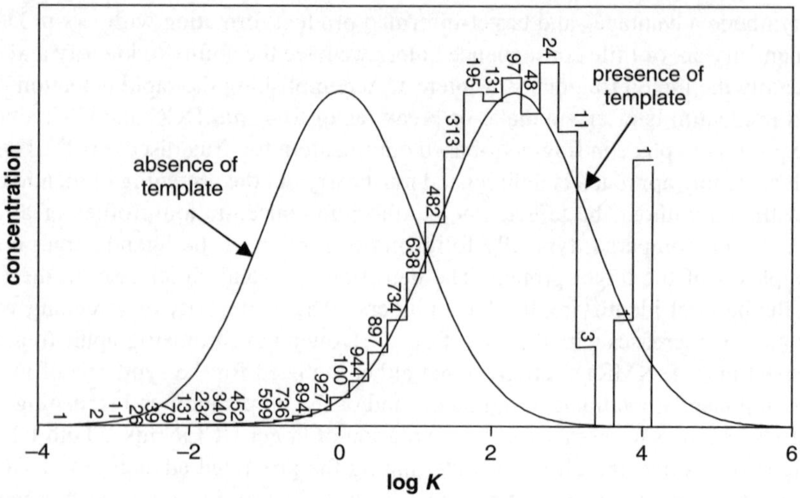

Figure 7.4 *Histograms representing the composition of a continuous DCL in the absence and presence of a template and a typical simulated DCL containing 10 000 compounds (bars). Bars are labelled with the number of compounds in the affinity class. Reprinted with permission from Corbett, P.T.,* et al. Org. Lett. *2004,* ***6****, 1825–1827. Copyright (2004) American Chemical Society.*

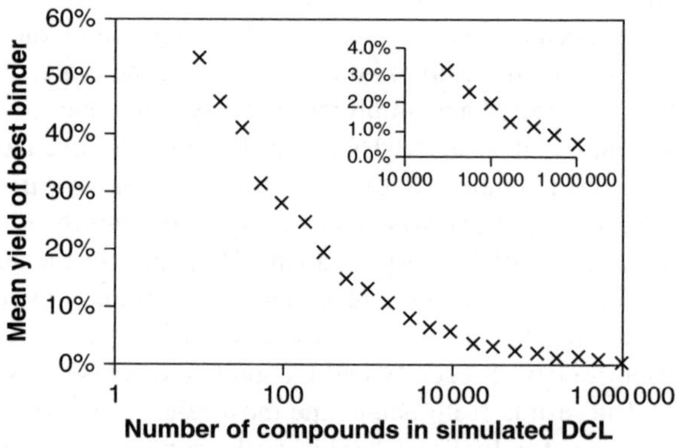

Figure 7.5 *Yields of the best binders as a function of library size. Each data point represents the average from 100 simulated DCLs. Reprinted with permission from Corbett, P.T.,* et al. Org. Lett. *2004,* ***6****, 1825–1827. Copyright (2004) American Chemical Society.*

the target biomolecule, aqueous buffer, fragment building blocks and noncovalent target–ligand complexes. A major difference is that *in situ* Click chemistry, unlike *in situ* DCC, has minimal background reaction in the bulk solution. *In situ* Click chemistry additionally presents a far tougher screening challenge owing to the very low level of conversion of fragments to potential binding ligands. At present there is no material in the primary literature that describes the theoretical level of product conversion to expect; it has, however, been qualitatively indicated that the quantities of triazole products generated are much lower than the concentration of the target biomolecule.[24, 38] In the examples of *in situ* Click chemistry that follow, we will attempt to make more transparent the level of product formation by scrutinizing the experimental data so far reported.

7.8 Electrospray Ionization Mass Spectrometry (ESI-MS)

The analytical demands of combinatorially generated compound libraries have directed the attention of the pharmaceutical industry towards mass spectrometry owing to its attributes of speed, intrinsic sensitivity, specificity, low sample consumption and the capability of resolving vast numbers of compounds in complex compound mixtures. Mass spectral analysis requires that the species of interest in solution be transferred to the gas phase as fully desolvated, charged species, for which the mass spectrometer then determines the mass-to-charge ratio (m/z). Using the well-established technique of ESI-MS, charged molecules in a solution are desolvated by a process that involves the formation of fine droplets generated by an applied high voltage; this process is usually assisted by pressure or ultrasonic nebulization.[48] The fine droplets successively decrease in size by evaporation of solvent and as the charge density in the droplet becomes higher, the ejection of yet smaller droplets occurs. This process continues until a final stage of desolvation by evaporation leaves fully desolvated, charged molecules in the gas phase which are then transferred into the mass analyzer of the mass spectrometer guided by a series of ion guides and ion optics. The mass analyzer of the mass spectrometer then measures the m/z of the ion; this is an information-rich parameter for the medicinal chemist, as will become evident in the following text.

7.8.1 ESI-MS and Proteins

ESI-MS is generally used to accomplish the transfer of intact biological macromolecules such as proteins from solution into the gas phase. Many commonly used mass analyzers such as quadrupoles and ion traps have optimal performance in the m/z range up to 3000 with a maximum typically of 4000–10 000. Therefore, in order to bring the m/z of a protein with a mass of 30 kDa into the optimal m/z range of a typical mass spectrometer, 10–30 charges must be incorporated on the protein in the gas phase. When using ESI-MS for molecular weight determination, samples are typically dissolved in a denaturing 1:1 water–organic solvent (acetonitrile or methanol) buffer system containing a few percent of acetic or formic acid from which a very stable electrospray can be produced. Under these conditions, the protein is unfolded, exposing many of the amino acid side-chains to the bulk solvent (Figure 7.6a). The trace acid ensures that many basic residues on the protein are protonated, thus maximizing the signal obtained from the protein when running the mass spectrometer in the positive ion mode. The mass spectrum for denatured carbonic anhydrase is shown in Figure 7.6b. The spectrum is characterized by a broad envelope of charge states from +35 to +20, corresponding to the family of differently protonated molecular species present in solution.

If the protein molecular weight is unknown, then the charge states are also not known. The process of determining the molecular weight for an unknown protein from such a mass spectrum is known as deconvolution and is based on simple algebra. If MW is the

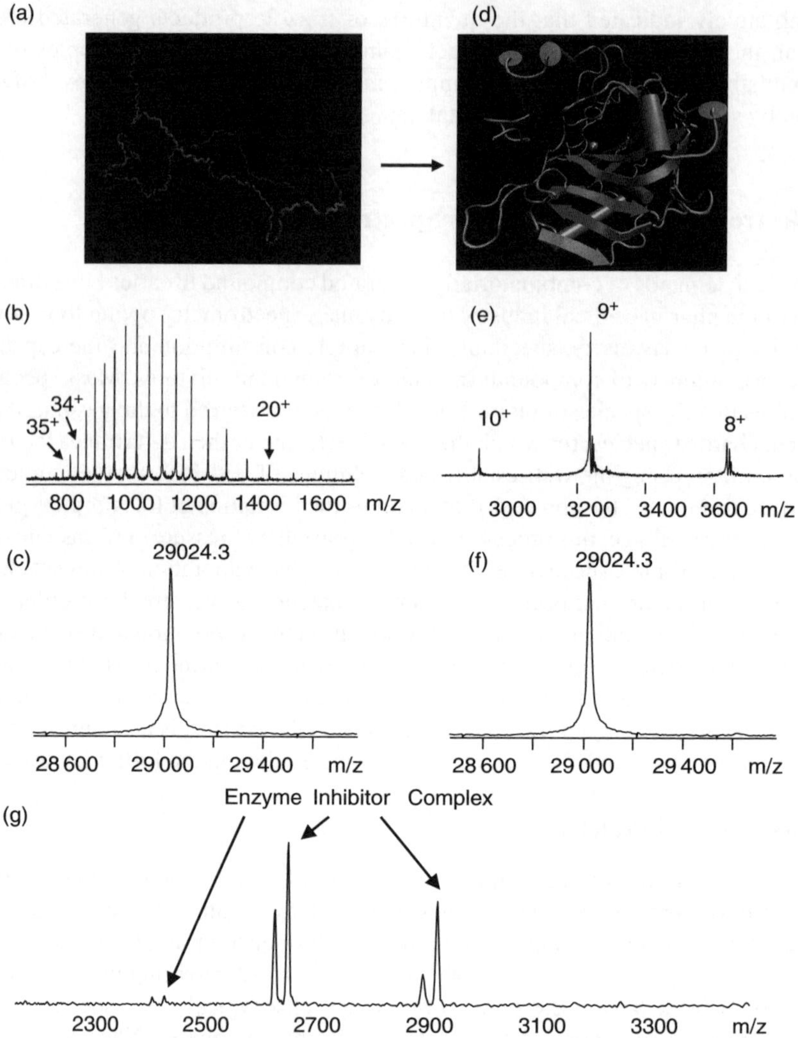

Figure 7.6 *ESI of zinc metalloenzyme carbonic anhydrase, (a)–(c) under acidic denaturing conditions, (d)–(f) native state conditions, and (g) native state conditions with a specific inhibitor. Structure of entry (d) is Protein Data Bank ID 1BN1. Boriack-Sjodin, P.A., Zeitlin, S., Chen, H.H., Crenshaw, L., Gross, S., Dantanarayana, A., Delgado, P., May, J.A., Dean, T., Christianson, D.W. Structural analysis of inhibitor binding to human carbonic anhydrase II.* Protein Sci., *1998, 7, 2483–2489.*

molecular weight of the protein, H is the mass of the proton charge carrier and z is the number of protons in a particular charge state, then the observed m/z for charge state z will be $(MW + zH)/z$. If we use the experimentally observed m/z values for two adjacent charge states z_1 and z_2, observed at mass-to-charge ratios $(m/z)_1$ and $(m/z)_2$, then we can write two equations that can easily be solved for the molecular weight of the protein:

$$(m/z)_1 = (MW + z_1H)/z_1 \tag{7.1}$$

$$(m/z)_2 = (MW + z_2H)/z_2 \tag{7.2}$$

As z_1 and z_2 are adjacent charge states then $z_2 = z_1 + 1$. This relationship allows Equations (7.1) and (7.2) to be combined to give

$$z_1 = [(m/z)_2 - H]/[(m/z)_1 - (m/z)_2] \tag{7.3}$$

and

$$MW = z_1(m/z)_1 - z_1H \tag{7.4}$$

Although it is trivial to solve these equations for simple peptides or a pure small protein, deconvoluting a mass spectrum to determine molecular weights and relative abundances of various components rapidly becomes intractable for manual calculations when working with larger proteins or mixtures of proteins. All modern ESI-equipped mass spectrometers include sophisticated deconvolution software derived from the above mathematical relationships to search automatically and routinely for peaks that are from the same molecular species but differ by charge state. The intensities of all related peaks are added together and converted to a single +1 or zero charge state to produce the 'deconvoluted' spectrum revealing the protein molecular weight (Figure 7.6c). Calculations generally require only a fraction of a second for even fairly complicated mass spectra.

7.8.2 ESI-MS and Noncovalent Ligand–Protein Interactions

The screening of targets against ligands by mass spectrometry requires that the experimental parameters of the ESI-MS process be adjusted such that the noncovalent interactions between the ligand and target are maintained during both the desolvation process and in the gas phase. ESI-MS has been used extensively to study proteins and complexes of proteins with naturally occurring substrates, inhibitors and drugs. An extensive literature on this topic (more than 300 publications and numerous review articles, as pointed out in a recent review[49]) indicates that although there are pitfalls, it is generally straightforward to adjust the parameters of the ESI-MS measurement so that the signals measured in the mass spectrometer for a protein, its ligands and the noncovalent complexes thereof reflect the equilibrium concentrations of these species in solution. The possibility that transient, nonspecific protein–ligand interactions may be maintained in the desolvation process and observed in the mass spectrometer (thus generating false positives) has been raised as an objection against the use of mass spectrometry in screening applications. However, there is now abundant evidence that the complexes that are observed in the mass spectrometer do indeed reflect the binding constants in solution as the conditions required for complete desolvation of the protein and specific protein–ligand complexes ensure that the weaker complexes resulting from nonspecific interactions are not maintained in the gas phase and are not observed in the mass spectrum.[49] To confirm correct ESI-MS parameter adjustment, parameters for the ESI-MS measurement are generally first optimized on a target with a ligand of known binding constant such that the known binding constant is observed in the results. Early work on screening of inhibitors against carbonic anhydrase showed that multiple inhibitors could be screened against a single target, with the results obtained by mass spectrometry correlating well with those obtained by conventional means.[50] Methods for simultaneously screening multiple binders require that the interaction between the target and any component in the screen be independent of the interactions with other components. This requirement can be met by using a sufficient excess of the target in each screen.[49] Thus, when screening ligand libraries with unknown binding affinities, it is also possible to include a small amount of compound with a known binding constant (K_d) to act as an internal calibrant for determining the K_ds for those library ligands for which noncovalent complexes with the target are observed.[49] Mass spectrometry has very high sensitivity and a wide dynamic range and previous results have shown that using these techniques it is possible to detect and determine K_ds for noncovalent complexes in the range 10 nM–1 mM.[49]

An important consideration in the success of direct observation of specific noncovalent complexes of native protein with small molecules using ESI MS is the nature of the sample itself, which is aqueous, near physiological pH and with an ionic strength capable of maintaining the native target conformation and specific noncovalent interactions thereof. These considerations will be discussed fully later. Under such conditions, the protein remains fully folded with many of the acidic and basic side-chains either involved in strong hydrogen bonding interactions (that help maintain the native state protein structure) or otherwise inhibited from charging by their position in the protein structure or pK_a value (Figure 7.6d). The fully folded protein typically shows a very narrow m/z distribution with just a few charge states observed and with a maximum in the distribution at much higher m/z than for the denatured protein. This is shown for carbonic anhydrase in Figure 7.6e, where only three charge states are observed (+8, +9 and +10) compared with the more extensive charge state distribution under denaturing conditions (Figure 7.6b). The mass spectrum can similarly be deconvoluted as already described and will result in the same protein molecular weight as determined under denaturing conditions (Figure 7.6f). If an active site inhibitor for carbonic anhydrase is added, the noncovalent complex of carbonic anhydrase + inhibitor is then observed in the ESI mass spectrum together with some unbound protein (Figure 7.6g). The mass difference between the peaks for the unbound protein and the protein–inhibitor complex ($\Delta m/z$) can be multiplied by the charge state to give directly the molecular weight of the binding inhibitor, i.e. $MW_{\text{inhibitor}} = \Delta m/z \times z$.

In general, the speed, simplicity and sensitivity of mass spectrometry-based screening of ligands against targets make it an excellent choice for a primary screen. This is especially true for *in situ* medicinal chemistry applications where the identity of the binding species is not initially known, except that any binding species must be composed of the building block fragments employed in the *in situ* experiment. The mass spectral detection of protein–ligand complexes readily allows the determination of the mass of the binding ligand, which can be used to ascertain the building block fragments comprising the ligand. On this basis, the integration of *in situ* medicinal chemistry synthetic approaches (both DCC and Click) with a mass spectrometry-based screen should fulfil the screening requirements and provide a very effective means for identifying the combination of fragments that bind to a given target under biologically relevant conditions. Ligands identified by mass spectrometric screening may then be verified by more traditional biological activity secondary screening tests.

7.8.3 Sample Preparation and Limits of Detection

Sample 'contaminants' from an ESI-MS viewpoint include nonvolatile buffers, salts and detergents, all of which may stabilize the target biomolecule sample but are incompatible with ESI. These sample additives, through various mechanisms, can reduce or suppress completely the ion abundance from the target of interest, leading to a poor or even null result in terms of observation of the mass spectrum. Volatile solution components that do not form gas-phase adducts with the target biomolecule are therefore essential and the most investigated and proven reliable volatile components for this purpose include the salts ammonium acetate and ammonium bicarbonate.[51] These salts can be purchased as ESI-MS grade, i.e. free of contaminants not compatible with ESI-MS. Proteins are generally dissolved in pure water with ammonium acetate or ammonium bicarbonate present in a concentration between 5 and 500 mM, depending on the protein. In-line techniques such as desalting and size-exclusion chromatography have been routinely employed as a sample cleanup directly prior to ESI-MS analysis. This permits target sample solutions to be first exchanged into ESI-compatible buffers, replacing unwanted nonvolatile solution components while preserving target–ligand noncovalent complexes. The importance of these sample preparation considerations cannot be overstated and are probably the single most common reason why, in the hands of the inexperienced MS operator, experiments with native state proteins fail.

Modern ESI-MS instruments have sufficient sensitivity to detect proteins present at low- and sub-micromolar concentrations. The most common types of ESI sources employed use a pressurized pneumatic assist gas to assist in the formation of fine droplets by pneumatic nebulization and operate with flow rates around 1 μL min^{-1}. Thus, a screen can be carried out with as little as 1 μL of a protein at a concentration of about 1 μM, meaning that as little as a few picomoles of protein are sufficient for a screen. A number of different types of ESI sources, known as nanospray sources, have been designed that can operate at even lower sample flow rates (10–200 nL min^{-1}). These generate smaller sample droplets and improve the signal intensity of the protein–ligand noncovalent complexes further, with the added benefit of reducing protein consumption up to 100-fold compared with standard electrospray flow rates. Nanospray has also been reported to be more tolerant of nonvolatile cations in solution.[52] Nanospray has long been associated with difficult manual work involving fine spray needles and is therefore avoided by many MS operators. Recently, an automated fabricated chip nanospray source has been developed. This chip-based device has improved the ease of use for nanospray and the design eliminates carryover effects as the spray is produced directly from an orifice on each sample well of the chip.[53] The application of nanospray to screening DCLs will be of interest and could allow screens to be carried out with sub-picomole quantities of target protein.

7.9 Advantages of FTMS for Screening Ligands for Protein Binding

A full discussion of the theory and performance of FTMS (Fourier transform mass spectrometry) is beyond the scope of this chapter; however, books and review articles are available for the interested reader.[54] For the purpose of this discussion, an outline of the basic principles and performance characteristics of FTMS which are important for the measurement of protein–ligand noncovalent complexes will be presented. The FTMS mass analyzer is typically a cylindrical-shaped cell capped by plates known as trapping plates; the cylinder walls are divided into four quadrants. Opposing pairs of plates in the four quadrants are known as the excite plates and detect plates (Figure 7.7). This cell is located within a vacuum chamber that is housed inside the homogeneous magnetic field region of a superconducting solenoid magnet. The basic equation of ion motion in FTMS is

$$\boldsymbol{F} = q\boldsymbol{v} \times \boldsymbol{B} \tag{7.5}$$

where $\boldsymbol{F}$ is the force on the ion, $\boldsymbol{v}$ is the velocity of the ion and $\boldsymbol{B}$ is the applied magnetic field from the superconducting magnet that lies along the axis of the cylindrical analyzer cell and q is the charge on the ion (variables in bold font are vector variables). The ion motion equation cross product is zero when the velocity vector of the ion is parallel to the magnetic field. In a physical sense, this means that the ions are unconstrained by the magnetic field as they move along the central axis of the cylindrical mass analyzer cell. For this reason, a 'trapping' voltage, typically about +1.0 V, is applied to the trapping plates (a positive trapping voltage for positive ions or a negative trapping voltage for negative ions) to hold ions in the analyzer cell. For velocity components of ions that are not parallel to the magnetic field, the ions experience a force perpendicular to both the magnetic field and the velocity vector of the ions, which causes them to move in a circular path around the magnetic field axis and is known as cyclotron motion (Figure 7.7).

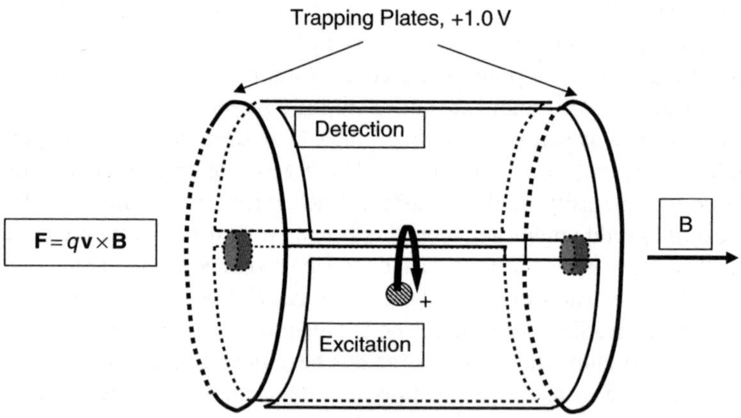

Figure 7.7 *An FTMS mass analyzer cell depicting the motion of trapped ions in the presence of the magnetic field.*

The radius of the cyclotron motion depends on the energy of the ions. Ions are typically injected into the mass analyzer cell with low energies, resulting in small initial cyclotron radii of 0.01–0.1 mm. The frequency f of the cyclotron motion is

$$f = qB/2\pi m \tag{7.6}$$

where B is the magnetic field strength in tesla, q is the charge on the ion and m is the mass of the ion [Equation (7.6) describes the ideal situation, i.e. with no applied electric fields; this is an approximation, however, as there are electric fields from both the trapping plates and other ions within the analyzer cell; calibration equations have been developed to correct for these electric field effects].Thus, if the frequency of the ion cyclotron motion (f) can be determined then the mass-to-charge ratio ($m/z = m/q$) can be derived based on the relationship of Equation (7.6). The cyclotron frequency f is measured by excitation of the cyclotron motion of the trapped ions; this is achieved through application of an RF field to the excite plates of the analyzer cell that is in resonance with the cyclotron motion of the ions. The ions absorb energy from this RF field and move to a larger radius, with all ions of a given mass-to-charge ratio moving coherently (Figure 7.8a). This collection of ions induces an image current that can be measured and amplified as shown in Figure 7.8b. The ions continue to move in a coherent packet at this larger radius until they are eventually knocked out of phase with each other, caused by collisions with molecules of the background neutral gas, and the image current thus decays with time.

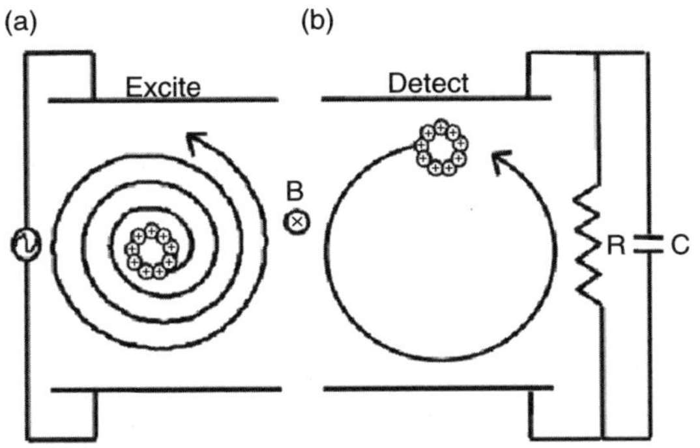

Figure 7.8 Excitation (a) and detection (b) of the ion cyclotron motion within an FTMS mass analyzer cell. Reprinted from Marshall, A.G. and Hendrickson, C.L., Fourier transform ion cyclotron resonance detection: principles and experimental configurations. International Journal of Mass Spectrometry, **215**, 59–75. Copyright (2002), with permission from Elsevier.

Some typical mass spectra for a peptide are shown in Figure 7.9b and d. Excitation and detection of ions of more than one mass-to-charge ratio leads to a time domain signal (i.e. signal intensity as a function of time) that consists of a superposition of sinusoidal waveforms for each mass-to-charge ratio present within the mass analyzer cell (Figure 7.9a and c). Fourier transformation of the time domain signal yields a spectrum that shows the signal intensity as a function frequency, also known as the frequency domain signal. As described above, the mass-to-charge ratio ($m/z = m/q$) can then be derived using Equation (7.6) to generate the corresponding mass spectrum (Figure 7.9). The time domain signal decay is directly dependent on the pressure in the mass analyzer cell. For slow signal decay, necessary to achieve both high resolution and high mass accuracy, ultra-high vacuum (UHV) is required in the analyzer cell of FTMS instruments. If the pressure is high in the mass analyzer cell then a more rapid decay of the signal occurs due to collisions with the abundant background gas and poor resolution and peak shape in the mass spectrum result. If lower pressure mass analyzer cell conditions are achieved, then the transient signal response

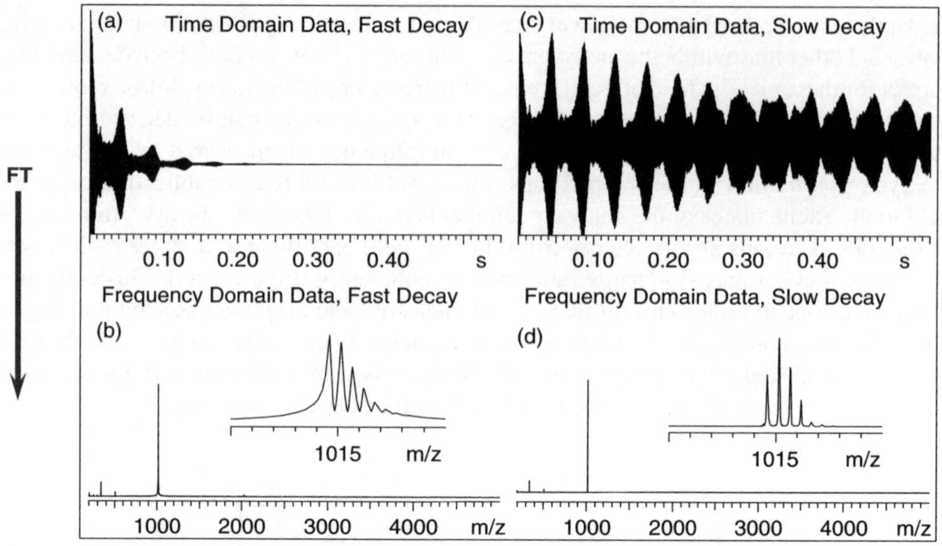

Figure 7.9 *Fourier transformation of the time domain signal (a and b) to the frequency domain signal and corresponding mass spectrum (c and d), showing the effects of signal decay time on spectrum resolution and mass accuracy.*

is much longer and excellent peak shapes and resolution can be achieve to generate low- or sub-ppm mass measurement accuracy (compare Figure 7.9 entry a with b and entry c with d).

A typical commercial FTMS system is pictured in Figure 7.10, with a schematic cutaway showing the regions of differential pumping required to achieve the UHV conditions in the mass analyzer cell. FTMS relies on high field superconducting magnets to achieve its high-performance characteristics. The ability to achieve high resolution and isotopically resolved spectra improves with increasing magnetic field strength so that higher field magnets are preferred for this work. Commercial systems are now available with field strengths up to 15.0 T, though 7.0, 9.4 and 12.0 T field strength systems are more common. Other important characteristics of FTMS include its versatility in performing MS–MS experiments and in obtaining highly accurate mass measurements. These last two features can be of great benefit in determining the identity of the unknown binding species in noncovalent protein–ligand complexes.[49] Using well-established methods, a single protein adduct (or several adducts) can be isolated in the FTMS mass analyzer cell and energized using either a laser or collisions with an added neutral gas, leading to dissociation of the noncovalent complex. The low mass ligands generated may then be observed and measured with sufficient mass accuracy (typically low ppm) to determine elemental composition and importantly the identity of library ligands. An example of the application of such high mass accuracy measurements of binding ligands from carbonic anhydrase has been published previously[50a] and an example of its application to a DCC study is shown later in this chapter.[15] Thus the suite of tools available from FTMS make it an ideal instrument for work with *in situ* medicinal chemistry as the fragments comprising the bound ligand can be readily determined; this will be illustrated with examples later.

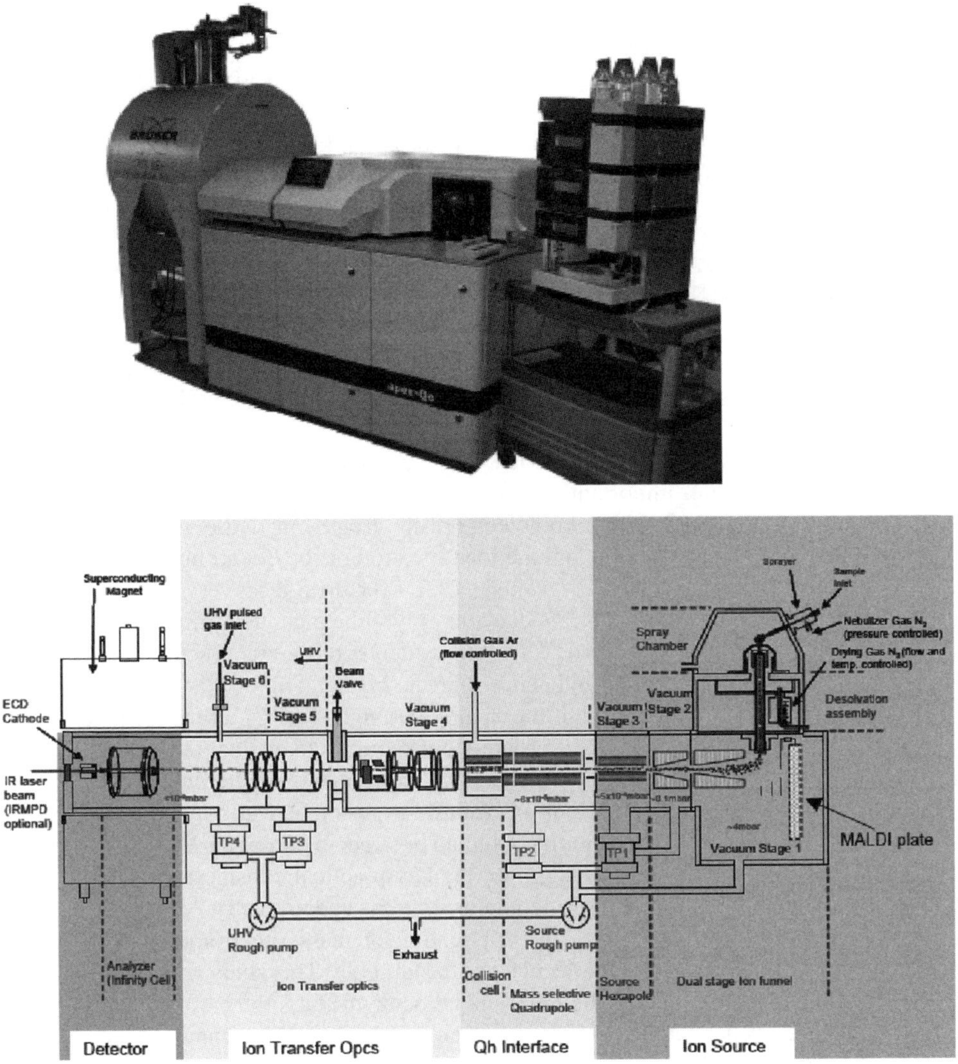

Figure 7.10 *A commercial FTMS system with a front-end nanoflow LC system. The schematic cutaway view depicts the sequential vacuum stages that are required to transfer ions in the ESI source (formed at atmospheric pressure) to the mass analyzer cell of the FTMS (10^{-10} mbar).*

7.10 Mass Spectrometry and *In Situ* DCC

In the early days of DCC, the integration of synthesis with screening was touted as a key advantage of this approach. What actually emerged as the standard *in situ* DCL screening approach has, probably owing to analytical constraints, deviated somewhat from this original conception and instead an indirect method that screens for ligand enrichment has evolved. This indirect screening approach carries out the analysis of identical DCLs twice, with and without the target biomolecule. The equilibrium concentration profiles of all library ligands when generated in both the presence and absence of the target protein (following disruption of the ligand–target complexes) are compared and the detection of ligand enrichment in the targeted DCL is the basis of identifying the 'best binders'. Only recently has the original promise to integrate DCC synthesis and screening occurred, as reported by Poulsen[15] and followed by Schofield and colleagues[28] using mass spectrometry and Congreve *et al.*[14] using X-ray crystallography. These direct screening approaches represent a major leap forward in addressing the inconvenient and somewhat impractical indirect analysis that has come to be associated with drug discovery applications of DCC. X-ray crystallography is without doubt unrivalled with regard to the resolution of structural information yielded; however, it is a relatively slow turn-around experiment with a high consumption of protein and so from a primary screening viewpoint is somewhat restricted. By describing some recently published examples we will demonstrate that mass spectrometry allows direct screening of DCLs, requiring only a minimal protein sample, and is capable of generating a rapid result that yields the mass of any ligands noncovalently bound to the target biomolecule. Importantly, when using ESI-FTMS this mass determination is of high accuracy and hence with a knowledge of the fragments and reversible chemistry employed this analysis permits the unique identification of the bound ligand and is therefore an excellent primary screen to evaluate DCLs for drug discovery applications. An exception to this unique identification will occur only if the mass determined for the bound ligand(s) can be attributed to isobaric DCL constituents; these are structural isomers that share the same molecular formula and hence share an identical molecular mass and isotope distribution fingerprint. The contribution of these ambiguous DCL constituents to a 'hit' identified in the MS screen can be differentiated if required to evaluate the identity of the fragments in question, for example by tandem MS experiments or alternatively 'knockout' MS experiments, both ready extensions of the basic experiment. Mass spectrometry has no requirement for modification or labelling of ligand or target to permit the analysis. Throughout this discussion, we will aim to highlight the key practical considerations involved to deliver a successful DCL screening outcome using mass spectrometry.

Poulsen was the first to describe the application of mass spectrometry to the direct screening of a DCL wherein a biomolecule was the target.[15] This study applied ESI-FTMS to the screening of a DCL against the Zn(II) metalloenzyme target of carbonic anhydrase II (CA II). Recent evidence implicates CA inhibition as a validated target and/or biomarker for a range of disease states.[55] The classical CA recognition moiety is an aromatic sulfonamide ($ArSO_2NH_2$); the sulfonamide group (deprotonated as $ArSO_2NH^-$) serves as a zinc binding function in the active site of CAs (Figure 7.11). When this privileged CA anchor fragment is derivatized with 'tail' fragments, CA inhibitors with optimized target affinity, selectivity and other pharmaceutical properties may be generated (Figure 7.11).[55, 56]

In this ESI-FTMS screening proof-of-concept experiment, a DCL was generated using the hydrazone exchange reaction (described earlier in this chapter) from two hydrazide fragments (**1** and **2**) and five aldehyde fragments (**A–E**) (Figure 7.12). Hydrazide **1** was designed as the CA anchor fragment and hence necessitated dual functionality: an $ArSO_2NH_2$ moiety for reliable CA affinity and a hydrazide moiety to take part in hydrazone exchange. Hydrazide **2** lacked the sulfonamide moiety of **1**, but was still able to participate in hydrazone exchange and thus functioned as a control compound through provision of hydrazone compounds expected to have *no* affinity for the target CA. Fragment aldehydes **A–E**, the

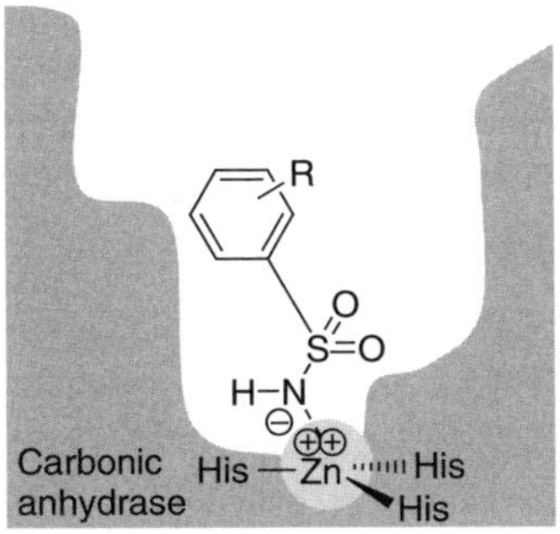

Figure 7.11 *Schematic of the carbonic anhydrase active site showing the Zn(II) cation coordinated to a benzenesulfonamide ($ArSO_2NH_2$) inhibitor molecule.*

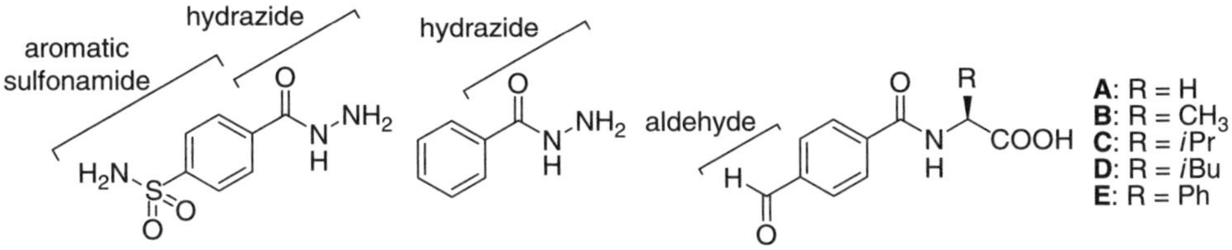

1 **2** **A-E**

Figure 7.12 *Fragments equipped with a hydrazide (**1** and **2**) or aldehyde (**A–E**) functional groups for DCC utilizing hydrazone exchange and targeting carbonic anhydrase.*

hydrazone exchange partners of **1** and **2**, were selected to introduce an array of 'tails' on to **1** to permit exploration of periphery recognition interactions with CA.

By monitoring a suite of control experiments using mass spectrometry, this study first confirmed that hydrazone exchange was dynamic in the aqueous buffer used in the subsequent *in situ* DCC mass spectrometry screening methodology. Then the *in situ* DCL of acyl hydrazones (**1A–E**, **2A–E**) was generated in the presence of CA II. The fragments and target were combined and incubated as follows: CA II (30 μM), **1** (15 μM), **2** (15 μM) and **A–E** (6 μM each), in 10 mM NH_4OAc (pH 7.2) with 1% DMSO to effect solubility, at 37 °C for 40 h (Scheme 7.3). The final DCL volume was 50 or 100 μL. The total accessible hydrazone concentration was equivalent to the concentration of CA II (30 μM), while individual hydrazone products could be formed to a maximum concentration of 6 μM or 20 mol% relative to CA II.

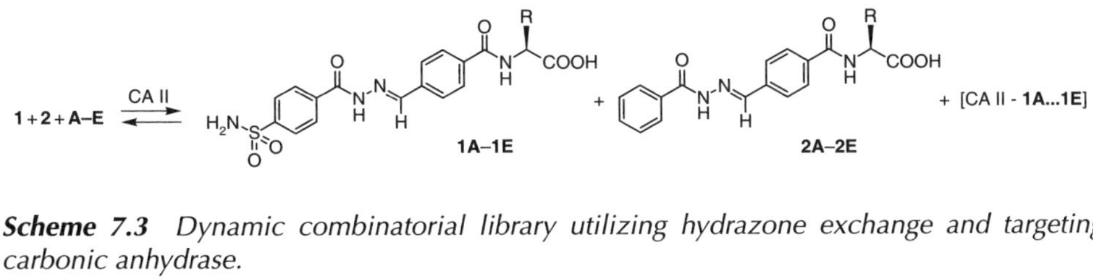

1A–1E **2A–2E**

Scheme 7.3 *Dynamic combinatorial library utilizing hydrazone exchange and targeting carbonic anhydrase.*

The mass spectrometry for this analysis was performed on an APEX III 4.7 T FTICR mass spectrometer (Bruker Daltonics, Billerica, MA, USA) fitted with an Apollo ESI source operated in the negative ion mode. Broadband excitation was used to analyze a mass range from *m/z* 100 to 4500, with instrument parameters tuned and optimized for detecting *m/z* ~3000. DCL samples were directly infused into the ESI source at 2 μL min^{-1} with an ESI source pressure of 6.2×10^{-7} mbar, a high-vacuum analyzer region pressure of 1.3×10^{-10} mbar and a hexapole ion accumulation time of 1 s. ESI-FTMS analysis of a solution containing only CA II (~29 kDa) yielded the ESI negative ion mass spectrum in Figure 7.13a. Peaks corresponding to the 8^- to 10^- charge states of CA II were observed, with the 9^- charge state predominating. This charge state envelope (low charge states and few charge states) is typical for CA II when in a compact, tightly folded native structure.[50, 57] The mass spectrum of the CA II-DCL solution (prepared from **1**, **2** and **A–E**) is presented in Figure 7.13b. The same charge state envelope as for free CA II (Figure 7.13a) was observed, but each charge state now consisted of a grouping of peaks: a peak that corresponded to native CA II and at higher *m/z* value a group of peaks that corresponded to the five different CA II–hydrazone noncovalent complexes CA II–**1A** . . . **1E** in addition to a small amount of CA II–**1**.

Owing to overlapping isotopic envelopes, a consequence of the molecular mass of CA and the broadband detection mode, the five complexes CA II–**1A** . . . **1E** were not completely resolved (Figure 7.13b, inset) and the MS–MS technique was employed to confirm the identity of the bound ligands from the DCL. MS–MS experiments were performed by sustained off-resonance irradiation collision-activated dissociation (SORI-CAD) using argon as the collision gas at an analyzer pressure of $\sim 10^{-8}$ mbar (argon inlet pressure 2.9×10^{-2} mbar). The parent ions bearing the 9^- charge state were selected by use of correlated sweep isolation. This was followed by SORI-CAD, resulting in dissociation of the noncovalent complexes. The collision energy for the experiment was tuned to cause dissociation of the noncovalent protein–ligand complexes. The result yielded free CA II (both 8^- and 9^- charge states) and, important for the application to DCC screening, singly charged negative ions for the hydrazone ligands **1A–E**, now well resolved by molecular mass (Figure 7.13c). The masses of these ions were consistent with the $[M - H]^-$ ions expected for the DCL sulfonamide hydrazone products **1A–E** (Table 7.2). No ions in the tandem mass spectrum could be attributed to hydrazones **2A–E** (lacking the sulfonamide moiety). The hydrazones identified from the DCL with affinity for CA II could only have been synthesized *in situ* and the result demonstrates that the MS screening approach was able to identify relevant combinations of fragments whilst in the presence of the target biomolecule. The sample quantity consumed for these MS–MS experiments was less than 100 μL, the initial ESI-FTMS experiment takes only minutes to perform, while the MS–MS experiment can be completed within 30 min. Confirmation of the results of this DCL experiment was then obtained by conducting a conventional solution-phase competitive binding assay for CA II to measure the equilibrium dissociation constants (K_i) for the compounds described in this study. The DCL products **1A–E** each exhibited increased affinity for the enzyme (K_i range 10.6–82.3 nM) compared with the scaffold building block **1** ($K_i = 150$ nM).

Follow-up ESI-FTMS experiments in our laboratory have been effective with a 10-fold reduction in CA II concentration (from 30 to 3 μM, ~4 μg protein/DCL based on a 50 μL reaction volume) while retaining the ability to detect protein–ligand noncovalent

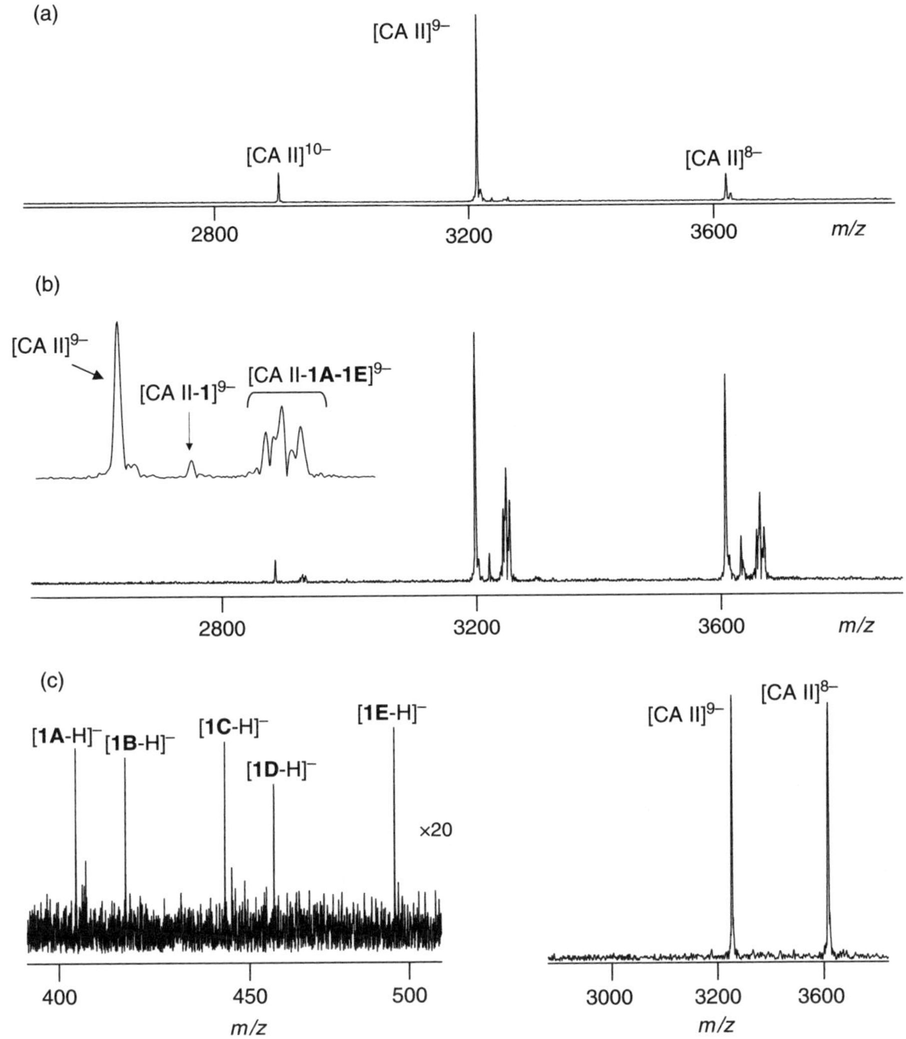

Figure 7.13 *(a) ESI-FTMS negative ion mass spectrum of CA II (30 μM) in 10 mM NH_4OAc solution, 1% DMSO; (b) ESI-FTMS negative ion mass spectrum of a mixture of CA II (30 μM) and DCL containing ten possible hydrazone products (**1A–E**, **2A–E**) in 10 mM NH_4OAc, 1% DMSO; (c) tandem mass spectrum obtained following collision-activated dissociation (CAD) of CA II–hydrazone noncovalent complexes to identify the hydrazones. Reprinted from Poulsen S.-A., Direct screening of a dynamic combinatorial library using mass spectrometry.* Journal of the American Society for Mass Spectrometry, ***17****, 1074–1080. Copyright (2006), with permission from Elsevier.*

complexes with good sensitivity down to 5 mol% ligand, whereas at 1 mol% ligand and below, the sensitivity of detection is at its limit and it is difficult to discern the protein–ligand complex confidently in the presence of the free protein (unpublished work). Nanoelectrospray ionization (nanoESI), as described earlier in this chapter, has superior capabilities

***Table 7.2** Molecular weights for CA II-bound ligands identified from* in situ *screening of the DCL by ESI-FTICR–MS.*

Molecular weight observed for $[M-H]^-$ (Da)[a]	Compound = [M]	Molecular weight calculated for $[M-H]^-$ (Da)
403.06	**1A**	403.07
417.07	**1B**	417.09
445.10	**1C**	445.12
459.11	**1D**	459.13
493.09	**1E**	493.12

[a] Observed molecular weights taken from mass spectrum in Figure 7.13c.

for investigating protein–ligand noncovalent complexes compared with standard electrospray; the lower sample flow rates (10–200 nL min^{-1}) and smaller sample droplets improve transfer of specific noncovalent complexes to the gas phase and hence increase the signal intensity with the added benefit of reducing protein consumption. As nanoESI sources become more user friendly and commonplace with FTMS, it is expected that more DCL applications will emerge that utilize MS for screening biomolecular targets to identify binding partners. The FTMS instrument that permitted these proof-of-concept experiments has a 4.7 T superconducting magnet and clearly with the higher field magnets now more commonly available more demanding DCLs could be analyzed with excellent mass accuracy results. In addition, coupling nanoESI and higher field magnets with automated sample handling could significantly reduce both the time and target quantity required to facilitate analysis; this is clearly desirable for drug discovery applications. In-line size-exclusion chromatography or dialysis techniques can be directly coupled to an ESI-FTMS instrument. Our preliminary investigations with such a system have permitted us to exchange DCLs from noncompatible ESI-MS buffers (e.g. phosphate-buffered saline, Tris) into compatible ESI-MS buffers directly prior to analysis (unpublished work); this simple procedure could undoubtedly further expand the range of biomolecular targets amenable to direct MS screening by permitting DCL experiments to be carried out in other buffers to fulfil stability requirements.

The application of mass spectrometry to the direct screening of a DCL has been adopted also by Schofield and colleagues, the biomolecular target for their study being the therapeutically relevant enzyme metallo-β-lactamase (BcII) from *Bacillus cereus*.[28] This enzyme (molecular weight ~25 kDa) catalyses the hydrolysis of a range of clinically used β-lactam antibiotics and is of interest as a medicinal chemistry target owing to its involvement in the resistance of bacteria to antibiotics.[58] This DCC investigation stemmed from some earlier work by the same group in which it was confirmed, also using mass spectrometry, that BcII contains two active site zinc cations and that simple thiol fragments formed BcII–inhibitor complexes through a zinc binding interaction.[59] A thiol fragment identified in this precursor study was adapted to give five anchor fragments (**F–J**) with suitable functionality for DCC using thiol–disulfide exchange as the reversible reaction. Each anchor fragment (**F–J**) had two thiol groups: one to facilitate Zn binding ($thiol_1$) and the second to participate in thiol–disulfide exchange for DCC ($thiol_2$). The DCC study examined 19 additional novel fragments: monothiols **(4–17, 19–22)** and dithiol **(18)** as the coupling partners for **F–J** (Scheme 7.4).

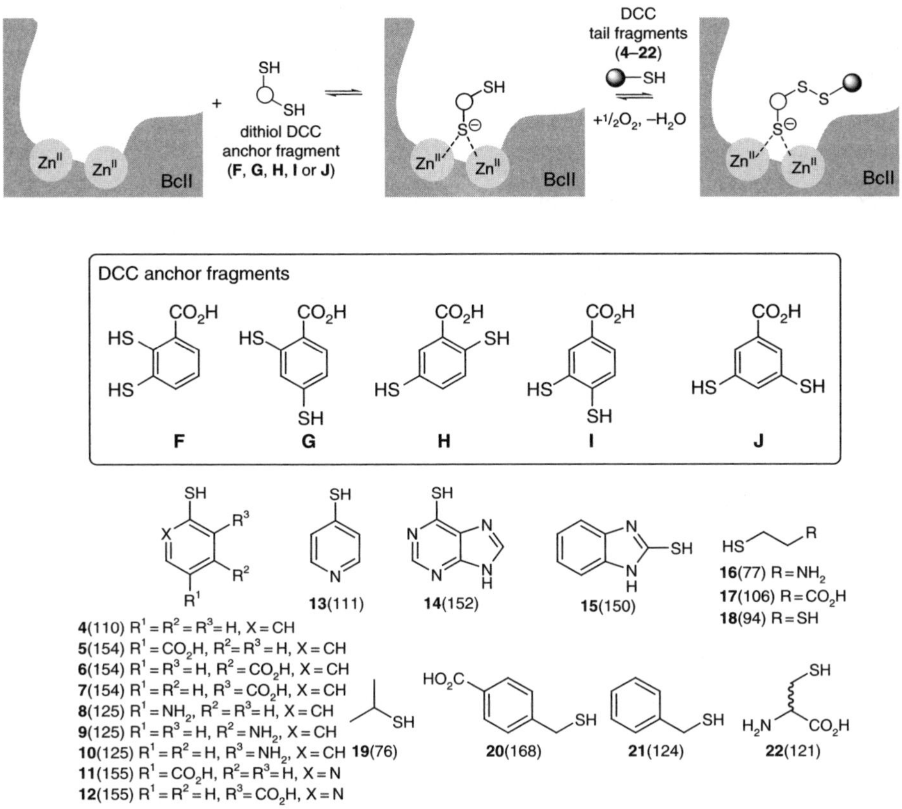

***Scheme 7.4** Anchor fragments equipped with two thiol moieties (**F–J**) and thiol partner fragments **(4–22)** for DCC utilizing thiol–disulfide exchange and targeting metallo-β-lactamase.*

The composition of the DCL accessible with thiol–disulfide exchange is complicated by the non-orthogonal nature of thiol/dithiol reactivity, which permits more permutations of reaction fragment combinations and also the potential for oligomerization compared with orthogonal reversible reactions such as hydrazone exchange. Schofield and co-workers minimized this complication by using controlled reaction conditions together with fast MS analysis as described next. The DCC experiments were prepared in an oxygen-free atmosphere (<1 ppm O_2) to limit the occurrence of thiol–disulfide exchange prior to the anchor fragments binding to the active site zinc cations of BcII. Individual thiols **(4–17, 19–22)** and dithiol **(18)** stock solutions (100 mM) were freshly dissolved in DMSO and then diluted to 75 μM with aqueous NH_4OAc (15 mM, pH 7.5). Similarly, solutions of the anchor fragments (**F–J**) were prepared at a final concentration of 100 μM in NH_4OAc (15 mM, pH 7.5). The fragment solutions were combined as appropriate with the target enzyme BcII to give a final *in situ* DCL that consisted of BcII (15 μM), anchor fragment dithiol (one each of **F–J**, 30 μM) and complementary fragments (**4–24** 10 μM each), in 15 mM NH_4OAc at pH 7.5 (Scheme 7.4). An aliquot of each DCL experiment was placed in a 96-well microtitre plate and sealed with adhesive aluminium foil and then removed from the oxygen-free environment. Analysis was performed by ESI-MS on a Q-TOF mass spectrometer (Q-TOFmicro, Micromass, Altrincham, UK) interfaced with a NanoMate chip-based nanoESI source (Advion Biosciences, Ithaca, NY, USA). Time courses were started when the NanoMate tip pierced the aluminium seal covering the 96-well microtitre plate and introduced O_2 into the system to initiate thiol–disulfide exchange. Samples were infused into the Q-TOF mass spectrometer through the ESI chip (estimated flow rate approximately 0.1 μL min^{-1}). Calibration and sample acquisitions were performed in the positive ion mode in the range *m/z* 500–5000. The results from this ESI-MS screen confirmed the preference of native BcII for certain disulfide heterodimers from all possible DCL constituents. A sample mass spectrum from this study is presented in Figure 7.14. Hit candidates (**H-5**, **H-8/9/10** and

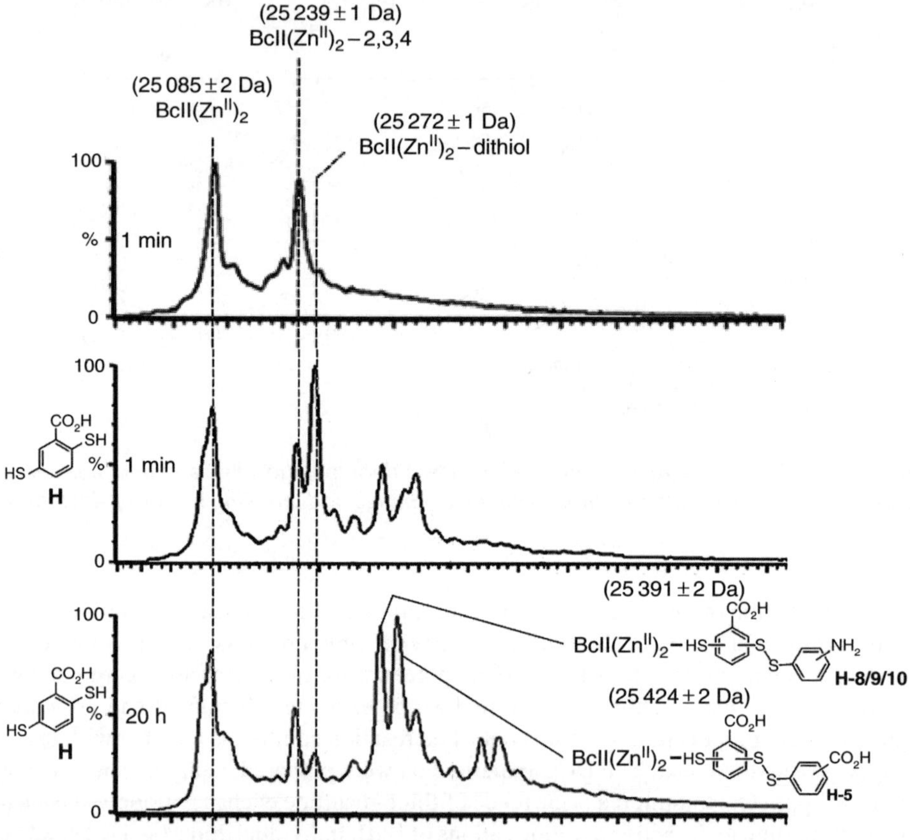

Figure 7.14 *Deconvoluted ESI mass spectra from an equimolar mixture of 19 thiols (10 μM each of* ***4–22****) and BcII (15 μM): (a) after 1 min; (b) plus anchor fragment dithiol* **H** *(30 μM) after 1 min of aerial exposure; (c) plus anchor fragment dithiol* **H** *(30 μM) after 20 h of aerial exposure. Reproduced with permission from Lienard, B.M.R., Selevsek, N., Oldham, N.J. and Schofield, C.J., Combined mass spectrometry and dynamic chemistry approach to identify metalloenzyme inhibitors.* ChemMedChem *2007,* **2**, *175–179. Copyright Wiley-VCH Verlag GmbH.*

J-20) were identified as having the potential to deliver improved potency over the lead compound **3**. These candidates were then used as the basis for a traditional medicinal chemistry investigation and optimization (Scheme 7.5). The overall outcome for this study was the identification of four novel thiol compounds with K_i 6–17 μM, 10–30-fold more potent than the lead compound **3** with a K_i of 185 μM.

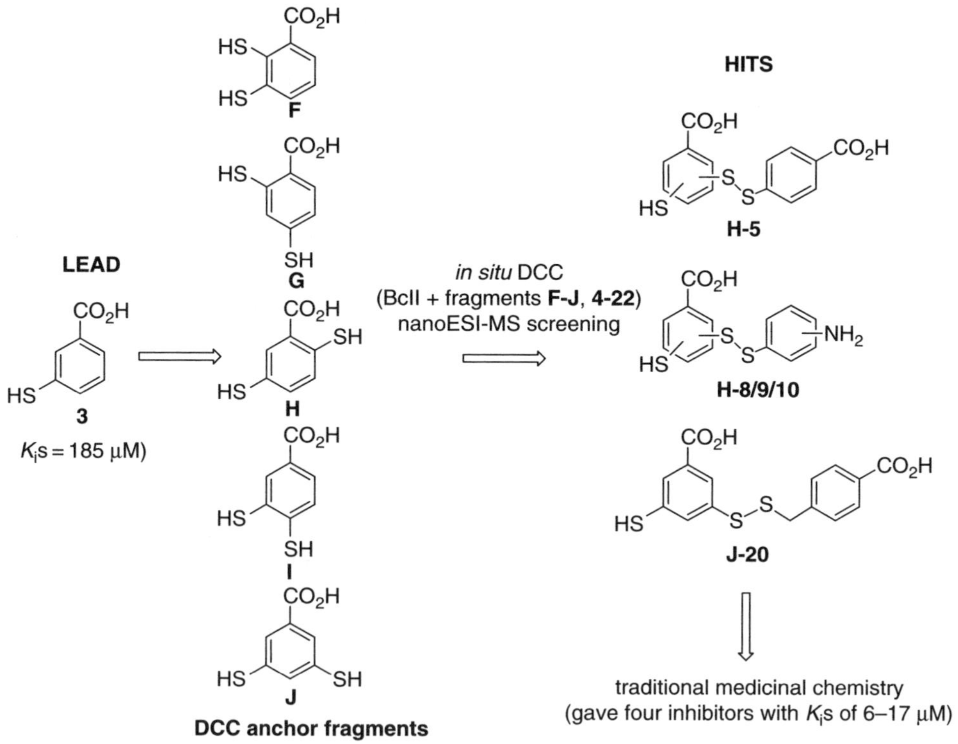

Scheme 7.5 *DCC enzyme inhibitor discovery and optimization protocol targeting metallo-β-lactamase.*

Both of the above examples demonstrate that there is enormous scope for ESI-MS as a direct and primary screening tool for the identification of small molecules formed by DCC in the presence of a biomolecular target. The ESI-MS screening has permitted concurrent identification of all ligands of interest through direct analysis. The approach distinguished the effective (from ineffective) combination of building blocks in the DCL by specific detection of the target protein–ligand noncovalent complexes and in both examples novel ligands were identified with improved enzyme inhibition properties compared with the lead fragments. When using the FTMS technique, it is reasonable to expect that increased DCL size need not increase the complexity of this screening protocol, owing to the sensitivity, high resolution and MS–MS capabilities, which should avoid the need for multiple sublibraries for deconvolution of larger DCLs.

7.11 Mass Spectrometry and *In Situ* Click Chemistry

The pioneering application of the 1,3-DCR for *in situ* Click chemistry to target a biomolecule was described by Sharpless, Finn and co-workers in 2002.[37] The enzyme acetylcholinesterase (AChE) was selected as the biological target for this proof-of-concept investigation and for subsequent optimization work in which the specificity, sensitivity and general capability of the analysis employed to screen with this target-guided synthetic strategy was enhanced.[37–39] Additional biomolecular targets have since been reported and include the enzymes carbonic anhydrase[40, 41, 44] and HIV-protease.[42] More recently, *in situ* Click chemistry was miniaturized by use of a microfluidic chemical reaction circuit in which the consumption of biomolecule and fragments alike was significantly reduced and together with automation this has undoubtedly improved the compatibility of this approach with drug discovery.[41] Each of the published examples has utilized mass spectrometry for the detection of the ligands formed by the *in situ* Click chemistry. Here we will describe the AChE proof-of-concept and optimization work and also the *in situ* microfluidic reactor-based work with a deliberate focus on the attributes of mass spectrometry for screening to identify lead compounds generated by *in situ* Click chemistry.

7.11.1 Proof-of-Concept

Acetylcholinesterase (AChE) is an enzyme that catalyses the hydrolysis of the neurotransmitter acetylcholine to give acetate and choline. This reaction terminates the neurotransmission process in both the central and peripheral nervous systems. It is already known that AChE has two distinct binding sites, a catalytic site and a peripheral site, and that these sites are in close proximity. The fragment library for the proof-of-concept study was derived from structural variants of known site-specific AChE inhibitors, namely tacrine (an active site ligand with a K_d of 18 nM) and phenylphenanthridinium (a peripheral site ligand with a K_d of 1.1 μM). These tacrine and phenylphenanthridinium fragments were thus complementary with respect to reactive functional groups (azide or acetylene) and binding site recognition (catalytic or peripheral). This investigation screened a total of 49 binary azide–acetylene fragment combinations with 98 potential triazole products (including the *syn*- and *anti*-triazole regioisomers) (Scheme 7.6). The *in situ* reactions consisted of solutions of AChE (~1 μM active enzyme), ammonium citrate buffer (2 mM, pH 7.3–7.5), tacrine fragment (**TZ2–6, TA1–3**; 30 μM) and complementary coupling partner (either 30 μM of tacrine fragment or 66 μM of phenanthridinium fragment **PZ6–8, PA2–6**). Each reaction mixture was allowed to stand at room temperature with analysis performed at daily intervals for 7 days.

As for DCC, a robust analytical method to identify fragment combinations that are assembled in the presence of the biological target is critical for the application of *in situ* Click chemistry to drug discovery. A variant of the MALDI (matrix-assisted laser desorption/ionization) MS technique was adopted for screening, wherein the desorption/ionization is performed on a silicon surface, known as desorption/ionization mass spectrometry (DIOS-MS). The advantage of the DIOS variant of the MALDI-MS technique is that it is unhampered by interfering background matrix peaks normally associated with MALDI. DIOS-MS was performed on a MALDI-TOF mass spectrometer, with just 0.25–0.5 μL of

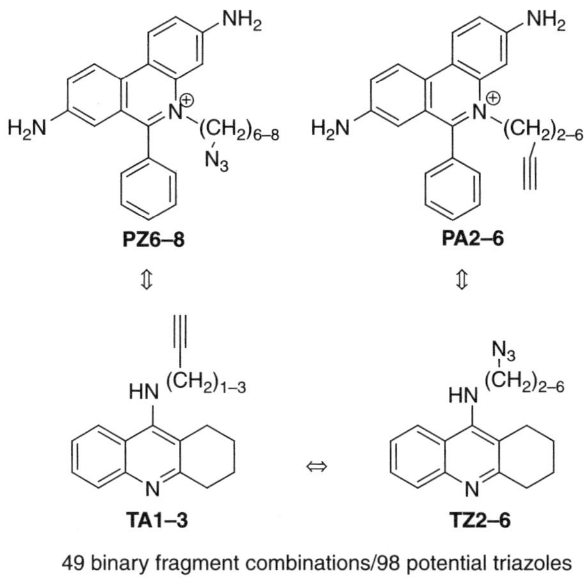

Scheme 7.6 *Binary azide–acetylene fragment combinations with 98 potential triazole products (including the* syn- *and* anti-*triazole regioisomers) for targeting AChE with* in situ *Click chemistry.*

sample being required for the analysis. This analysis applied MALDI laser energies that spanned from sub-threshold levels to above the threshold intensity for appearance of starting material ions. The laser intensity applied was higher than used in conventional MALDI or DIOS analyzes due to the presence of large amounts of protein in the samples. A single hit compound was identified in this proof-of-concept study, namely triazole *syn*-**TZ2PA6** (molecular weight = 661 Da) formed by the reaction of fragments **TZ2** and **PA6** and the presence of AChE. The DIOS mass spectrum demonstrating the formation of **TZ2PA6** is shown in Figure 7.15. The assignment of this hit as the *syn*-regioisomer required authentic synthesis of *syn*-**TZ2PA6** and *anti*-**TZ2PA6** and determination of K_d values. The results demonstrated that only *syn*-**TZ2PA6** was formed in the presence of AChE. A remarkable finding is that this compound was the most potent noncovalent AChE inhibitor reported to date, with K_d values of 77 fM (eel AChE) and 410 fM (murine AChE). In contrast, the *anti*-**TZ2PA6** isomer is not formed by the enzyme and is less active by two orders of magnitude. These first experiments also demonstrate that the *in situ* experiment avoided the requirement to synthesize all 98 possible triazoles; instead, the synthesis of just two triazoles was required.

It was clear from this early work that mass spectrometric analysis would be the most appropriate technique to detect hit compounds for *in situ* Click chemistry applications. Although the DIOS-MS method was able to directly detect the low quantity of triazole product conversion in the presence of large amounts of protein and parent fragments, the sensitivity for this measurement was very low, with a poor signal-to-noise ratio. It was therefore both logical and desirable to optimize the sensitivity and selectivity of the MS

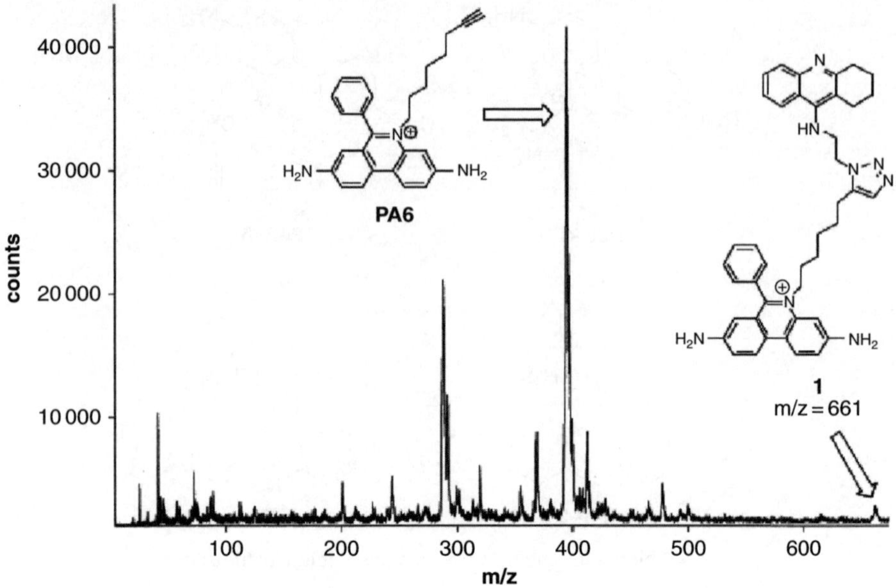

Figure 7.15 *DIOS mass spectrum of the* in situ *Click chemistry experiment of tacrine fragment* **TZ2**, *phenanthridinium fragment* **PA6** *targeting AChE. Unreacted* **TZ2** *is detected with low sensitivity under these conditions (independent measurements confirmed). Reproduced with permission from Lewis, W.G., Green, L.G., Grynszpan, F., Radic, Z., Carlier, P.R., Taylor, P., Finn, M.G. and Sharpless, K.B., Click chemistry* in situ*: acetylcholinesterase as a reaction vessel for the selective assembly of a femtomolar inhibitor from an array of building blocks.* Angewandte Chemie, International Edition *2002,* ***41****, 1053–1057; Supporting Information. Copyright Wiley-VCH Verlag GmbH.*

analysis to accommodate better the low product conversion rates of fragments to triazoles and also to enable the capability to evaluate complex multi-fragment libraries rather than only binary fragment combinations. Follow-up work also targeting AChE delivered the breakthrough advance in screening performance by using selected ion monitoring mass spectrometry (SIM-MS), in which the mass spectrometer is set to scan over a very small *m*/*z* range (typically 1 Da) rather than a broad *m*/*z* range. This narrow mass range permits greater specificity of the analysis as only compounds with the selected mass of interest are detected. The SIM mode also increases the sensitivity of detection of the triazole molecule by decreasing the contribution of other components to the recorded MS signal. In addition, SIM-MS was coupled with HPLC, which allowed the triazole products to be characterized by both their mass and their retention time. The retention time is of special significance for this application as it permits the *syn*-/*anti*-regioisomer assignment of the Click triazoles formed, which are identical by mass analysis but typically have different retention times. The other key advantage of the HPLC-SIM-MS technique is that it allowed the analysis of *in situ* Click chemistry performed with multi-fragment mixtures that consisted of the target and up to 11 fragments. Additionally, chromatographic removal of the molecules that may otherwise obscure the mass spectrum of the triazole product allowed a decrease in the reaction time of binary fragment combinations from 6 days to 6 h and for the multi-fragment combinations to 24 h. Fragment concentrations could also be lowered considerably (azide fragments, 4.6 μM; acetylene fragments, 24 μM) with 15 μL of reaction mixture required for HPLC-SIM-MS analysis.

With AChE as the target for *in situ* Click chemistry, a total of 58 fragments have now been screened, with eight fragment combinations undergoing conversion to triazoles in the presence of AChE (Figure 7.16). Interestingly, the regioisomer contribution of these fragment combinations has so far exclusively been attributed to the *syn*-triazoles as determined by comparison of HPLC retention times against authentic samples. The inhibition properties for all hit triazoles have also been determined and have proven to be the most potent noncovalent inhibitors of AChE yet reported, with K_is in the femtomole range (Figure 7.16).

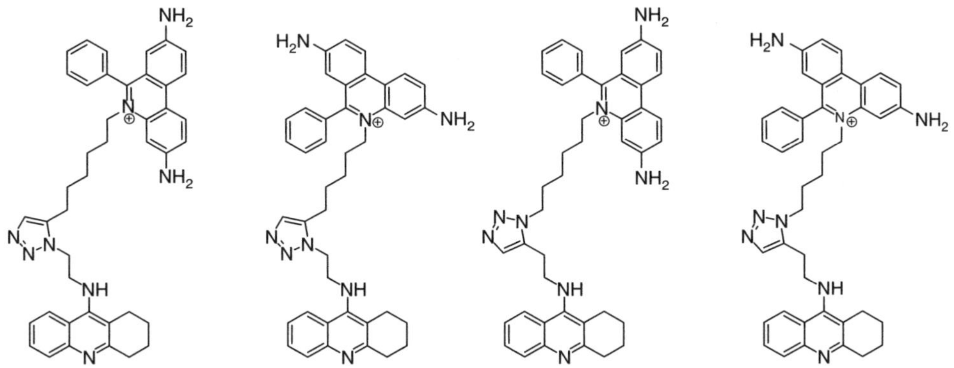

syn-TZ2PA6 K_d(eAChE) 99 fM K_d(mAChE) 410 fM

syn-TZ2PA5 K_d(eAChE) 100 fM K_d(mAChE) 2.3 pM

syn-TA2PZ6 K_d(eAChE) 830 fM K_d(mAChE) 610 fM

syn-TA2PZ5 K_d(eAChE) 540 fM K_d(mAChE) 3.0 pM

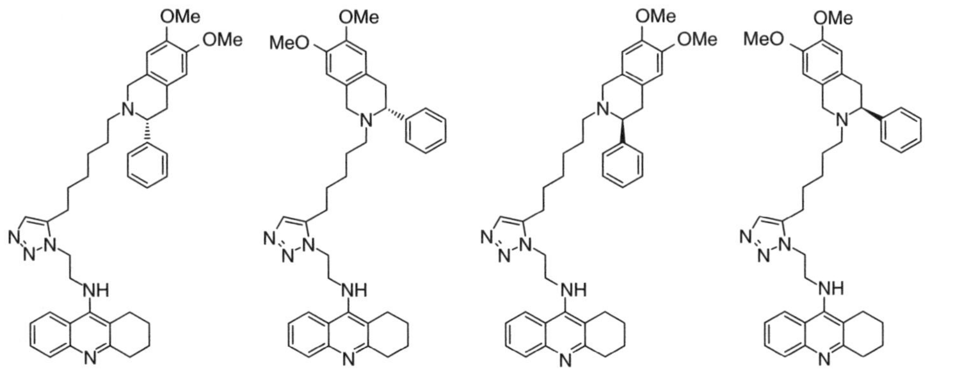

syn-(S)-TZ2PIQA6 K_d(eAChE) 96 fM K_d(mAChE) 1.1 pM

syn-(S)-TZ2PIQA5 K_d(eAChE) 33 fM K_d(mAChE) 500 fM

syn-(R)-TZ2PIQA6 K_d(eAChE) 360 fM K_d(mAChE) 1.7 fM

syn-(R)-TZ2PIQA5 K_d(eAChE) 36 fM K_d(mAChE) 100 fM

Figure 7.16 *Summary of* in situ *Click chemistry hits targeting AChE.*

The hit compound detection limit for the HPLC-SIM-MS technique was investigated and it was shown that triazole molecules could be detected down to 0.4 mol% of the active AChE concentration. The rate of any background 1,3-DCR between azide and acetylene fragments was also assessed by monitoring for triazole formation over 2 weeks. A plot of the integrated second-order kinetic equations appropriate to each tested binary combination gave the observed rate constants and demonstrated that the background 1,3-DCRs proceeded to less than 0.5% triazole formation over the 2 weeks. The experimental details reported in these papers demonstrate that the level of triazole formation with *in situ* chemistry is clearly very small, certainly $<<2\%$, likely $<1\%$, but necessarily greater than the background reaction if a 'hit' is to be claimed. Ample evidence is included in these proof-of-concept studies to confirm that the observed triazole formation is not a consequence of background reaction but is indeed induced by the presence of AChE. More specifically, this evidence includes a number of results from control experiments, all of which gave no detectable amounts of triazoles (analysis by either DIOS-MS or HPLC-SIM-MS): (i) a mixture of azides and acetylenes in the absence of AChE; (ii) a mixture of azides and acetylenes in the absence of AChE and in the presence of 10 μM bovine serum albumin; (iii) pretreatment of AChE with 100 μM tacrine to saturate the AChE active site, followed by incubation with azides and acetylenes; and (iv) prior inactivation of AChE by covalent phosphorylation of the active site serine followed by incubation with azides and acetylenes. In this last experiment, AChE was subsequently reactivated by treatment with pralidoxime chloride, and the reactivated enzyme then induced formation of the triazole from fragments.

7.11.2 Integrated Microfluidics

In 2006, Kolb and co-workers reported the application of an integrated microfluidic device for parallel screening of an *in situ* Click chemistry library.[41] The intention in this proof-of-concept study was to make lead discovery through *in situ* Click chemistry more convenient, more reliable, less expensive and more diverse compared with earlier efforts in which the Click chemistry was carried out in parallel in 96-well microtitre plates using largely manual operation. The microfluidics-based study targeted the already successful Click chemistry system with carbonic anhydrase (CA) and utilized 4-ethynylbenzenesulfonamide as the CA anchor fragment, with 20 complementary azide fragments (Scheme 7.7). The microfluidic chemical reaction circuit consisted of four major components (Figure 7.17). These components were utilized as follows:

1. A nanolitre-level rotary mixer – used to mix nanolitre quantities of reagents in a precise manner; this mixer has a total volume of 250 nL.
2. A microliter-level chaotic mixer – used to combine reagent solutions from the rotary mixer with microlitre quantities of enzyme solution in buffer to give a homogeneous reaction mixture (CA in PBS, pH 7.4);
3. A microfluidic multiplexer – used to deliver the reaction mixture to one of the 32 microvessels
4. Microvessels – 32 in total, used to store the reaction mixture for the *in situ* chemistry. The microvessels are 1.3 mm in diameter and 6 mm in depth, with a volume of ~8 μL.

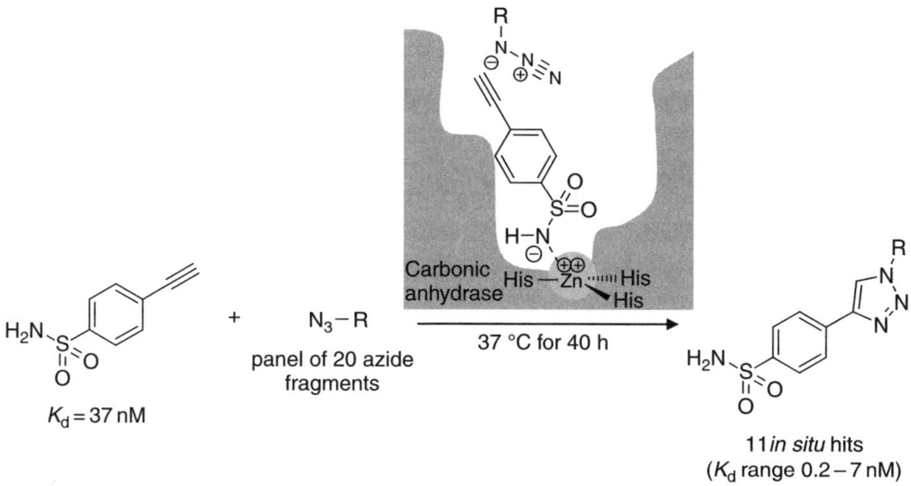

Scheme 7.7 In situ *Click chemistry targeting carbonic anhydrase.*

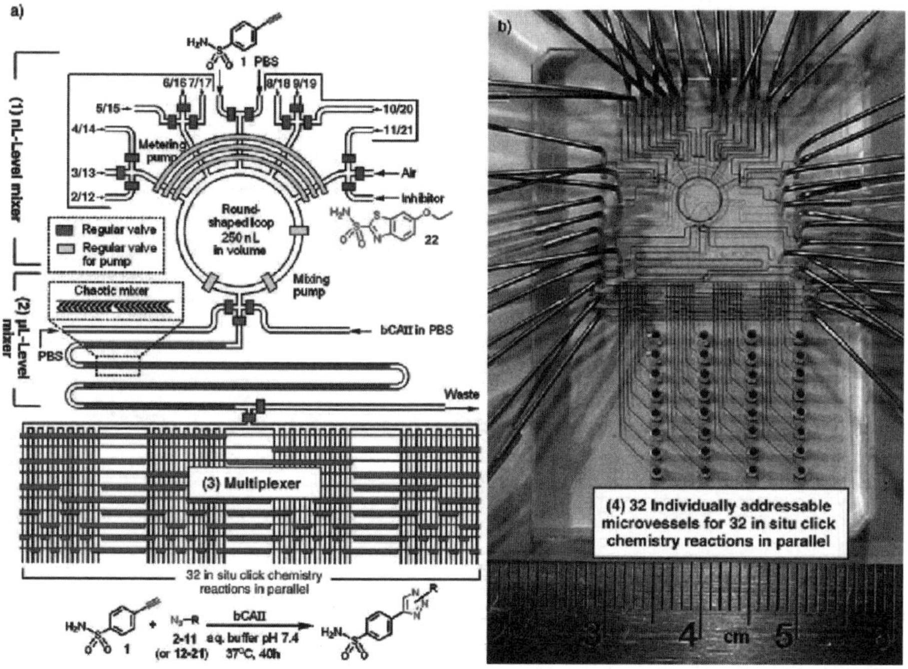

Figure 7.17 *Schematic representation of microfluidic chemical reaction circuit used for parallel screening with* in situ *Click chemistry. Reproduced with permission from Wang, J., Sui, G., Mocharla, V.P., Lin, R.J., Phelps, M.E., Kolb, H.C. and Tseng, H.-R., Integrated microfluidics for parallel screening of an* in situ *click chemistry library.* Angewandte Chemie International Edition *2006,* ***45****, 5276–5281. Copyright Wiley-VCH Verlag GmbH.*

The *in situ* experimental protocol using the microfluidic device allowed parallel screening and controls for 10 azide–acetylene fragment combinations per batch. Each screening campaign consisted of 32 individual reaction mixtures comprising the following format:

(i) acetylene + azide (×10, individually) + CA
(ii) acetylene + azide (×10, individually) + CA + CA inhibitor
(iii) acetylene + azide (×10, individually), no CA
(iv) CA only (×2).

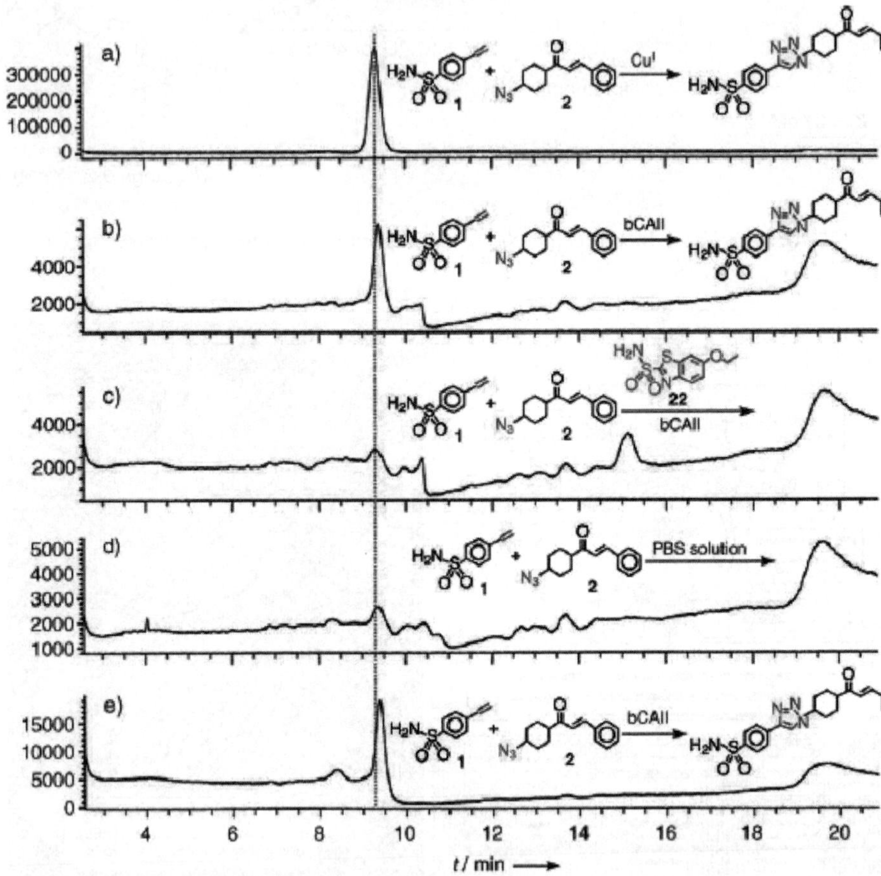

Figure 7.18 *Representative* in situ *Click chemistry reaction analysis by HPLC-SIM-MS; the triazole product is indicated by the dashed line. (a) Triazole products obtained by conventional synthesis with Cu(I) catalysis. (b)–(d) Microfuidic device reactions performed at 37 °C for 40 h: (b) in the presence of CA II, (c) in the presence of CA II and potent CA inhibitor to block the CA active site and (d) in the absence of CA II. (e) Reaction performed at 37 °C for 40 h in microtitre plate. Reproduced with permission from Wang, J., Sui, G., Mocharla, V.P., Lin, R.J., Phelps, M.E., Kolb, H.C. and Tseng, H.-R., Integrated microfluidics for parallel screening of an* in situ *click chemistry library.* Angewandte Chemie International Edition *2006, **45**, 5276–5281. Copyright Wiley-VCH Verlag GmbH.*

The individual reagent quantities per microvessel were acetylene, 80 nL of 30 mM stock = 0.514 μg or 2.4 nmol; azide, 120 nL of 30 mM stock = 3.6 nmol; CA, 3.8 μL of 5 mg mL^{-1} in PBS stock = 19 μg or 0.65 nmol; and CA inhibitor, 40 nL of 11 mM stock = 4 nmol. The interested reader is directed to the original paper for specific protocol details.[41] Significant for drug discovery applications is that the 32 *in situ* reactions were prepared in ~30 min (~57 s per reaction cycle); this is a remarkably short operation time compared with the microtitre plate reaction format and also uses substantially less protein (3.8 μL of 5 mg mL^{-1} protein compared with ~94 μL of 1 mg mL^{-1} protein for microtitre plate). The reaction circuit was incubated at 37 °C for 40 h prior to analysis by HPLC-SIM-MS using electrospray ionization as described above for AChE (Figure 7.18). The major determinant of the reagent and target quantities for these microfluidics experiments was the need to detect the triazole product formed, and in principle a more sensitive MS instrument than that used could facilitate even smaller reagent quantities.

7.12 Summary and Outlook

The *in situ* chemistry examples presented here are testament to a synthetic achievement that, although a niche component of medicinal chemistry, has the potential to impact on and influence the direction of future drug discovery campaigns. Typically, the reported *in situ* medicinal chemistry successes have stemmed from prior knowledge of a reliable target recognition fragment or 'anchoring' fragment. This anchor fragment is then furnished with the necessary functional group(s) to participate in either DCC or Click chemistry, typically with a much larger panel of complementary functionalized fragments. The *in situ* chemistry is free to scan the active site architecture of the target to link those fragments that can best exploit additional molecular recognition interactions with the target. As these approaches continue to develop, the field should aim to deliver results driven entirely from novel anchor fragments identified independently and without this prior knowledge. Thus fragments discovered through fragment-based screening are well poised to lead into either *in situ* DCC or Click chemistry campaigns to underpin target-guided fragment optimization, and it is expected that the currently small overlapping footprint of fragment-based drug discovery with these synthetic methodologies will expand. *In situ* DCC and Click chemistry can in principle be applied to a multitude of biomolecular targets for which small-molecules drug are sought. So far enzymes predominate in published examples, but receptor proteins, carbohydrate-binding proteins, DNA, RNA and whole cells have been targeted successfully. The real challenge that now presents itself is to demonstrate practicality of these synthetic methods in a drug discovery setting.

This research has progressed only because the science was willing to allow Nature's biomolecules to intervene earlier than is usual for medicinal chemistry protocols. The role of biomolecules has been elevated from spectator to key player, to guide and instruct the synthesis of potent inhibitor molecules, while the medicinal chemist acts to facilitate this process. The *in situ* selection and assembly of fragments could in principle be screened by an activity assay or by any number of analytical techniques; for drug discovery, however, it is a combination of speed, miniaturization and ultimately the nature of the information sought that drives the choice of analysis. With *in situ* medicinal chemistry, mass spectrometry is without doubt the screening method of choice as it allows the rapid identification of hit ligands from the complex reaction environment, with speed, sensitivity and minimal sample cleanup or consumption required. The role of mass spectrometry in drug discovery has changed dramatically; it is now a valued technique prominently integrated into drug discovery settings rather than overlooked and placed in the obscurity of the basement.

Professor K. Barry Sharpless, Nobel Laureate for Chemistry (2001), has provided an elegant description that captures the integral role of mass spectrometry leading to success with *in situ* Click chemistry, and this sentiment is similarly valid for *in situ* DCC:

> *'This stealthy trickery would have been for naught but for the awesome POWER of modern mass spectrometry to 'see' the tiny amount of triazole made 'accidentally' by the enzyme itself. In effect the system found itself with the opportunity to shepherd a reaction step along that meshed so perfectly that the new product fit its mould like a fine glove. Fitting so well in fact the enzymes natural function was strongly interfered with.'*
>
> Personal communication, March 2007.

References

[1] (a) Payne, D.J., Gwynn, M.N., Holmes, D.J., Pompliano, D.L. Drugs for bad bugs:confronting the challenges of antibacterial discovery. *Nat. Rev. Drug Discov*. 2007, **6**, 29–40; (b) Frantz, S. Pipeline problems are increasing the urge to merge. *Nat. Rev. Drug Discov*. 2006, **5**, 977–979; (c) Gray, N.S. Drug discovery through industry–academic partnerships. *Nat. Chem. Biol*. 2006, **2**, 649–653.

[2] Huc, I., Lehn, J.-M. Virtual combinatorial libraries: dynamic generation of molecular and supramolecular diversity by self-assembly. *Proc. Natl. Acad. Sci. USA* 1997, **94**, 2106–2110.

[3] Lewis, W.G., Green, L.G., Grynszpan, F., Radic, Z., Carlier, P.R., Taylor, P., Finn, M.G., Sharpless, K.B. Click chemistry *in situ*: acetylcholinesterase as a reaction vessel for the selective assembly of a femtomolar inhibitor from an array of building blocks. *Angew. Chem. Int. Ed*. 2002, **41**, 1053–1057.

[4] Griffey, R.H., Swayze, E.E. Ligand SAR using electrospray ionization mass spectrometry. In *Fragment-based Approaches in Drug Discovery*, ed. Jahnke, W., Erlanson, D.A., Wiley-VCH Verlag GmbH, Weinheim, 2006, pp. 267–283.

[5] Min, W., English, B.P., Luo, G., Binny, J. Cherayil, B.J., Kou, S.C., Xie, X.S. Fluctuating enzymes: lessons from single-molecule studies. *Acc. Chem. Res*. 2005, **38**, 923–931.

[6] Kubinyi, H. Drug Research: myths, hype and reality. *Nat. Rev. Drug Discov*. 2003, **2**, 665–668.

[7] Corbett, P.T., Leclaire, J., Vial, L., West, K.R., Wietor, J.-L., Sanders, J.K.M., Otto, S. Dynamic combinatorial chemistry. *Chem. Rev*. 2006, **106**, 3652–3711.

[8] Cousins, G.R.L., Poulsen, S.-A., Sanders, J.K.M. Molecular evolution: Dynamic combinatorial libraries, autocatalytic networks and the quest for molecular function. *Curr. Opin. Chem. Biol*. 2000, **4**, 270–279.

[9] Ramström, O., Lehn, J.-M. Dynamic ligand assembly. In *Comprehensive Medicinal Chemistry II*, Vol. 3, ed. Taylor, J.B., Triggle, D.J., Elsevier, Oxford, 2006, pp. 959–976.

[10] Rowan, S.J., Cantrill, S.J., Cousins, G.R.L., Sanders, J.K.M., Stoddart, J.F. Dynamic covalent chemistry. *Angew. Chem. Int. Ed*. 2002, **41**, 898–952.

[11] Ramström, O., Lohmann, S., Bunyapaiboonsri, T., Lehn, J.-M. Dynamic combinatorial carbohydrate libraries: probing the binding site of the concanavalin A lectin. *Chem. Eur. J.* 2004, **10**, 1711–1715.
[12] Bunyapaiboonsri, T., Ramström, O., Lohmann, S., Lehn, J.-M., Peng, L., Goeldner, M. Dynamic deconvolution of a pre-equilibrated dynamic combinatorial library of acetylcholinesterase inhibitors. *ChemBioChem* 2001, **2**, 438–444.
[13] Bunyapaiboonsri, T., Ramström, H., Ramström, O., Haiech, J., Lehn, J.-M. Generation of bis-cationic heterocyclic inhibitors of *Bacillus subtilis* HPr kinase/phosphatase from a ditopic dynamic combinatorial library. *J. Med. Chem.* 2003, **46**, 5803–5811.
[14] Congreve, M.S., Davis, D.J., Devine, L., Granata, C., O'Reilly, M., Wyatt, P.G., Jhoti, H. Detection of ligands from a synamic combinatorial library by X-ray crystallography. *Angew. Chem. Int. Ed.* 2003, **42**, 4479–4482.
[15] Poulsen, S.-A. Direct screening of a dynamic combinatorial library using mass spectrometry. *J. Am. Soc. Mass Spectrom.* 2006, **17**, 1074–1080.
[16] Polyakov, V.A., Nelen, M.I., Nazarpack-Kandously, N., Ryabov, A.D., Eliseev, A.V. Imine exchange in *O*-aryl and *O*-alkyl oximes as a base reaction for aqueous 'dynamic' combinatorial libraries. A kinetic and thermodynamic study. *J. Phys. Org. Chem.* 1999, **12**, 357–363.
[17] Cousins, G.R.L., Poulsen, S.-A., Sanders, J.K.M. Dynamic combinatorial libraries of pseudopeptide hydrazone macrocycles. *Chem. Commun.* 1999, 1575–1576.
[18] Giuseppone, N., Schmitt, J.-L., Schwartz, E., Lehn, J.-M. Scandium(III) catalysis of transimination reactions. Independent and constitutionally coupled reversible processes. *J. Am. Chem. Soc.* 2005, **127**, 5528–5539.
[19] Huc, I., Nguyen, R. Optimizing the reversibility of hydrazone formation for dynamic combinatorial chemistry. *Chem. Commun.* 2003, 942–943.
[20] Dirksen, A., Dirksen, S., Hackeng, T.M., Dawson, P. E. Nucleophilic catalysis of hydrazone formation and transimination: implications for dynamic covalent chemistry. *J. Am. Chem. Soc.* 2006, **128**, 15602–15603.
[21] Huisgen, R., Szeimies, G., Moebius, L. 1,3-Dipolar cycloadditions. XXXII. Kinetics of the addition of organic azides to carbon-carbon multiple bonds. *Chem. Ber.* 1967, **100**, 2494–2507.
[22] Tornøe, C.W., Christensen, C., Meldal, M. Peptidotriazoles on solid phase: 1,2,3-triazoles by regiospecific copper(I)-catalyzed 1,3-dipolar cycloadditions of terminal alkynes to azides. *J. Org. Chem.* 2002, **67**, 3057–3064.
[23] Rostovtsev, V.V., Green, L.G., Fokin, V.V., Sharpless, K.B. *Angew. Chem. Int. Ed.* 2002, **41**, 2596–2599.
[24] Sharpless, K.B., Manetsch, R. *In situ* click chemistry: a powerful means for lead discovery. *Expert Opin. Drug Discov.* 2006, **1**, 525–538.
[25] Krasinski, A., Fokin, V.V., Sharples, K.B. Direct synthesis of 1,5-disubstituted-4-magnesio-1,2,3-triazoles, revisited. *Org. Lett.* 2004, **6**, 1237–1240.
[26] Zhang, L., Chen, X., Xue, P., Sun, H.H.Y., Williams, I.D., Sharpless, K.B., Fokin, V.V., Jia, G. Ruthenium-catalyzed cycloaddition of alkynes and organic azides. *J. Am. Chem. Soc.* 2005, **127**, 15998–15999.
[27] Kolb, H.C., Finn, M.G., Sharpless, K.B. Click chemistry: diverse chemical function from a few good reactions. *Angew. Chem. Int. Ed.* 2001, **40**, 2004–2021.
[28] Lienard, B.M.R., Selevsek, N., Oldham, N.J., Schofield, C.J. Combined mass spectrometry and dynamic chemistry approach to identify metalloenzyme inhibitors. *ChemMedChem* 2007, **2**, 175–179.
[29] Bugaut, A., Toulme, J.-J., Rayner, B. SELEX and dynamic combinatorial chemistry interplay for the selection of conjugated RNA aptamers. *Org. Biomol. Chem.* 2006, **4**, 4082–4088.
[30] Valade, A., Urban, D., Beau, J.-M. Two galactosyltransferases' selection of different binders from the same uridine-based dynamic combinatorial library. *J. Comb. Chem.* 2007, **9**, 1–4.

[31] Valade, A., Urban, D., Beau, J.-M. Target-assisted selection of galactosyltransferase binders from dynamic combinatorial libraries. An unexpected solution with restricted amounts of enzymes. *ChemBioChem* 2006, **7**, 1023–1027.

[32] Pei, Z., Larsson, R., Aastrup, T., Anderson, H., Lehn, J.-M., Ramström, O. Quartz crystal microbalance bioaffinity sensor for rapid identification of glycosyldisulfide lectin inhibitors from a dynamic combinatorial library. *Biosens. Bioelectron.* 2006, **22**, 42–48.

[33] Andre, S., Pei, Z., Siebert, H.-C., Ramström, O., Gabius, H.-J. Glycosyl-disulfides from dynamic combinatorial libraries as *O*-glycoside mimetics for plant and endogenous lectins: their reactivities in solid-phase and cell assays and conformational analysis by molecular dynamics simulations. *Bioorg. Med. Chem.* 2006, **14**, 6314–6326.

[34] Shi, B., Stevenson, R., Campopiano, D.J., Greaney, M.F. Discovery of glutathione *S*-transferase inhibitors using dynamic combinatorial chemistry. *J. Am. Chem. Soc.* 2006, **128**, 8459–8467.

[35] Milanesi, L., Hunter, C.A., Sedelnikova, S.E., Waltho, J.P. Amplification of bifunctional ligands for calmodulin from a dynamic combinatorial library. *Chem. Eur. J.* 2006, **12**, 1081–1087.

[36] Danieli, B., Giardini, A., Lesma, G., Passarella, D., Peretto, B., Sacchetti, A., Silvani, A., Pratesi, G., Zunino, F. Thiocolchicine–podophyllotoxin conjugates: dynamic libraries based on disulfide exchange reaction. *J. Org. Chem.* 2006, **71**, 2848–2853.

[37] Lewis, W.G., Green, L.G., Grynszpan, F., Radic, Z., Carlier, P.R., Taylor, P., Finn, M.G., Sharpless, K.B. Click chemistry *in situ*: acetylcholinesterase as a reaction vessel for the selective assembly of a femtomolar inhibitor from an array of building blocks. *Angew. Chem. Int. Ed.* 2002, **41**, 1053–1057.

[38] Manetsch, R., Krasinski, A., Radic, Z., Raushel, J., Taylor, P., Sharpless, K.B., Kolb, H.C. *In situ* click chemistry: enzyme inhibitors made to their own specifications. *J. Am. Chem. Soc.* 2004, **126**, 12809–12818.

[39] Krasinski, A., Radic, Z., Manetsch, R., Raushel, J., Taylor, P., Sharpless, K.B., Kolb, H.C. *In situ* selection of lead compounds by click chemistry: target-guided optimization of acetylcholinesterase inhibitors. *J. Am. Chem. Soc.* 2005, **127**, 6686–6692.

[40] Mocharla, V.P., Colasson, B., Lee, L.V., Roeper, S., Sharpless, K.B., Wong, C.-H., Kolb, H.C. *In situ* click chemistry: enzyme-generated inhibitors of carbonic anhydrase II. *Angew. Chem. Int. Ed.* 2005, **44**, 116–120.

[41] Wang, J., Sui, G., Mocharla, V.P., Lin, R.J., Phelps, M.E., Kolb, H.C., Tseng, H.-R. Integrated microfluidics for parallel screening of an *in situ* click chemistry library. *Angew. Chem. Int. Ed.* 2006, **45**, 5276–5281.

[42] Whiting, M., Muldoon, J., Lin, Y.-C., Silverman, S.M., Lindstrom, W., Olson, A.J., Kolb, H.C., Finn, MG., Sharpless, K.B., Elder, J.H., Fokin, V.V. Inhibitors of HIV-1 protease by using *in situ* click chemistry. *Angew. Chem. Int. Ed.* 2006, **45**, 1435–1439.

[43] Pitram, S.M., Druzina, Z., Fokin, V.V., Schimmel, P., Sharpless, K.B. Inhibitors of tryptophanyl-tRNA synthetase from *Plasmodium falciparum* via *in situ* click chemistry. Abstracts of Papers, 232nd ACS National Meeting, San Francisco, CA, 10–14 September 2006.

[44] Kolb, H., Mocharla, V.P., Walsh, J.C., Padgett, H.C., Tanpure, R.T., Toyokuni, T., Su, H., Weber, W.A., Czernin, J., Jain, N., Ishikawa, T.-O, Herschman, H. Application of click chemistry to the development of COX-2 and CA-II inhibitors. Abstracts of Papers, 231st ACS National Meeting, Atlanta, GA, 26–30 March 2006.

[45] Weber, L. *In vitro* combinatorial chemistry to create drug candidates. *Drug Discov. Today: Technol.* 2004, **1**, 261–267.

[46] Corbett, P.T., Otto, S., Sanders, J.K.M. What are the limits to the size of effective dynamic combinatorial libraries? *Org. Lett.* 2004, **6**, 1825–1827.

[47] Moore, J.S., Zimmerman, N.W. 'Masterpiece' copolymer sequences by targeted equilibrium-shifting. *Org. Lett.* 2000, **2**, 915–918.

[48] Fenn, J.B., Mann, M., Meng, C.K., Wong, S.F., Whitehouse, C.M. Electrospray ionization for mass spectrometry of large biomolecules. *Science* 1989, **246**, 64–71.

[49] Hofstadler, S.A., Sannes-Lowery, K.A. Applications of ESI-MS in drug discovery: interrogation of noncovalent complexes. *Nat. Rev. Drug Discov.* 2006, **5**, 585–595.

[50] (a) Cheng X, Chen R, Bruce J.E., Schwartz B.L., Anderson, G.A., Hofstadler, S.A., Gale, D.C., Smith, R.D., Gao, J., Sigal, G.B., Mammen, M., Whitesides, G.M. Using electrospray ionization FTICR mass spectrometry to study competitive binding of inhibitors to carbonic anhydrase. *J. Am. Chem. Soc.* 1995, **117**, 8859–8860; (b) Gao, J., Cheng, X., Chen, R., Sigal, G.B., Bruce, J.E., Schwartz, B.L., Hofstadler, S.A., Anderson, G.A., Smith, R.D., Whitesides, G.M. Screening derivatized peptide libraries for tight binding inhibitors to carbonic anhydrase II by electrospray ionization-mass spectrometry. *J. Med. Chem.* 1996, **39**, 1949–1955; (c) Gao, J., Wu Q., Carbeck, J., Lei, Q.P., Smith, R.D., Whitesides, G.M. Probing the energetics of dissociation of carbonic anhydrase–igand complexes in the gas phase. *Biophys. J.* 1999, **76**, 3253–3260.

[51] Loo, J.A. Electrospray ionization mass spectrometry: a technology for studying noncovalent macromolecular complexes. *Int. J. Mass Spectrom.* 2000, **200**, 175–186.

[52] Benkestock, K., Sundqvist, G., Edlund, P.O., Roeraade, J. Influence of droplet size, capillary-cone distance and selected instrumental parameters for the analysis of noncovalent protein–ligand complexes by nanoelectrospray ionization mass spectrometry. *J. Mass Spectrom.* 2004, **39**, 1059–1067.

[53] (a) Keetch, C.A., Hernánndez, H., Sterling, A., Baumert, M., Allen, M.H., Robinson, C.V. Use of a microchip device coupled with mass spectrometry for ligand screening of a multi-protein target. *Anal. Chem.* 2003, **75**, 4937–4941; (b) Zhang, S., Van Pelt, C.K., Wilson, W.D. Quantitative determination of noncovalent binding interactions using automated nanoelectrospray mass spectrometry. *Anal. Chem.* 2003, **75**, 3010–3018; (c) Benkestock, K., Van Pelt, C.K., Akerud, T., Sterling, A., Edlund, P.O., Roeraade, J. Automated nanoelectrospray mass spectrometry for protein–ligand screening by noncovalent interaction applied to human H-FABP and A-FABP. *J. Biomol. Screen.* 2003, **8**, 247–256.

[54] (a) Asamoto, B., ed., *FT-ICR/MS: Analytical Applications of Fourier Transform Ion Cyclotron Resonance Mass Spectrometry*, VCH, New York, 1991; (b) Marshall, A.G., Verdun, F.R., eds, *Fourier Transforms in NMR, Optical and Mass spectrometry: a User's Handbook*, Elsevier, Amsterdam, 1990; (c) Marshall, A.G., Hendrickson, C.L., Jackson, G.S. Fourier transform ion cyclotron resonance mass spectrometry: a primer. *Mass Spectrom Rev.* 1998, **17**, 1–35; (d) Marshall, A.G., Hendrickson, C.L., Emmett, M.R., Rodgers, R.P., Blakney, G.T., Nilsson, C.L. Fourier transform ion cyclotron resonance: state of the art. *Eur. J. Mass Spectrom.* 2007, **13**, 57–59.

[55] For a series of comprehensive reviews on CA expression, distribution and therapeutic potential of CA inhibition and activation, see: (a) Thiry, A., Dogne, J.-A., Masereel, B., Supuran, C.T. Targeting tumour-associated carbonic anhydrase IX in cancer therapy. *Trends Pharmacol. Sci.* 2006, **27**, 566–573; (b) Pastorekova, S., Parkkila, S., Pastorek, J., Supuran, C.T. Carbonic anhydrases: current state of the art, therapeutic applications and future prospects. *J. Enzyme Inhib. Med. Chem.* 2004, **19**, 199–229; (c) Supuran, C.T. Carbonic anhydrases as drug targets – an overview. *Curr. Top. Med. Chem.* 2007, **7**, 825–833; (d) Supuran, C.T., Scozzafava, A. Carbonic anhydrases as targets for medicinal chemistry. *Bioorg. Med. Chem.* 2007, **15**, 4336–4350.

[56] Supuran, C.T. Carbonic anhydrases: catalytic and inhibition mechanism, distribution and physiological roles. In *Carbonic Anhydrase: Its Inhibitors and Activators*, ed. Supuran, C.T., Scozzafava, A., Conway, J., CRC Press: Boca Raton, FL, 2004, pp. 1–24.

[57] Smith, R.D., Bruce, J.E., Wu, Q., Lei, P. New mass spectrometric methods for the study of noncovalent associations of biopolymers. *Chem. Soc. Rev.* 1997, **26**, 191–202, and references therein.

[58] Bush, K., Jacoby, G.A., Medeiros, A.A. A functional classification scheme for β-lactamases and its correlation with molecular structure. *Antimicrob. Agents Chemother*. 1995, **39**, 1211–1233.

[59] Selevsek, N., Tholey, A., Heinzle, E., Lienard, B.M.R., Oldham, N.J., Schofield, C.J., Heinz, U., Adolph, H.-W., Frere, J.-M. Studies on ternary metallo- β-lactamase inhibitor complexes using electrospray ionization mass spectrometry. *J. Am. Soc. Mass Spectrom*. 2006, **17**, 1000–1004.

8

Computational Approaches to Fragment and Substructure Discovery and Evaluation

Eelke van der Horst and Adriaan P. IJzerman

8.1 Introduction

Nowadays, large molecular databases are easily accessible to the research community. This is illustrated by the advent of free online resources such as PubChem[1] and eMolecules.[2] These publicly available databases consist of structure and property data for millions of small molecules. Both databases are accessible through web-based search tools and are therefore an unprecedented source of small-molecule data. Outside the public domain, similar progress is taking place. Large molecular databases are becoming available that include bioactivity data, for example WOMBAT[3] (WOrld of Molecular BioAcTivity). Molecular data from these sources may be used to construct predictive models, such as structure–activity/property relationships (SARs/SPRs) or classification models. These models can be based on molecular properties, such as lipophilicity, solubility and molar weight, but also on molecular structures *per se*. *In silico* fragmentation of molecular structures is often used to provide a dataset of structural elements of the intact molecule. Analysis of the resulting fragments is useful to derive novel classifiers, e.g. for predicting the activity of new molecules. What is meant by the term *fragment* depends on the context. In the chemical sense, a fragment is a small, low molecular weight substance with weak affinity often used to 'build' a higher affinity lead compound. This is different from the computational sense. In the computational context, the term fragment or substructure denotes some structural part

Fragment-Based Drug Discovery: A Practical Approach Edited by Edward R. Zartler and Michael J. Shapiro

of the 2D structure of a molecule. It is the result of fragmentation of the molecule according to some 'breaking rules'. This chapter focuses on the *computational fragment*. We review fragment discovery and evaluation in the context of large molecular databases as described in the recent literature. Definitions, use and applications of fragments are addressed in addition to fragmentation methods. Fragmentation of 3D molecular structures will not be discussed.[4] In Section 8.2, we discuss the ways in which fragments can be derived. In the Section 8.3, a few examples of what can be learned from such fragmentation methods are presented together with their applications.

8.2 Fragmentation Methods

What is considered a fragment depends on the definition. A 'ring' could be a fragment or a particular chain of carbon atoms could be a fragment. The definition follows from the breaking rules that are used. To find structural patterns in a database, molecules should be broken into manageable parts that are readily analyzed. Graph theory is extensively used to this end (see Section 8.2.1). There are two approaches to molecule fragmentation. The first approach is to find all possible fragments that form some part of the molecular structure; the second is to dissect the molecule into fragments according to predefined (breaking) rules. The first approach allows a complete analysis of the fragments that exist in the set. However, the number of substructures for a single structure may then become very large, even for a moderately sized molecule. Several methods allow consideration of all (potential) fragments for analysis without generation of the full substructure set. The substructure approach will be the subject of the subsections *Frequent subgraph mining* and *Common substructures* in Section 8.2.2. The second fragmentation approach generally has a lower yield of fragments per molecule. Fragments result from 'breaking' the molecular structure into nonoverlapping, predefined parts. Thus, 'ring structures' may be defined in addition to functional groups. Fragmentation into molecular building blocks according to predefined rules follows in the subsections *Molecular building blocks* and *Virtual retro-synthesis* in Section 8.2.3.

8.2.1 Graph Representation

Graph theory plays an important role in fragmentation. The 2D structure of a molecule and its fragments are often represented as graphs.[5] A graph is a mathematical object that consists of a set of vertices or nodes and a set of edges that connect these nodes. The molecular structure conveniently translates into a graph, where vertices represent the atoms and edges represent the bonds.[5] This abstraction enables the use of generic methods that are under study in graph theory, such as the discovery of rings (cycles).

To illustrate the representation of molecules as graph, let us consider the sample structure in Figure 8.1 (taken from the PubChem compound database,[1] accession number CID9959891). Figure 8.2 shows the graph representation of the molecule in Figure 8.1 Hydrogen atoms, even when connected to heteroatoms, are omitted. Note that with standard graphs, representation of the molecule is limited to reproducing the connection pattern (connectivity) between the atoms. Any other information such as atom type or bond order is disregarded.

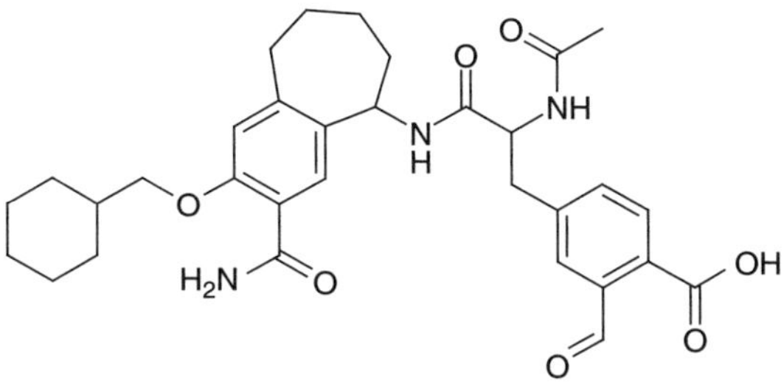

Figure 8.1 Example structure taken from the PubChem compound database.[1] IUPAC name 4-{(2S)-2-acetamido-2-[[(2S)-10-carbamoyl-9-(cyclohexylmethoxy)-2-bicyclo[5.4.0]undeca-7,9,11-trienyl]carbamoyl]ethyl}-2-formylbenzoic acid, PubChem CID 9959891.

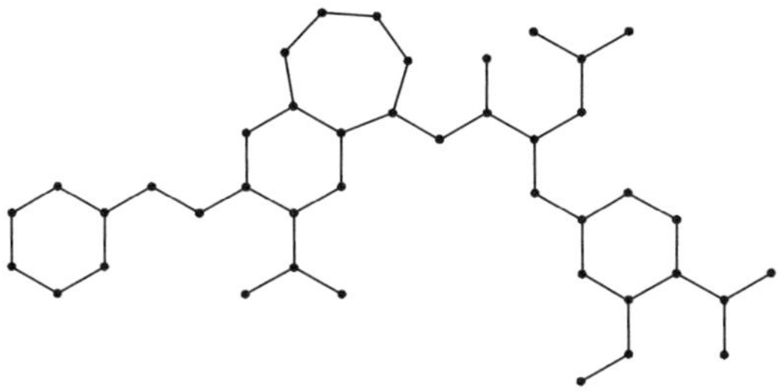

Figure 8.2 Graph representation of the example structure in Figure 8.1. Nodes (black dots) represent the atoms and edges (solid lines) represent the bonds. Note that standard graph representation disregards any extra information such atom type or bond order.

8.2.2 Substructure Methods

Frequent subgraph mining. Graph-based data mining aims to find interesting patterns in graph data. It has a variety of applications, such as analysis of literature citation networks, weblogs and web searches. Frequent subgraph mining is the process of finding all frequently recurring topological patterns in a database. In drug discovery, frequent subgraph mining, or fragment mining in the context of molecular databases, can be used to find structural patterns that are frequent in one class of compounds and infrequent in another. First, the general procedure of subgraph mining will be described. After that, a number of algorithms and tools for molecular fragment mining will be presented.

To find the frequently occurring fragments in a set of graphs, a typical algorithm would enumerate all possible fragments that exist in the set and find for each fragment the graphs in which it occurs. The frequency of a fragment is the number of graphs in which it occurs. The process of testing whether a fragment is part of a graph is called subgraph isomorphism testing. It searches the graph for a subgraph that is isomorphic to the fragment. A typical example is the ethyl fragment (C–C) in *n*-propane (C–C–C); it occurs twice and the one is 'isomorphic' to the other. In terms of computing steps, graph/subgraph isomorphism tests are relatively costly. This translates to prolonged computing time or memory requirements. It is one of the key issues in graph mining since there currently exist no efficient algorithms for isomorphism testing on general graphs. In the worst case, the number of computing steps is exponentially proportional to graph size, which contributes to the inefficiency of an algorithm. Therefore, most algorithms seek ways to avoid graph/subgraph isomorphism tests as much as possible.

Starting from an empty fragment, all possible fragment extensions (refinements) are generated, a process that will be explained below for the simple amino acid alanine. This is done by recursively adding edges and nodes to already generated fragments. In case of a ring closure, only an edge is added. Generated fragments are compared against the graphs in the database to check whether they occur. New refinements can only appear in those graphs that already hold the original fragment. Accordingly, the algorithm keeps appearance lists to restrict isomorphism testing to the graphs in the lists only. The support for a fragment is the proportion, or percentage, of graphs in the database in which it occurs. Obviously, found fragments are more relevant if they occur in at least a given minimum number/fraction of molecules. This minimum is called the minimum support value. Fragments are discarded if they occur in fewer molecules than allowed for the minimum support value, which is related to the significance of the found fragments. In general, lower minimum support values will yield higher numbers of fragments. Choosing a sufficiently high minimum support value will result in a comprehensible number of fragments while mining is completed within a reasonable time-scale. By definition, the support value of a fragment never exceeds the support values of the fragments it contains. This restricts refinement generation further, starting only from fragments with sufficient support (cf. *A priori* rule[6]). To focus isomorphism testing, fragment-mining algorithms may keep a mapping of the nodes and edges of a fragment to the corresponding nodes and edges of the graph in which it occurs. This is known as an embedding.

To illustrate the process, let us consider a graph mining experiment on a molecule database with alanine (Figure 8.3). For a single molecule in the database, such as alanine, a search tree can be constructed of all possible fragments. Figure 8.4 shows all these fragments for alanine with hydrogen atoms omitted as discussed before. On top is an empty fragment and each following fragment is a substructure of its descendants below. Fragments on the same level (six in total) have the same number of bonds (edges). For instance, the first level contains the elements N, O and C, since these are the constituents of the molecule. The C–C fragment on the second level forms the common core for the C–C–N, C–C–C, C–C=O and C–C–O fragment on the third level. The arrows indicate the paths leading from an empty fragment to the complete structure, yielding one extension at a time.

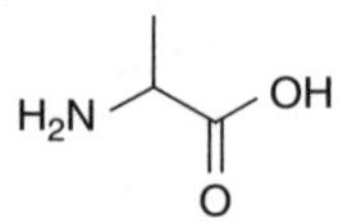

Figure 8.3 *Chemical structure of alanine. Implicit hydrogens are omitted.*

There are two ways to travel the subgraph lattice as the entire scheme in Figure 8.4 is called: breadth-first and depth-first. A breadth-first search considers all refinements at the same level before advancing to the next. For Figure 8.4, this means stepping through the lattice one row of fragments at a time. Storage requirements are proportional to the maximal

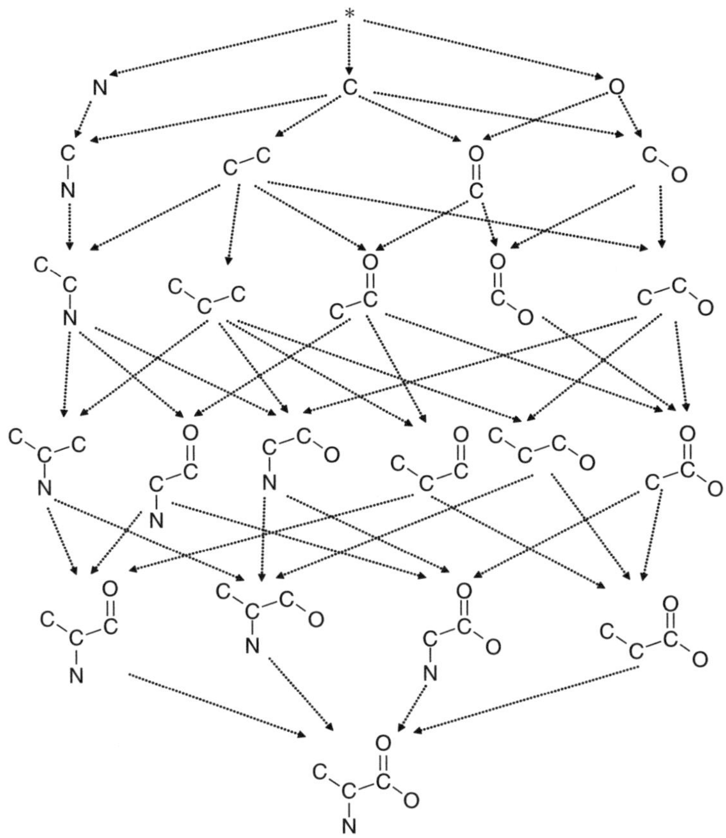

Figure 8.4 *The complete lattice of substructures for alanine (bottom). Level numbers in the lattice increase with fragment size until the final structure of alanine is reached.*

number of subgraphs at one level. Depth-first searching requires less storage, since a graph is completely searched before advancing to the next. Therefore, it is proportional to the size of the biggest graph. Modern graph mining algorithms, such as the ones described below, work in a depth-first manner.

There are three problems central to frequent subgraph mining; the difference between algorithms lies in how they address these problems. First, as was mentioned, subgraph isomorphism tests are expensive in terms of computation steps needed to perform the search. Second, the generation of refinements should be restricted. Third, since generated duplicates require isomorphism tests, their number should be kept to a minimum, e.g. by using a unique graph representation for testing.

The fragment miner MoFa (Molecule Fragment Miner)[7] was made especially for the purpose of molecule mining. All embeddings are stored and used for isomorphism testing and for restriction of fragment extensions to refinements that actually exist in the database. To reduce the number of generated refinements, MoFa sorts all nodes and edges of a fragment in the order in which they were added. Refinements may only occur at the same or newer nodes. Nonetheless, many duplicates are generated, with time-consuming isomorphism tests as a consequence. Two extensions exist for MoFa; the first treats rings as single units and the other treats chains of arbitrary length as a single unit. One of the advantages of treating rings as single units becomes clear when fragmenting steroid structures. Normally, MoFa considers more than 300 000 fragments per steroid, whereas the ring extension generates only 93 fragments. Another advantage is that the ambiguity of aromatic bond representations in rings, either single or double, is circumvented.

In gSpan (Graph-based Substructure Pattern),[8] a canonical graph representation is used, which is constructed from the concatenation of edge representations in the order in which they are visited. To generate unique representations, the algorithm dictates a strict, depth-first traversal of the subgraph lattice, hence the name 'depth-first search' code (dfs-code). Since the string of concatenated edge/vector representations resembles the sequence of letters in a word, graph representations are compared in the same way, that is, lexicographically. The elements of the string are sequentially compared until a mismatch is found or if one string ends. Lower edge/vector labels precede higher ones; if all labels match, the shorter string precedes the longer. The dfs-code of a fragment determines which nodes can be extended, thereby restricting the number of refinements for that fragment. Only those refinements are generated that have the smallest dfs-code. Appearance lists are used instead of embeddings; hence subgraph isomorphism tests are still necessary for the graphs in these appearance lists.

FFSM (Fast Frequent Subgraph Mining)[9] uses a canonical code, the Canonical Adjacency Matrix (CAM) code, to identify isomorphic graphs and to restrict refinement generation. It is based on a matrix representation of the graph. By concatenating all entries of the matrix, a string is formed that is used for lexicographic ordering of the graphs. FFSM stores embeddings for the nodes only. In this way, embeddings are rapidly created for new fragments made from joining or extension.

Gaston (GrAph/Sequence/Tree extractiON)[10] exploits the fact that various types of substructures are contained in each other and that for the simple types more efficient algorithms exist. First, only paths are considered in a substructure search. After that, paths are transformed to trees and trees are searched. Finally, trees are transformed to general graphs with cycles. This type of graph requires the most advanced and time consuming algorithms. As stated before, finding subgraph isomorphisms is a laborious task compared with other search problems and therefore time consuming. Therefore, they are only used, when they are really needed. Gaston stores all embeddings, in order to restrict the generation of fragment refinements to those that actually appear in the database and for isomorphism testing.

The tools have recently been compared and evaluated in the context of molecule mining.[11] Wörlein *et al.* reimplemented all four methods (same code base, programming expertise and optimization effort). Benchmarks were carried out on a comprehensive set of graph databases, including molecular databases. The molecular databases used were the IC93 (1283 compounds),[12] the HIV assays 1999 (42 689 compounds)[13] and the NCI (237 771 compounds).[14]

The IC93 database served to investigate how the algorithms behaved when the number of found fragments and the fragments themselves become large. For example, a support value of 4% resulted in 37 727 fragments, of which the largest had 22 bonds. The HIV database served to measure performance, whereas the NCI was used to test how the algorithms scale with increasing database size. For this, molecules were randomly divided into sets of various

Table 8.1 *Comparison of four frequent substructure-mining algorithms in terms of performance.*[a]

Algorithm	Runtime (minimum support value)				Memory usage (GB) (minimum support value)	
	5%		20%		5%	20%
	Total (min)	Per fragment found (s)	Total (min)	Per fragment found (s)		
Gaston[10]	7.4	0.1	2.5	0.4	1.3	0.9
gSpan[8]	19	0.3	4.5	0.8	0.3	0.2
FFSM[9]	19	0.3	8.3	1.5	1.2	0.9
MoFa[7]	80	1.1	11.8	2.2	0.6	0.6

[a] Performance was measured by applying each algorithm to the NCI HIV database (42 689 compounds). Runtime and memory usage are provided for two support thresholds: 5 and 20%. The runtime per fragment found is also provided to correct for the runtime overhead due to the higher number of fragments at lower support values. Data taken from performance charts from Wörlein et al.[11]

sizes. Sample measurements are provided for illustrating the quantitative comparison of the algorithms. Table 8.1 lists the performance measurements for the algorithms applied to the HIV data. The runtime of the algorithms increases with lower support values. Gaston was the fastest and MoFa the slowest algorithm. However, Gaston used the largest amount of memory, whereas MoFa needed less. gSpan had the lowest memory requirements. Note that these figures may differ for other data sets. Size and contents of the database, the minimum support value, and also implementation details and even the underlying hardware architecture may influence the performance of the algorithm. The data in Table 8.1 are indicative of the overall outcome of the quantitative comparison. For all algorithms, lower support values resulted in an exponential increase in runtime. This is probably due to the runtime overhead caused by the exponential rise in found fragments at lower support values. The benchmark results permitted a ranking of methods. gSpan needed the least memory, since it does not use embedding lists. MoFa, which stores only one subgraph embedding per node in the search tree, was also memory efficient. FFSM required more memory than gSpan and MoFa, probably because it stores the main subgraphs together in a node in the search tree. Gaston needed most memory, since with this method embedding lists for new fragments are based on those of 'parent' fragments. Extensions to the parent's list are stored with the 'children'. The size of embedding lists also depends on the number of children per fragment.

In terms of runtime, Gaston was always the fastest algorithm, except at lower support values on the complete NCI. The gSpan algorithm was faster than FFSM for the large datasets, although FFSM was faster than gSpan for the IC93 dataset. Embedding lists are not used in gSpan, which, in fact, speed up testing, especially for larger fragments. MoFa was always the slowest algorithm. The authors suggested that the slowdown of Gaston at lower support values on the complete NCI was due to the large amount of bookkeeping related to the vast number of embeddings. This results in a slowdown due to memory operations. However, the authors found that this effect varies for different systems. Some memory architectures penalize the memory-intensive operations of Gaston. Although MoFa was the slowest in all tests, it offers more functionality for molecular databases, e.g. there is an extension for treating rings as single entities, as mentioned above.[15] Another extension offers finding fragments with carbon chains of varying length. This can be useful for the exploration of biochemical reactions where this length is less important.[16]

Interestingly, the four fragment miners mentioned above have been made available as a single package named ParMol (Parallel Molecular Mining).[17] In addition to uniform access to MoFa, gSpan, FFSM and Gaston, the authors included a 2D viewer for molecular structures, parallel (multiprocessor) search and support for several file formats such as SMILES and SDF and a number of options to customize mining.

Other algorithms for frequent fragment mining that are more database-centric include Molfea[18] and Warmr.[19] Molfea (Molecular Feature Miner)[18] is in essence an inductive database framework. It finds patterns based on first-order logic. Molecules are encoded as basic facts and queries result in a combination of facts. The fragments that can be searched for or result from queries are linear sequences of non-hydrogen atoms and bonds. The fact that Molfea only finds chains of atoms limits its usefulness since almost all molecules have rings or branching points. Warmr[19] is a general-purpose Inductive Logic Programming (ILP) data-mining tool for finding frequently occurring patters in relational data.[20] [ILP is a machine learning technique used for knowledge discovery. The purpose of ILP is hypothesis generation, given some background knowledge and a set of positive and negative examples. Examples and background knowledge are encoded as a facts and rules in a relational database. From this, possible hypotheses are generated through inductive learning. Logic programming is used to represent examples, background knowledge and hypotheses, in a uniform way.] ILP has been successfully applied to chemical data, for instance to find frequent substructures in carcinogenic compounds. First, molecules are described in a relational language. Atoms are related to molecules and to other atoms through bonds. Algorithms such as Warmr perform multi-relational data mining, which means they are capable of finding patterns that span across multiple relations. Warmr searches the available patterns in a breadth-first manner, starting from the most general relations and gradually increasing the level of complexity, to find patterns that are more specific. Candidates that are more specific are generated by pruning nonfrequent patterns from the next level. Several meaningful relationships were reported for application of ILP on toxicity data. Although Warmr should be able to produce identical results compared with the fragment miners, it inherits some of the drawbacks related to ILP. First, a high level of expertise is required to encode the molecules, i.e. the graph and their properties, into relations that can be mined. Second, the complexity of relations queried, places high demands on computing resources [19]

Common substructures. Fragments are also derived by comparing molecular structures. For a pair of molecules, a number of substructures/fragments may exist that occur in both structures. A 'common substructure' is a set of atoms that two molecules have in common. Corresponding atoms should have the same atom type and the same topological distance to other common atoms, in both molecules. The topological distance is the number of bonds that form the shortest path between two atoms. The 'maximum common substructure' (MCS) is a continuously bonded substructure that has the highest number of common atoms.[21] Note that there may be multiple MCSs for a pair of molecules. Figure 8.5 shows an example of the MCS of two molecules, of which the largest is the molecule from Figure 8.1 The 'highest scoring common substructure' (HSCS)[21] is similar to the MCS, but also allows discontinuous common substructures. Scores are based on the number of common atoms and are corrected with a penalty for discontinuous pieces. In Figure 8.5, the HSCS and MCS are equal. Common substructure methods, such as the MCS and HSCS, are used to detect and visualize structural similarities between molecules.[21] In addition, the HSCS has been applied for discovery of common chemical replacements and to find fragments associated with multiple biological activities.[22, 23] These applications are described in Section 8.3.3.

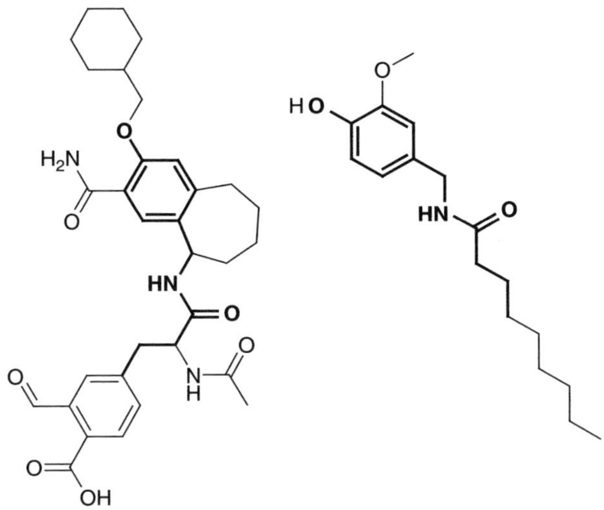

Figure 8.5 *Maximum Common Substructure of two molecules (MCS, drawn in bold). The structure on the left is the example structure from Figure 8.1. The structure on the right is vanillyl-N-nonylamide, IUPAC name N-[(4-hydroxy-3-methoxy-phenyl)methyl]nonanamide, PubChem CID 2998.*

8.2.3 Building Blocks

The fragmentation methods described in the previous section all use the 'full substructure set'. Despite the high level of detail of these approaches, exhaustive study of all possible fragments can be costly, however. A more restrictive, but still sensible, approach may be to focus on chemically meaningful fragments only, instead of including every single fragment in a study.

Molecular building blocks. To accomplish this, compounds are dissected into molecular building blocks. This method splits molecules into nonoverlapping structural parts according to a predefined set of breaking rules. These rules follow from the definition of individual building blocks. This approach yields (chemically) more intuitive fragments such as rings/ring systems, linkers, side-chains, functional groups, etc. Figure 8.6 illustrates the derivation of building blocks. A typical compound (Figure 8.6a) is fragmented into molecular parts, according to the method described by Bemis and Murcko.[28] Three ring systems (Figure 8.6d) are at the core of this compound, which are connected by two linkers (Figure 8.6e). Together, ring systems and linkers form the molecular framework (Figure 8.6c). Attached to this framework are the five side-chains (Figure 8.6b), yielding the complete molecule. There are many variations to this method; most methods differ in the precise definition of building blocks.

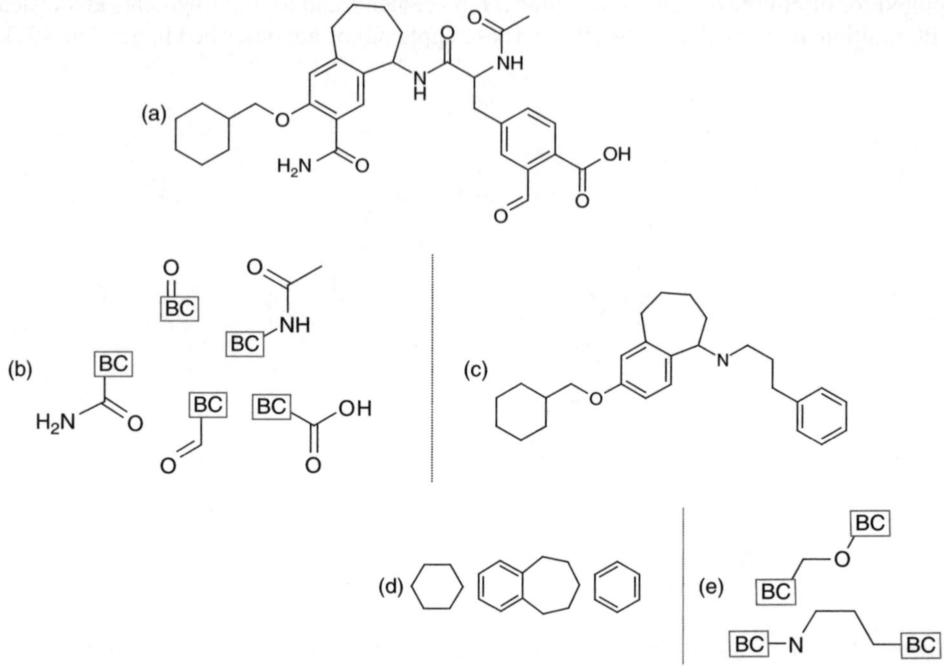

Figure 8.6 *Molecular building blocks, according to Bemis and Murcko.*[28] *(a) The structure that will be fragmented (CID9959891, see Figure 8.1). By removing (b) the side-chains from this structure, (c) the molecular framework is revealed. The framework consists of one or more (d) ring systems connected by (e) linkers. The connection point to the framework or rings is indicated by a rectangular label composed of the letter B and the atom type that it is connected to. For instance, the BC label means a carbon connection point in the framework.*

Virtual retro-synthesis. Another way to split a molecule into smaller parts is by virtual retro-synthesis. This method applies a set of breaking rules based on chemical reactions. Bonds that are typically formed by one of these reactions, are cleaved, essentially reversing synthesis. The resulting fragments are precursors from which the molecule can be synthesized using the set of chemical reactions. Although this approach might seem useful from a chemical point of view, it is not so appropriate for precise analysis. A different choice of synthesis rules may result in a different set of fragments. In addition, rules may conflict or the derived fragments may overlap. Moreover, there are indications that actual synthesis may not be reflected very well (as indicated e.g. by Vinkers *et al.*[24]). For a general overview of retro-synthesis, the reader is referred to a review by Todd.[25] Furthermore, a recent application of this synthetic approach was described by Vieth and Siegel.[26] They investigated four sets of bioactive molecules, fragmented these and analyzed the fragment distribution within a single set and between the four sets. An interesting example is the distribution of the β-lactam framework within antibiotics. This framework was prevalent in the older marketed drugs and absent in new ones. This may reflect the problem of the developing resistance observed against older antibiotics. Another example is the absence of amino acid scaffolds and side-chains in marketed oral drugs. Likewise, the majority of amino acid scaffolds are exclusive to injectable drugs.

8.3 Learning from Existing Databases

There is a lot to be learned from existing (drug) compound databases in terms of fragments: which fragments exist, how frequent they are and how the occurrence of one fragment is related to the occurrence of another, nonoverlapping fragment.[27] For instance, one can find single fragments that occur extremely often (e.g. a phenyl ring) or chemical templates on which some drug classes are based (e.g. benzodiazepines). Fragments which have low abundance might indicate barely explored parts of chemical space,[27] potentially interesting for designing new compounds. Insight can be obtained into preferences regarding chemistry and also in differences among databases. In the following sections, we will further expand on this, discussing analysis and evaluation of such databases (Sections 8.3.1 and 8.3.2) and applications of the findings thereof (Sections 8.3.3, 8.3.4 and 8.3.5).

8.3.1 Analysis of a Single Database

In an effort to identify the common features present in drug molecules, Bemis and Murcko[28] analyzed the structures of 5120 drugs extracted from the Comprehensive Medicinal Chemistry (CMC) database.[29] Two types of representation were used, in order to analyze structures at different levels of detail. At a more general level, properties of the molecular graphs were analyzed. Since the same graph may represent multiple molecules of similar shape, the common structure classes are revealed. For example, benzene, hexane and pyridine are all represented by the same hexagonal graph. In a more detailed analysis, the authors also considered atomic properties such as atom type, hybridization and bond order. They defined four nonoverlapping structural units that form a hierarchical description of the molecule: ring systems, linkers, frameworks and side-chains, as discussed in the subsection *Molecular building blocks* in Section 8.2.3. The authors justified their choice of this classification scheme by highlighting its useful features. For example, most frequent frameworks are easily identified, which may guide future drug design. Moreover, ring systems and linkers can serve as input for combinatorial library generation. In addition, the simple building blocks in existing drugs are already useful for checking the overlap between compound libraries.

The graph theoretical approach, as outlined in Section 8.2 and in Figure 8.2, identified a set of 1179 different frameworks, of which the six-membered ring was the most common one found. Of all these frameworks, 783 (66%) were unique, i.e. they occurred only once in the database. However, a small set of only 32 frameworks accounted for 50% of the drug molecules in the database. Analysis that also considered atomic properties logically resulted in a more diverse set of frameworks. There were 2506 different frameworks, of which 1908 (76%) were unique. Not surprisingly, a small set of 41 frameworks accounted for 1235 drug molecules (24%) in the database. Benzene was the most common framework found (8.5%). When we think of molecules as a common framework decorated with side-chains, phenyl and other small rings may be considered side-chains just as well, as in peptides. In this study, however, they were not; the few rings present in a small molecule are needed to derive a reasonable framework. In a continuation of this work, Bemis and Murcko focused on the various side-chains found in drugs.[30] Additional information was included in the side-chain description, i.e. the connection point and type of framework atom to which the side-chain was bonded. Side-chains consisting of a single (heavy) atom other than hydrogen, e.g. chlorine, were also considered. The set of molecules extracted from the CMC database was slightly smaller now, 5090 molecules in size. From this set, 4689 had side-chains. The total number of side-chains was 18 664, on average four side-chains per scaffold. The average length of a side-chain was two atoms. Side-chains of one heavy atom in length were found most (66%). Since oxygen atoms double-bonded to a ring system have a profound effect on the ring's electronic properties, it may be reasonable to consider these as part of the ring. In this case, the number of side-chains was reduced to 57%.

Lameijer *et al.* explored the possibility of gaining new insights solely from the structures that exist in the database.[27] For this, the NCI database[14] was mined. The authors reasoned that the substructures, and the combinations they occur in, provide insight into synthetic feasibility and 'chemical habits'. These habits emerge from an analysis of compound types that are made frequently or substructures that are often found together. The most frequently occurring fragments and fragment combinations were denoted 'chemical clichés'. Graph splitting was used to break the molecules into parts suitable for mining. For this, the method described by Bemis and Murcko[28] was adopted, with the extension that frameworks were further split into ring systems and linkers. Another difference was that only side-chains connected to a ring counted as side-chains. Side-chains attached to a linker were part of the linker. Figure 8.7 shows an example of a molecule split into molecular parts according to Lameijer *et al.*[27]

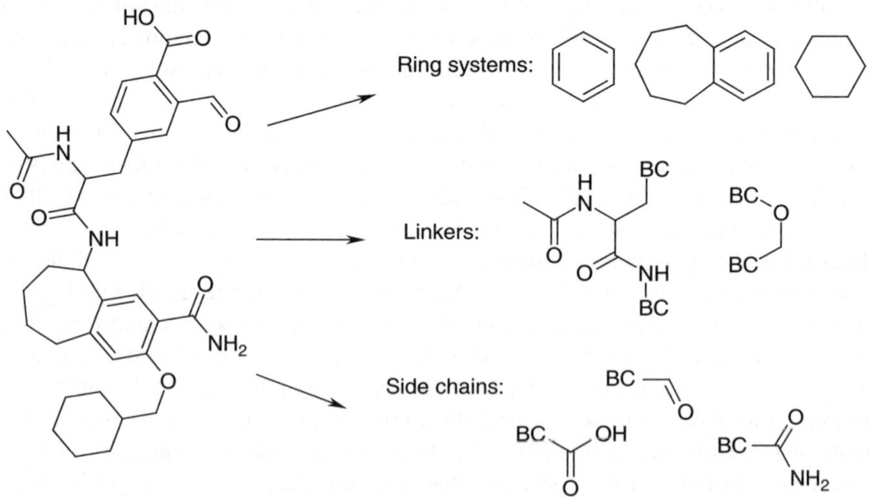

Figure 8.7 *Example structure (see also Figure 8.1) split into ring systems, linkers and side-chains according to the algorithm of Lameijer* et al.[27] *In contrast to Figure 8.6, side-chains in this figure stem from rings only. Side-chains connected to a linker are considered part of the linker. Again, boxed 'B and atom type' labels are used to indicate a connection point to a ring.*

By fragmenting 250 251 compounds from the NCI database, they found 65 612 fragments of the three different types of ring systems, side-chains and linkers. This already yielded useful information, for instance which ring systems occur and which do not, i.e. finding an N_6-ring to be nonexistent may complement some chemical commonsense. In total, 13 509 ring systems were found, 18 015 side-chains, 9675 linkers with two ring systems, 2531 linkers with three ring systems and 2280 linkers with four or more ring systems (up to 18 ring systems). In general, larger ring systems or branches occurred less frequently. Almost 70% of the three types of fragments occurred only once in the database. Branches with a higher number of attachment points seemed to have lower abundance. An exception to this rule was formed by linkers with six, or multiples of six, attachment points. These linkers occurred much more frequently than their neighbors did. Inspection revealed that these linkers were symmetrical.

The co-occurrence of fragments was also analyzed, to see whether the occurrence of one fragment in a molecule is related to the occurrence of another. This type of analysis can be compared to studying the contents of a shopping basket in a supermarket, a so-called Market Basket Analysis. Wine and olives may be frequently bought together, as are beer and potato chips, whereas beer and olives might be rarely observed together. Market Basket Analysis is a data-mining tool for finding regularities in the shopping behavior of customers of supermarkets, online shops, etc. A stochastic experiment was conducted first, since for frequently occurring fragments the chance is higher that a relationship is found, even if there is none. A new 'NCI' database was simulated using fragments that occurred in 20 or more molecules. Each fragment was used as many times as it occurred in molecules of the real NCI. Fragments were randomly divided over virtual molecules in the new database and each combination was counted. This process was repeated 1000 times, after which the expected occurrence of each fragment pair was calculated, together with the standard deviation of the occurrence. The expected occurrences were compared with actual co-occurrences in the NCI. A significant difference between the simulated/expected and the real co-occurrence implies that the fragments are correlated. The z-values were calculated and compared to detect that correlation.

Table 8.2 presents some examples of fragment pairs that occur in the same molecule much more or much less frequently than expected. In the first row of Table 8.2, tetrahydrofuran and a CH_2OH group are together; they were expected to occur 122 times together, but do so much more frequently in 2292 molecules. This is 19 (2292/122) times more than expected and very significantly different (z-value of 206) from the simulated database. The explanation is that the combination is found in (substituted) nucleosides that have been tested for anti-tumor activity. The second row presents another example of frequently co-occurring fragments that present a single structure class, viz. dihydrocholesterol analogues.

Interestingly, the situation is the opposite for the combination of a tetrahydrofuran and a phenyl group expected to occur in 2653 molecules. However, in the NCI there are only 270 of such instances, a factor of ~0.10 (270/2653). Apparently, this combination is underrepresented. A possible explanation for this effect might be that the 'avoiding' fragments belong to different compound classes with little overlap. Typical members from one class will be abundant in that class and scarce in others, adding to an overall reduction in co-occurrence frequency. Similarly, typical members from the same class are prone to be found together. Tetrahydrofuran-containing compounds generally differ in origin from

***Table 8.2** Some fragment pairs that occurred much more and much less often together than expected.*[a]

z-Value	Fragment 1	Fragment 2	Expected occurrence	Real occurrence	Multiple
206	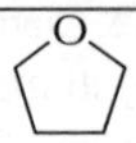		122	2292	19
117	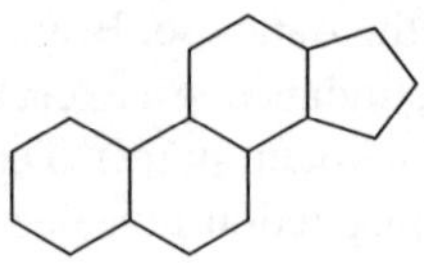	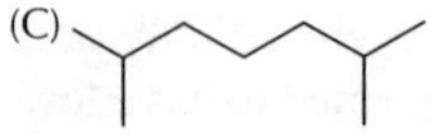	2.3	206	88
−19	CH_3–(C)	CF_3–(C)	544	139	0.26
−67			2653	270	0.10

[a] The first row, consisting of the tetrahydrofuran and the $-CH_2OH$ group, would be expected to occur 122 times together, but the pair appears in 2292 molecules leading to a multiple of 19 (see also text; data taken from Lameijer et al.[27]).

phenyl-containing compounds. The tetrahydrofuran ring often stems from the ribose moiety of nucleosides, either natural or chemically modified, whereas the phenyl ring is often found in industrial chemicals.

The authors suggest that the derived fragment and co-occurrence lists are useful in creating new chemistry. For instance, these listings provide insight into the most popular and therefore most commonly used side-chains and ring systems for synthesis. Rarer fragments also come forward through these lists, indicating less explored parts of chemical space. Finally, by looking at the fragments that do not occur together, new chemical space can be explored. The co-occurrences may be used to find a replacement for a structural feature. Examples of fragment pairs that are replacements of one another are chlorine and bromine or naphthalene and benzene.[27] These fragment pairs rarely occur together,[27] possibly because of their comparable physicochemical properties.

8.3.2 Analysis of Multiple Databases

To facilitate the design of libraries for high-throughput screening, Xue and Bajorath extracted scaffolds and side-chains and analyzed the distributions.[31] A 'scaffold' was defined as a molecular fragment without side-chains, essentially identical with the definition of frameworks (Figure 8.6). A 'side-chain' was defined as any acyclic chain or functional group with a single connection point to the rest of the molecule. As a source, the authors used Optiverse (OV),[32] a combinatorial screening library designed for diversity, and the Maybridge collection (MB),[33] a library of compounds used in medicinal chemistry. Acyclic structures were removed prior to screening (1214 from OV and 1060 from MB). The remaining sets were 116 762 (OV) and 58 239 (MB) compounds in size. To isolate scaffolds and side-chains, ring structures were detected first. Starting from these rings, all connected fragments were inspected. Acyclic fragments were removed from the structure and stored as side-chains. The remaining structure was stored as a scaffold. Using this algorithm, the authors extracted 52 529 unique scaffolds and 4486 side-chains from OV and 15 690 scaffolds and 2851 side-chains from MB. Only a minor overlap was observed: 2945 scaffolds and 407 side-chains occurred in both sets.

The ratios between the number of unique scaffolds and database size suggest that on average one scaffold is found in 2.2 (OV) and 3.7 (MB) molecules, respectively. However, the authors observed an unequal distribution of scaffolds: 8% (OV) and 7% (MB) of scaffolds occurred in 50% of the molecules. Moreover, more than 90% of the scaffolds occurred only once or twice. Aromatic structures and heterocycles were found most. The distribution of side-chains was similarly imbalanced. The 10 most frequent side-chains accounted for almost 75% of occurrences, whereas the majority occurred only once. Among the top 10 were classic substitutions such as halogens, the nitro group, the hydroxy group and organic functional groups such as the methoxy group. The methyl group accounted for 25% (OV) and 20% (MB) of occurrences, respectively.

Xu[34] derived molecular scaffolds to evaluate chemical compound libraries in terms of diversity, distribution in chemical space and differences/similarities with respect to existing drugs. The author used a Scaffold-based Classification Approach (SCA) that groups compounds into the same class if they share the same topological scaffold or so-called class center. The rationale behind this approach was that medicinal chemists intuitively group compounds based on scaffolds and functional groups and not so much on structural descriptors that most classification algorithms use. Scaffolds were derived similar to Xue and Bajorath[31] and Bemis and Murcko.[28] However, unsaturated bonds connected to a ring were considered part of the scaffold, since they change the chemical behavior of the ring system. Normally, scaffold analysis overlooks aliphatic compounds, since scaffolds are defined to consist of at least one ring. To overcome this, an extended definition of scaffold was adopted that also covered the aliphatic compounds. Double and triple bonds of acyclic compounds were treated as ring bonds, hence part of the scaffold. For saturated acyclic compounds, the scaffold consisted of the heteroatoms and carbon atoms that connect them. In all other cases, the carbon backbone formed the scaffold. Although the purpose of this extended definition is to extract scaffolds from all possible compound classes, some compounds from the same class may appear unrelated. For instance, amino acids that possess a cyclic side-chain are separated from those with an aliphatic chain. The structural scaffold derived will be the ring system in the first case and the characteristic amino/carboxyl group core in the second case.

First, a list of unique scaffolds was derived and sorted by complexity. The complexity was calculated from four structural descriptors, namely number of rings in the smallest set of smallest rings, number of heavy atoms, number of bonds and the sum of heavy atomic numbers in the scaffold. Each scaffold or class center in the list was assigned an ID that corresponded to its position in the list. How much a molecule resembled its class center was determined by the number of side-chains attached to the scaffold. Fewer side-chains will give a closer resemblance to the class center. The similarity of a drug with the class center was reflected in the *membership value*. The membership value was based on the sum of heavy atomic numbers, the number of rotating bonds, the number of one and two nodes and the number of double and triple bonds in a molecule compared with its scaffold. Since the membership value indicated the contribution of rings in the class center for a certain molecule, this term was called *cyclicity*. The four databases ACD,[35] NCI,[14] CMC[29] and MDDR[36] were analyzed according to this scaffold-based classification approach. Only the orally available drugs of CMC and MDDR were used. A diversity map was constructed that mapped complexity values against cyclicity values for each compound. Libraries that are more diverse have a wider spread on this map. An interesting outcome was the ranking of the four libraries according to chemical diversity. The ACD was most diverse, followed by the NCI, then the CMC and finally the MDDR. Two factors contribute to the low diversity of the MDDR: the majority of compounds are analogues and all compounds comply with the 'drug-likeness' property values. Molecules contributing to the high diversity of the ACD included RNAs/DNAs and fullerene C_{60}. Another interesting finding was that the orally active drugs from the CMC and MDDR were distributed in a narrower region than the other libraries.

8.3.3 Biological Activity

Sheridan[22] used common substructures to find fragment replacements in (drug-like) molecules. For this, 98 445 drug-like molecules from the MDL Drug Data Report (MDDR)[36] database were clustered according to similar biological activity, resulting in 556 clusters. Compounds from the same cluster were compared to find the 'highest-scoring common substructure' (HSCS).[21] Only compounds with an HSCS significantly larger than two randomly selected molecules of the same size were used to extract the fragments pairs that differed. Two different methods were used to extract replacement fragment pairs. The first method used atom-wise comparison of fragments, i.e. based on element and hybridization of atoms. The second method also considered possible rings that the atoms were in and adjacent functional groups, such as $-NO_2$, $-CO$, $-SO_2$ or $-PO_3$. Many of the classical replacements in medicinal chemistry were found.[22] With atom type, substitution of C with N in an aromatic ring (e.g. phenyl versus pyridine) was the most common. The next most common was replacement of –O– with –S– in both rings and chains, followed by –N– with –O– in rings, chains and esters versus amides. Another interesting commonly found replacement was the change between a five- a six-membered ring. Also considering the context of atoms in the comparison, e.g. a ring or functional group yielded a qualitatively similar fragment list. For a more complete list of replacements, the reader is referred to Sheridan.[22]

In a subsequent study, Sheridan[23] utilized the HSCS to identify fragments that are associated with multiple biological activities. Sheridan considered activity in the widest sense, ranging from *in vivo* biological effects (e.g. anti-hypertensive) to *in vitro* measures (e.g. affinity for a receptor). Since high specificity is very much desired for new drugs, knowledge about multi-activity fragments may be useful to avoid chemical classes likely to have unwanted side-effects. On the other hand, scaffolds that are active on a variety of receptors may form an attractive starting point in combinatorial library design. Pairs of molecules with similar structure and dissimilar activity were identified first. For each pair, the highest scoring common substructure (HSCS) was derived.[21] Again, only those HSCSs were kept that were significantly larger than would be expected for two randomly selected molecules. A 'consensus substructure' was generated from each molecule and its HSCS. It consists of atoms that are considered to be 'conserved', i.e. atoms that appeared relatively often in the set of HSCSs for that molecule. The most interesting consensus substructures are those that are found in many molecules and have many unique activities. Therefore, the generated consensus substructures were ranked according to both frequency of occurrence and number of unique activities. In case of structurally similar consensus substructures, only the highest in rank was kept. The steroid skeleton was found as a fine example of a multi-activity structure due to the many physiological processes in which steroid hormones are involved. Similarly highly ranked were tricyclic structures as in imipramine and doxepine. They bind to many G-protein-coupled receptors and transport proteins.

8.3.4 Predictive Models

In an attempt to organize available data in mutagenicity databases, Kho *et al.* described an automated approach to extract and organize ring systems occurring in a mutagenicity dataset.[37] They suggested that this method can be applied to any other set of molecules classified by some property, e.g. biological activity. A common assay for mutagenicity prediction is the Ames test, in which Ames-positive compounds are suspected to have mutagenic characteristics, whereas Ames-negatives are not. The database[38] was searched for the occurrence of ring types and their frequency in the Ames-positive and -negative categories. Emphasis was not so much on the development of predictive algorithms, but more on organizing the available data for use by chemists. Simple scaffolds were identified using a program that finds scaffolds by comparing all molecules in a set.[39] The results were presented as a hierarchy according to complexity. In this approach, simple rings are placed at the highest level and more complex ring systems that contain the parent rings as descendants. An example hierarchy is presented in Figure 8.8 Note that the

	Complexity→		Confidence interval of proportions (%)
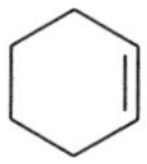 Equal odds			50–64
↳	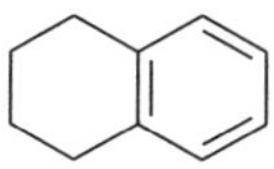 Equal odds		30–50
	↳	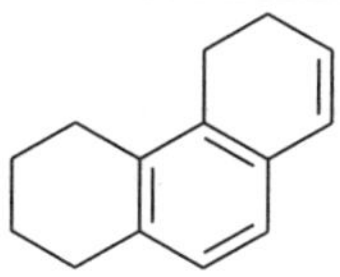 Ames+	6–33
	↳	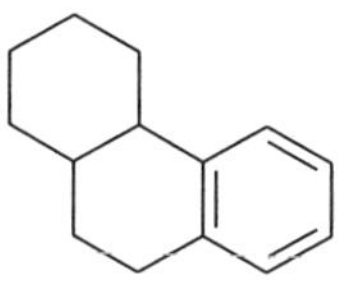 Ames–	100

Figure 8.8 *Scaffold (cyclohexene) hierarchy derived from mutagenicity data.*[38] *The proportion of Ames negative to Ames positive counts is qualitatively indicated below each scaffold. The confidence interval of the proportions is shown on the right of the scaffolds. Data taken from Kho* et al.[37]

Table 8.3 *Pairwise comparison of bicyclic rings (taken from Kho* et al.[37]*).*[a]

	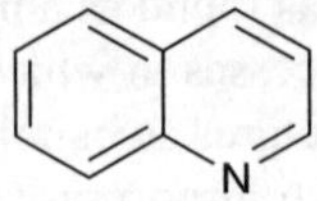	…	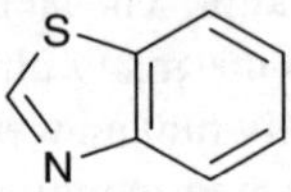	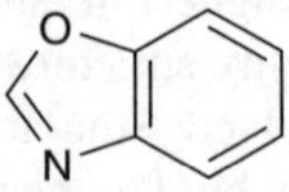
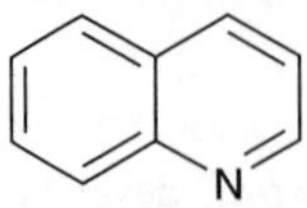	0.000	…	↑ −1.0182	↑ −3.0331
…			…	…
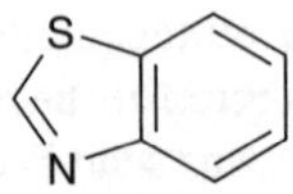 Benzothiazole			0.000	↑ −2.0149
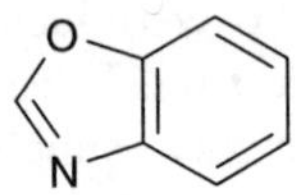 Benzoxazole				0.000

[a] The numbers are the logarithm of the odds ratio and indicate the preference in terms of mutagenic potential of one ring system relative to the other. For instance, a value of −1.0182 (second row, second column from the right) means that the left ring system has higher odds of being found in Ames positive compounds, so the top ring system is preferred. The arrow points to the fragment that is more likely to be found in the Ames-negative class. Many more ring systems were considered, indicated by the (empty) third column

tetrahydronaphthalene branch (first child), having equal odds of being found in either set, leads to an Ames-positive and an Ames-negative scaffold. A selection of the bicyclic rings found is presented in Table 8.3 Such a two-way entry table may be useful for selection of (bio)isosteric replacements with higher odds in the Ames-negative set. Similar tables can be constructed for other properties. A general finding from these data was that an increase in aromaticity or extension of conjugation enhances the odds for mutagenic compounds. An increase in the aliphatic character of rings decreases the mutagenic potential. To evaluate the usefulness of the mutagenicity dataset (with a total of 6039 compounds), the authors compiled a reference dataset consisting of 3882 commercially available drugs. Analysis revealed that the chemical diversity within the mutagenicity dataset was significantly less than the diversity of the marketed drugs. For the smaller drug set, 750 ring systems were found, in contrast to the 427 ring systems found in the Ames-test dataset. The two sets had 199 ring systems in common.

Instead of studying a limited set of structural features such as ring systems,[37] others have taken a more exhaustive approach. In such a scenario, all possible fragments are examined to find those discriminative for a certain property, e.g. toxicity. Kazius *et al.* used frequent fragment mining in order to derive toxicophores.[40] Similarly to Kho *et al.*,[37] structural elements were arranged according to mutagenic potential, thereby forming a decision list. Most substructure mining methods use only part of the chemical information in a molecule, viz. connectivity of the molecular graph (Figure 8.2), atom type labels and bond order (sometimes including aromaticity). To increase the level of chemical detail that is considered, Kazius *et al.* used an extended chemical representation.[40] Figure 8.9 shows a typical compound in standard chemical notation and two types of elaborate chemical representation. Elaborate chemical representation uses atomic hierarchies in addition to atom type labels, thereby including both general and more specific information. Atomic hierarchies are tree-like structures that consist of a root of a general atom label representing an atomic property and branches of more atom-specific labels (specifiers). Aliphatic nitrogen and oxygen atoms were labeled as 'small heteroatom' with specifiers for the atom type and number of connected hydrogens, as shown in Figure 8.9 Aliphatic sulfur and phosphorus atoms were labeled 'large heteroatom' with an additional specifier for the atom type. Chlorine, bromine and iodine atoms were labeled 'halogen' with atom type specifiers (Figure 8.9). For rings, two types of elaborate chemical representation were used. The 'aromatic' setting used a special atom label and bond type to represent aromatic atoms and bonds and attached a type specifier to aromatic heteroatom. Examples of aromatic atoms and bonds are shown in chemical representation **I** in Figure 8.9 The 'planar' setting used a

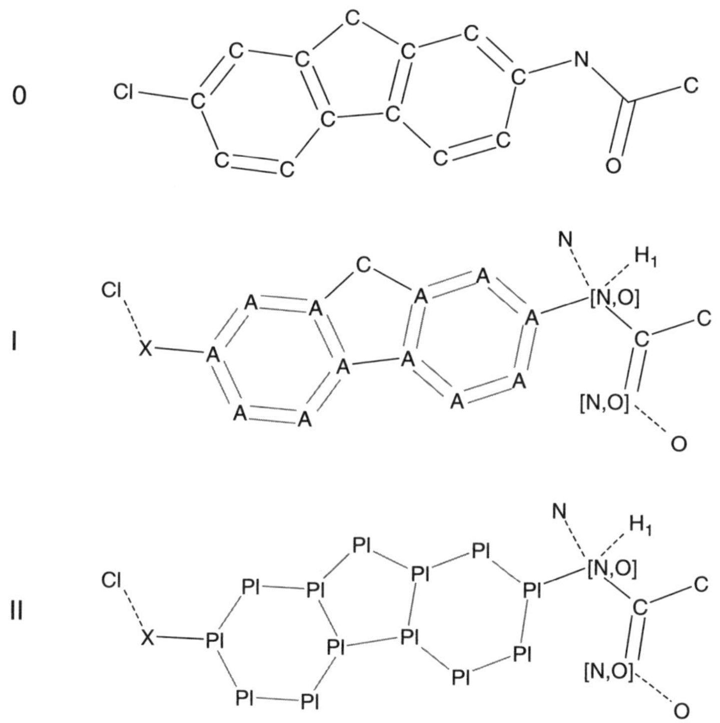

Figure 8.9 *A typical compound (PubChem[1] CID78776) in standard chemical notation* (**0**) *and two types of elaborate chemical representation, viz. the 'aromatic' setting* (**I**) *and the 'planar' setting* (**II**)*. Bonds are either single, double, aromatic (gray double bonds in* **I**) *or planar (gray single bonds in* **II**)*. Additional information is attached using the dashed bonds. Atom labels are carbon* (C)*, nitrogen* (N)*, oxygen* (O)*, small heteroatom* ([N,O])*, halogen* (X)*, chlorine* (Cl)*, aromatic atom* (A)*, planar atom* (Pl) *and number of implicit hydrogens* (H_1).

special atom label and bond type for atoms and bonds in aliphatic five- and six-membered rings or aromatic rings, including atom type specifiers. Planar atoms and bonds are shown in chemical representation **II** in Figure 8.9 All other atoms were labeled with the atom type. An additional atom specifier for the atom type was connected to heteroatoms and halogens and a specifier for implicit hydrogens was connected to heteroatom. Standard and elaborate chemical representations were used to extract substructures from mutagenicity data, both with and without considering nonlinear fragments. The dataset consisted of 4069 compounds from the Chemical Carcinogenesis Research Information System database.[41] Compounds were categorized as nonmutagens if all mutagenicity tests had a negative outcome. This resulted in 2294 compounds classified as mutagens.

Fragments from all methods were used together to find nonredundant substructures that are discriminative for mutagenicity. Only those substructures that occurred in more than 70 mutagens were considered. A decision list was constructed (Figure 8.10) by using the fragment with lowest p-value to split the set into two subsets (one that contained the fragment and one that did not). The p-value of a fragment was defined as the probability of finding a statistical association with mutagenicity based on chance alone. It was calculated from the amount of mutagens versus nonmutagens that are detected using that fragment. For the subset that did not contain the fragment, p-values were recomputed and the next most mutagenic fragment was used to split this set. In the case of multiple fragments with the lowest p-value, the largest fragment was used. The process was repeated as long as the new set had more than 60% mutagenic compounds. If the best-selected fragment had a p-value of more than 10^{-20}, no further splits were made. From all methods, the use of elaborate chemical representation combined with detection of nonlinear fragments proved best: mutagens were detected with a sensitivity of 84%. The resulting decision list (Figure 8.10) consisted of six nonredundant discriminating substructures, starting with a polycyclic planar system that described at least three rings and consisted of 11 planar atoms connected by planar bonds. The next most discriminating fragment was a nitrogen atom double-bonded to a nitrogen or oxygen, followed by a three-membered heterocycle (aliphatic epoxides and aziridines) and then an aliphatic halogen (chlorine, bromine and iodine). The second-last fragment was an aromatic primary amine and the list ended with a heteroatom bonded to a heteroatom fragment. Some of these substructures proved to be very similar to the general toxicophores derived previously by the authors in a laborious approach.[42] These results emphasize the benefit of elaborate chemical representation. For instance, the most discriminative fragment for mutagenicity would not have been detected by other methods, since the planar atom notation proved essential. Moreover, the importance of wildcards is underlined by their presence in all six substructures. Since the list contained two branched and one cyclic substructure, all possible graphs must be considered in substructure mining.

8.3.5 Ligand Design

The ring–linker frameworks approach described by Bemis and Murcko[28] was used to design new scaffold classes based on experimental structural information and to guide the optimization of modestly active ligands.[43] A set of 119 kinase inhibitors for at least 18 different targets was fragmented into ring systems and linkers and frequencies of occurrence were analyzed. Since bi- and tricyclic ring systems were relatively rare in the fragmented set, only monocyclic rings were considered. The authors found that the four rings benzene,

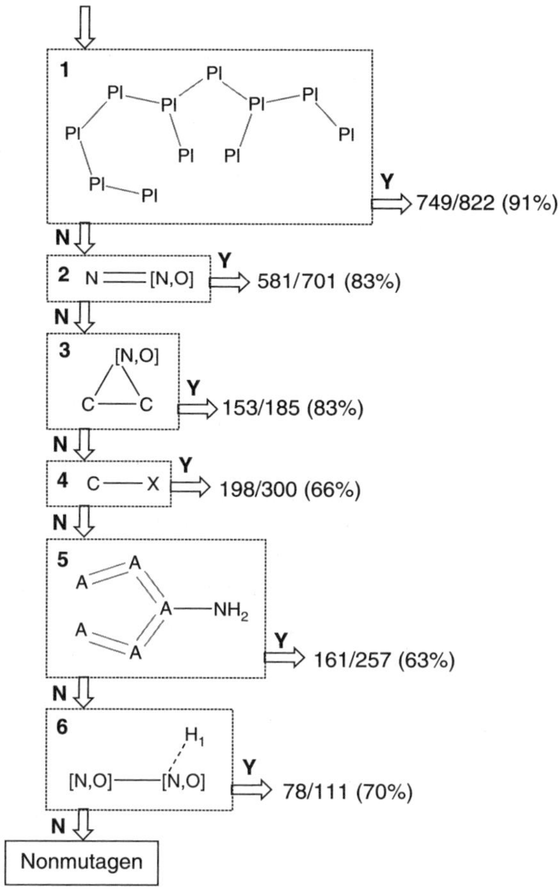

Figure 8.10 *Decision list derived from mutagenicity data.*[40] *Arrows indicate the direction to follow if a substructure is* (**Y**) *or is not* (**N**) *present in a compound. The number of mutagens, the total number of compounds and the percentage of mutagens is indicated for each subset (right).*

pyridine, pyrimidine and pyrrole comprise almost 90% of monocyclic ring occurrences in the fragmented data set. In addition, eight of the most abundant linkers were responsible for 90% of all linker occurrences in the set. From the four rings and eight linkers, a virtual library of kinase inhibitor scaffolds was constructed. Fragments known to form a critical interaction with the binding site of a kinase, served as a starting anchor. New scaffolds were generated by linking one of the rings to the anchor fragment, using one of the linkers. This was repeated for all ring–linker combinations and for each attachment point on the rings and anchor fragment. The newly designed scaffolds were docked against their targets, using the placement of the anchor fragment as constraint. A fit-based score was calculated and the highest scoring scaffolds were clustered according to the connection point at the anchor fragment. Using this method, the authors were able to reproduce the predominant structural motifs for known kinase inhibitors. In addition, they were able to suggest a number of alternative variations for these ligand cores.

Lameijer *et al.* developed a software tool to design drug-like molecules, the 'Molecule Evoluator'.[44] In this tool, both atom- and fragment-based evolutionary approaches were implemented. Fragments were taken from the analysis of the NCI database (ref. 27 and reviewed in Section 8.3.1). Through *interactive* evolution, a new principle in which the user acts as a fitness function, the authors suggested a number of simple yet novel molecules, eight of which were subsequently synthesized. Four compounds showed affinity for biogenic amine targets (receptor, ion channel and transport protein).[45]

8.4 Conclusion

In this chapter, we have compiled a number of computational strategies to dissect molecules into sets of constituting atoms, leading to fragments of different nature. Such fragments may also consist of elaborate atom representations, including wildcards. The reason for doing these, often computationally intensive, operations is found in the wealth of information that can be gleaned from such analyses. Virtual and real-world compound libraries can be mined for their diversity and/or similarity. In addition, the 'synthetic habits' of medicinal chemists can be explored. Furthermore, occurrence and co-occurrence of fragments may suggest new directions into chemical space. Fragments that appear linked to side-effects, via either multiple activities or straight toxicity, have been identified. This may help the medicinal chemist in designing safer or more selective lead compounds. Conversely, desired activities can be linked to fragments and such information may be a decisive factor in a successful medicinal chemistry program. With both the large number of HTS campaigns being performed and the resulting data increasingly being made available in the public domain, it is expected that steadily more dedicated datasets will become available for fragment mining. Rule- and knowledge-based design efforts will certainly benefit from this.

References

[1] PubChem database, pubchem.ncbi.nlm.nih.gov.

[2] eMolecules, www.emolecules.com.

[3] T.I. Oprea and J.M. Blaney, Cheminformatics approaches to fragment-based lead discovery, in *Fragment-based Approaches in Drug Discovery*, ed. W. Jahnke and D.A. Erlanson, Methods and Principles in Medicinal Chemistry, Vol. 34, Wiley-VCH Verlag GmbH, Weinheim, 2006, pp. 91–111.

[4] R.D. Cramer, R.J. Jilek and K.M. Andrews, Dbtop: topomer similarity searching of conventional structure databases, *J. Mol. Graph. Model.*, **20**, 447–462 (2002).

[5] P.J. Hansen and P.C. Jurs, Chemical applications of graph theory, *J. Chem. Educ.*, **65**, 574–580 (1988).

[6] R. Agrawal, T. Imielinski and A. Swami, Mining association rules between sets of items in large databases, in *Proceedings of the International Conference on Management of Data*, ACM Press, New York, 1993.

[7] C. Borgelt and M.R. Berthold, Mining molecular fragments: finding relevant substructures of molecules, in *Proceedings of the International Conference on Data Mining (ICDM)*, 2002, pp. 51–58.

[8] X. Yan and J. Han, gSpan: graph-based substructure pattern mining, in *Proceedings of the International Conference on Data Mining (ICDM)*, Maebashi City, Japan, 2002.

[9] J. Huang, W. Wang and J. Prins, Efficient mining of frequent subgraphs in the presence of isomorphism, *in Proceedings of the 3rd IEEE International. Conference on Data Mining (ICDM)*, IEEE Press, Piscataway, NJ, 2004.

[10] S. Nijssen and J.N. Kok, A quickstart in frequent structure mining can make a difference, in *Proceedings of the Tenth ACM SIGKDD International Conference on Knowledge Discovery and Data Mining*, ed. R. Kohavi, J. Gehrke, W. DuMouchel and J. Ghosh, ACM Press, New York, 2004.

[11] M. Wörlein, T. Meinl, I. Fischer and M. Philippsen, A quantitative comparison of the subgraph miners MoFa, gSpan, FFSM and Gaston, in *Proceedings of the 3rd European Conference of Principles of Knowledge Discovery and Data Mining (PKDD)*, 2005, pp. 392–403. http://www.springerlink.com/content/f510121050061j54/fulltext.pdf.

[12] *Index Chemicus*, Institute for Scientific Information (ISI), Philadelphia, PA, subset from 1993.

[13] HIV assay, dtp.nci.nih.gov/docs/aids/aids_data.html.

[14] National Cancer Institute database (NCI 3D), dtp.nci.nih.gov (Developmental Therapeutics Program NCI/NIH).

[15] H. Hofer, C. Borgelt and M.R. Berthold, Large scale mining of molecular fragments with wildcards, in *Intelligent Data Analysis* **8**, 495–504 (2004).

[16] T. Meinl, C. Borgelt and M.R. Berthold, Mining fragments with fuzzy chains in molecular databases, in *Proceedings of the Workshop W7 on Mining Graphs, Trees and Sequences (MGTS '04)*, ed. J.N. Kok and T. Washio, Pisa, 2004.

[17] T. Meinl, M. Wörlein, O. Urzova, I. Fischer and M. Philippsen, The ParMol package for frequent subgraph mining, in *Proceedings of the 3rd International Workshop on Graph Based Tools*, ed. T. Margaria-Steffen, J. Padberg and G. Taentzer, Electronic Communications of EASST 1, European Association of Software Science and Technology, Berlin, 2006; http://www2.informatik.uni-erlangen.de/Forschung/Projekte/ParMol/?language=en.

[18] C. Helma, S. Kramer and L. De Raedt, The molecular feature Miner MOLFEA, in *Proceedings of the Beilstein-Institut Workshop*, ed. M.G. Hicks and C. Kettner, Molecular Informatics: Confronting Complexity, Logos Verlag, Berlin, 2003.

[19] R.D. King, A. Srinivasan and L. Dehaspe, Warmr: a data mining tool for chemical data, *J. Comput.-Aided Mol. Des.*, **15**, 173–181 (2001).

[20] L. Dehaspe and H. Toivonen, Discovery of frequent DATALOG patterns, *Data Min. Knowl. Discov.*, **3**, 7–36 (1999).

[21] R.P. Sheridan and M.D. Miller, A method for visualizing recurrent topological substructures in sets of active molecules, *J. Chem. Inf. Comput. Sci.*, **38**, 915–924 (1998).

[22] R.P. Sheridan, The most common chemical replacements in drug-like compounds, *J. Chem. Inf. Comput. Sci.*, **42**, 103–108 (2002).

[23] R.P. Sheridan, Finding multiactivity substructures by mining databases of drug-like compounds, *J. Chem. Inf. Comput. Sci.*, **43**, 1037–1059 (2003).

[24] M. Vinkers, M. De Jonge, F. Daeyaert, J. Heeres, L. Koymans, J. Van Lenthe, P. Lewi, H. Timmerman, K. Van Aken and P. Janssen, SYNOPSIS: SYNthesize and OPtimize System in Silico, *J. Med. Chem.*, **46**, 2765–2773 (1998).

[25] M.H. Todd, Computer-aided organic synthesis, *Chem. Soc. Rev.*, **34**, 247–266 (2005).

[26] M. Vieth and M. Siegel, Structural fragments in marketed oral drugs, in *Fragment-based Approaches in Drug Discovery*, ed. W. Jahnke and D.A. Erlanson, Methods and Principles in Medicinal Chemistry, Vol. 34, Wiley-VCH Verlag GmbH, Weinheim, 2006, pp. 113–124.

[27] E.-W. Lameijer, J.N. Kok, T. Bäck and A.P. IJzerman, Mining a chemical database for fragment co-occurrence: discovery of 'chemical clichés', *J. Chem. Inf. Model.*, **46**, 553–562 (2006).

[28] G.W. Bemis and M.A. Murcko, The properties of known drugs. 1. Molecular frameworks, *J. Med. Chem.*, **39**, 2887–2893 (1996).

[29] Comprehensive Medicinal Chemistry (CMC-3D) Release 94.1, MDL Information Systems, San Leandro, CA.
[30] G.W. Bemis and M.A. Murcko, Properties of known drugs. 2. Side chains, *J. Med. Chem.*, **42**, 5095–5099 (1999).
[31] L. Xue and J. Bajorath, Distribution of molecular scaffolds and R-groups isolated from large compound databases, *J. Mol. Model.*, **5**, 97–102 (1999).
[32] C.D. Garr, J.R. Peterson, L. Schultz, A.R. Oliver, T.L. Underiner, R.D. Cramer, A.M. Ferguson, M.S. Lawless and D.E. Patterson, Solution phase synthesis of chemical libraries for lead discovery, *J. Biomol. Screen.*, **1**, 179–186 (1996).
[33] Maybridge, Maybridge Chemical Company, Trevillet, Cornwall.
[34] J. Xu, A new approach to finding natural chemical substructure classes, *J. Med. Chem.*, **45**, 5311–5320 (2002).
[35] *Available Chemicals Directory*, MDL Information Systems, San Leandro, CA.
[36] MD*L Drug Data Report, Version 99.1*, MDL Information Systems, San Leandro, CA.
[37] R. Kho, J.A. Hodges, M.R. Hansen and H.O. Villar, Ring systems in mutagenicity databases, *J. Med. Chem.*, **48**, 6671–6678 (2005).
[38] Ames-test data set, www.altoris.com.
[39] *SARvisonPlus 1.5*, ChemApps, San Diego, CA, www.chemapps.com.
[40] J. Kazius, S. Nijssen, J. Kok, T. Bäck and A.P. IJzerman, Substructure mining using elaborate chemical representation, *J. Chem. Inf. Model.*, **46**, 597–605 (2006).
[41] Chemical Carcinogenesis Research Information System, TOXNET, http://toxnet.nlm.nih.gov.
[42] J. Kazius, R. McGuire and R. Bursi, Derivation and validation of toxicophores for mutagenicity prediction, *J. Med. Chem.*, **48**, 312–320 (2005).
[43] A.M. Aronov and G.W. Bemis, A minimalist approach to fragment-based ligand design using common rings and linkers: application to kinase inhibitors, *Proteins*, **57**, 36–50 (2004).
[44] E.-W. Lameijer, J.N. Kok, T. Back and A.P. IJzerman, The Molecule Evoluator. an interactive evolutionary algorithm for the design of drug-like molecules, *J. Chem. Inf. Model.*, **46**, 545–552 (2006).
[45] E.-W. Lameijer, R.A. Tromp, R.F. Spanjersberg, J. Brussee and A.P. IJzerman, Designing active template molecules by combining computational *de novo* design and human chemists' expertise, *J. Med. Chem.*, **50**, 1925–1932 (2007).

9

Virtual Fragment Scanning: Current Trends, Applications and Web-based Tools

Bradley Feuston, M. Katharine Holloway, Georgia McGaughey and J. Christopher Culberson

9.1 Introduction

Significant resources are currently devoted to screening fragment databases with the goal of impacting lead finding and optimization in drug discovery. To facilitate this approach, it is clear that molecular modeling plays an important role in interpreting results and providing insights to help guide subsequent efforts.

While there are numerous references delineating the successes of fragment-based screening through technologies such as NMR, X-ray and surface plasmon resonance (SPR), which are discussed elsewhere in this book, the present chapter focuses on various aspects of virtual screening of fragment libraries. Crystal structures of proteins with bound small fragments have in general identified a number of conformations and binding sites for each ligand.[1] Such a lack of specificity in binding modes for fragments presents a difficult challenge for virtual screening. From a modeling perspective, small fragments are much weaker binders than typical drug-like molecules and are more challenging to score.[2] To normalize the binding energy of ligands by size, the concepts of binding efficiency index and ligand efficiency have evolved to choose preferentially weak binders with low molecular weight or low non-hydrogen atom counts, respectively.[3, 4] The same problem exists with virtual screening and scoring functions. A universal scoring function for bound ligands remains a

Fragment-Based Drug Discovery: A Practical Approach Edited by Edward R. Zartler and Michael J. Shapiro

challenge for larger molecules[5] and scoring functions specific for fragments have yet to be developed. However, software vendors are known to be actively pursuing suitable docking and scoring functions explicitly for fragment screening.[6] Virtual screening of fragment libraries has been most successful in applications where the active site is well characterized and degrees of freedom for molecular conformations and binding modes have been substantially reduced.

An important ingredient to both empirical and virtual screening is the design of the fragment libraries. Although this topic is covered in detail in other chapters, a short discussion will be given in the context of virtual screening. Design criteria and comparison of the physical properties distribution for various vendor offerings are presented in Section 9.2.

9.2 Fragment Databases

Due to the present popularity of fragment screening, vendors are now offering fragment libraries for general screening. The prevailing hypothesis is that small fragments, e.g. compounds of less than 300 Da, can be used to probe efficiently the functionality and shape of a protein binding site. The chemical space spanned by all the combinations of merging together small fragments in the library, i.e. *de novo* design of drug-like molecules, can be much larger than can be achieved by corporate compound databases. At present, ActiveSight, ChemBridge and Maybridge in particular offer fragment libraries for screening, although they differ significantly in property distributions and sizes.[7–9] Using the connection (SDF) files provided by the vendors, 384, 5213 and 848 molecules were obtained for ActiveSight, ChemBridge and Maybridge fragment libraries, respectively. In Figures 9.1–9.7, distributions for molecular weight, AlogP98,[10] polar surface area (PSA)[11], hydrogen bond acceptors (HBA), hydrogen bond donors (HBD), net charge

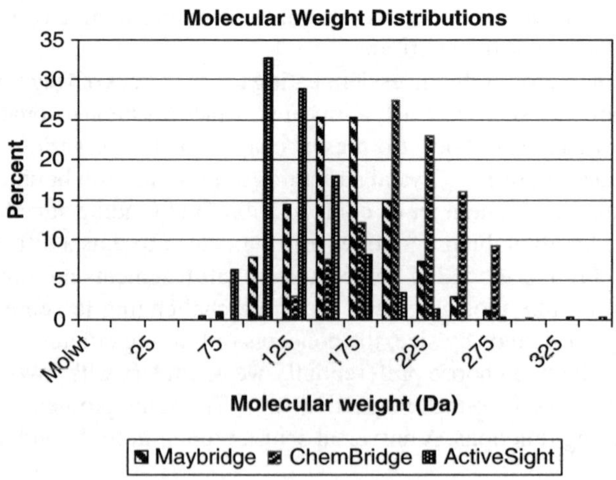

Figure 9.1 Molecular weight distributions for ActiveSight, ChemBridge and Maybridge fragment libraries.

and a measure for the number of rotatable bonds (NormBondFlx) are shown. The normalized bond flexibility is calculated by dividing the number of rotatable bonds by the total number of bonds, ranging from zero for a completely rigid structure to one for a chain of single bonds. The libraries were designed with similar selection criteria, e.g. small, drug-like, soluble fragments with both free and protected functionality, but differ significantly in their calculated property distributions. In addition, all vendor libraries utilized some form of diversity selection.

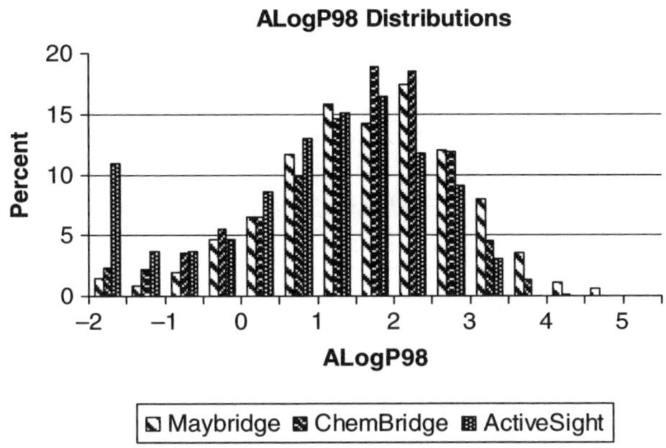

Figure 9.2 *Calculated AlogP98[10] distributions for ActiveSight, ChemBridge and Maybridge fragment libraries.*

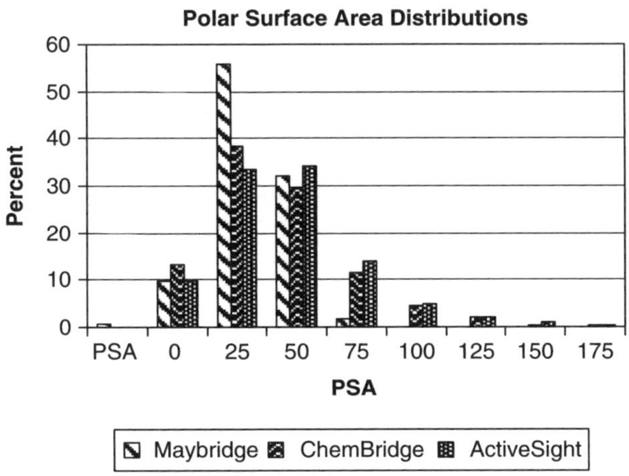

Figure 9.3 *Polar surface area (PSA) distributions for ActiveSight, ChemBridge and Maybridge fragment libraries.*

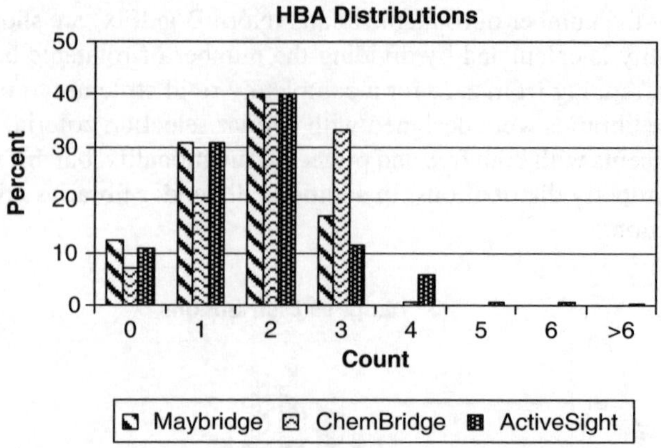

Figure 9.4 Hydrogen bond acceptor (HBA) distributions for ActiveSight, ChemBridge and Maybridge fragment libraries.

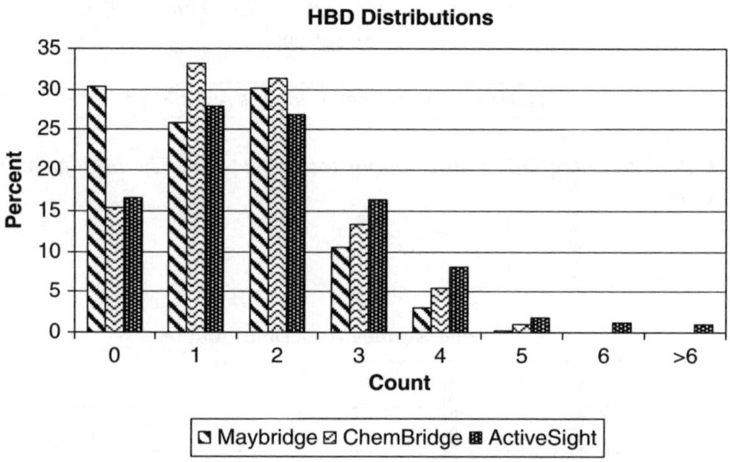

Figure 9.5 Hydrogen bond donor (HBD) distributions for ActiveSight, ChemBridge and Maybridge fragment libraries.

ActiveSight indicates that their small compound library is obtained from fragments of 'drug-like' molecules.[7] As seen by the molecular weight (Figure 9.1) and flexibility (Figure 9.7) distributions, these compounds are very small and rigid compared with the other two vendors. Most of the ActiveSight fragments consist of a single rigid core decorated with small side-chains. While nearly 95% of their molecules are less than 200 Da with cores of benzenes, benzimidazoles, cyclohexanes, furans, indoles, indazoles, naphthalenes, piperidines, pteridines, purines, pyrazoles, pyridines, pyrimidines, oxazoles, isoxazoles, quinolines, isoquinolines, quinazolines and thiazoles, the fragments provide a fairly large range of functionality, as witnessed by the distribution of HBAs and HBDs, shown in Figures 9.4 and 9.5, but with 75% of the fragments having a zero net charge (Figure 9.6).

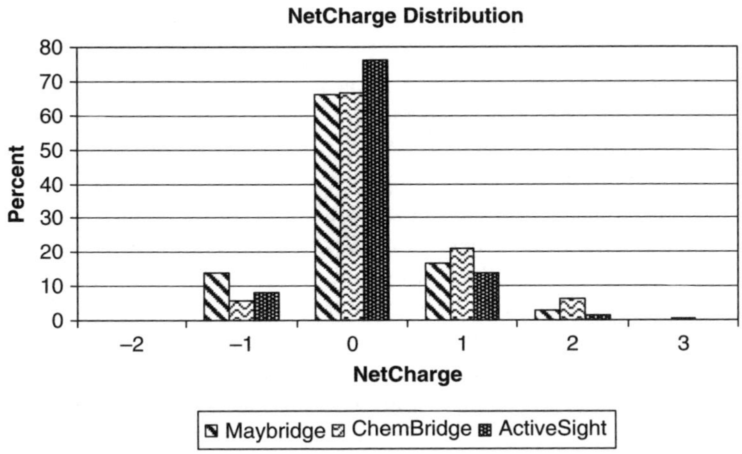

Figure 9.6 Net charge distributions for ActiveSight, ChemBridge and Maybridge fragment libraries.

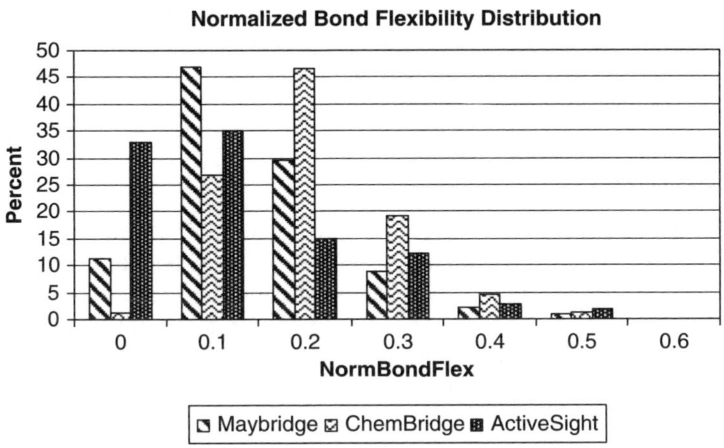

Figure 9.7 Normalized bond flexibility (number of rotatable-bonds/total bonds) distributions for ActiveSight, ChemBridge and Maybridge fragment libraries.

It was interesting to note that the ActiveSight fragments are reminiscent of early internal modeling efforts to construct a database of scaffolds for virtual design. While the cores are fairly flat, typically an undesirable feature for a diverse set of drug-like molecules, it still remains to be seen whether such rigid flat fragments benefit fragment screening and *de novo* design.

The ChemBridge collection of 5213 compounds was rationally selected from their EXPRESS-Pick Collection,[8] utilizing Astex's rule of three considerations ($MW \leq 300$, $HBD \leq 3$, $HBA \leq 3$, $CLogP \leq 3$)[12] and measures of diversity. ChemBridge has used a predicted aqueous solubility of greater than 3 mM for all selected compounds and can be seen in Figure 9.2 to have an AlogP98[10] distribution lying between the ActiveSight and Maybridge libraries.

Maybridge's fragment library, referred to as Ro3 since it strictly adheres to the rule of three,[12] has been constructed to be highly diverse. The Maybridge compounds are predominantly heterocycles that have been clustered for diversity while retaining reactive or 'linker-friendly' side-chains to promote their use as reagents in future syntheses of larger drug-like molecules. In addition, calculated LogS[13] values greater than 3 are used to filter out potential library members. Still, from the AlogP98 (Figure 9.2) and PSA (Figure 9.3) distributions, Maybridge's fragments have higher values than the ActiveSight set, an indicator of better solubility. Using Maybridge's Reactive Intermediates, uniquely capped fragments such as carboxamides and sulfonamides have also been included as suitable building blocks for quick follow-up of fragment hits.

It is clear that similar design criteria can yield very different distributions of physical properties and topologies. While these libraries are predominantly designed for crystallography screening, even greater differences in fragment libraries may be found when targeting NMR or SPR detection methodologies where solubility constraints may differ. More importantly from a virtual screening perspective, the physical limitations imposed by the various empirical approaches need not restrict which fragments are considered. Thus, all fragments are possible and only in the context of the design and target are constraints introduced.

9.3 Virtual Screening Using Fragments

Over the past decade, numerous companies have focused on fragment-based screening using X-ray crystallography,[1, 14] virtual screening,[15] SAR by NMR,[16] high concentration bioassay,[17] mass spectrometry,[18] and the automated ligand identification system (ALIS) technology developed at NeoGenesis.[19] The term 'fragment' itself has had various definitions, referring to compounds typically between ~150 and 450 Da. In the present discussion, 'fragments' will usually refer to small molecular compounds of less than 300 Da.

Computational chemists play a significant role in the lead optimization stage of drug discovery. In this position, computational chemists can aid in fragment identification through at least two avenues. First, fragments may be identified through the use of chemoinformatic tools such as screening and filtering of candidates by calculated physical properties, which is the more commonly applied involvement of computational design in fragment selection. Second, modeling may aid in selecting fragments to couple with a scaffold to enhance binding. A scaffold may be defined as a single structural motif which has demonstrated binding efficacy to a specific target or provided support for critical pharmacophoric elements. It is possible that the scaffold itself does not have measurable binding until it is part of a fully elaborated molecule. The scaffold could be any size fragment with sufficient functionality and/or proper shape and size to be accommodated by the protein's active site. The scaffold may be a fragment whose binding poses have been empirically identified through biological screening of a fragment library. Typically, chemists test different adornments of the scaffold to increase binding and specificity by synthetically combining a scaffold with a fragment. Molecular modeling can assist this lead optimization effort by selecting potential fragments (i.e. reagents), virtually enumerating these fragments with their target scaffold and, if a protein active site is known, scoring these virtual compounds in the protein. In such an application, the conformational space and binding modes of the fragments are substantially reduced to those most likely to positively impact protein–ligand interactions. Below, two in-house examples of such virtual screening of fragments have identified active molecules of interest to Merck's BACE-1 (Beta-site Amyloid precursor protein-Cleaving Enzyme) inhibitor program.

Even though there has been a paucity of papers specifically focused on virtual screening of small fragments, there is an increase in papers and presentations by scientists focusing on identification, scoring and docking of fragments.[20, 21] Virtual screening is the process of utilizing a hypothesis generated from existing biological or biophysical data on a given, or homologous, target and through the application of a computational method generating a list of compounds for screening in a validated biological assay. The choice of computational method could range from a simple 2D methodology such as a similarity-to-substructure method to a more sophisticated algorithm such as 3D docking of compounds and subsequent scoring of them in the active site. Hence virtual screening may include docking and scoring, but there are many other approaches, including ligand-based similarity searches and quantitative structure–activity relationship (QSAR) models.[22] Researchers are also combining virtual screening techniques with experimental methods, such as crystallography, to aid in hit identification.[23] In addition, independent software vendors (ISVs), such as Schrödinger, are specifically refining their scoring functions to describe fragment-protein interactions better.[6]

9.4 BACE-1: Identification of Binding Fragments

The identification of a small, brain penetrant BACE-1 inhibitor has been sought after for nearly a decade. There are patents and publications of brain penetrant BACE-1 inhibitors, and for a recent review of such compounds, the interested reader is referred to the report by Hills and Vacca.[24]

A couple of recent papers on BACE-1 fragments highlight complementary approaches used by Murray *et al.*[23] The first utilizes a crystallographic approach to identify the placement of the fragment. Even though computational chemistry was not used in the role of virtual screening to identify these small molecular inhibitors, it was applied in the generation of the fragment library which was biologically screened.

This particular fragment library was constructed to contain scaffolds and side-chains which often appear in drugs.[25] After screening the library, direct binding to the BACE-1 active site was determined by X-ray crystallography, yielding a hit rate around 0.6%. Two aminopyridine fragments hits are depicted in their binding mode in 2D format in Figure 9.8.

It is important to point out that when examining small fragments and trying to prioritize them, scientists will often utilize simple metrics such as binding efficiency index (BEI)[3] and ligand efficiency (LE).[4] The BEI is calculated using the equation

$$\mathrm{BEI} = \mathrm{IC}_{50}/\mathrm{MW} \tag{9.1}$$

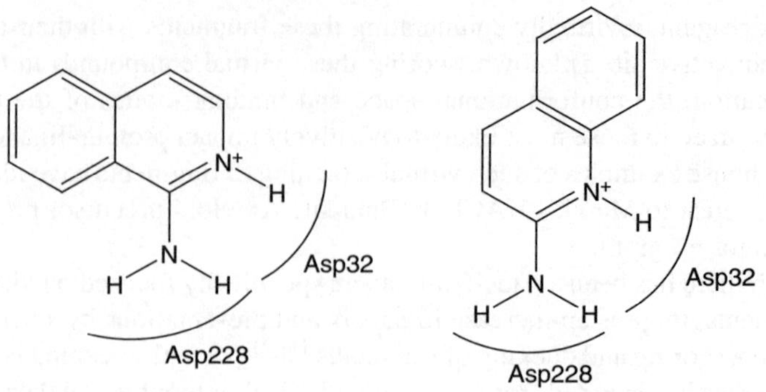

Figure 9.8 *Two aminopyridine fragments depicted in their binding modes with BACE-1 active site.*

where IC_{50} could also be pK_i or pK_d and the assay results are normalized by molecular weight (MW). LE is calculated by normalizing the free energy by heavy atom count (HAC) via the following equation:

$$LE = -\Delta G/HAC \approx -RT\ln(IC_{50})/HAC \tag{9.2}$$

The calculated LEs for the fragments in Figure 9.8 are 0.33. The Astex fragments are assumed to have an IC_{50} of ~2 mM and with molecular weights of ~145 Da their calculated BEI values are ~18. When compared with other targets, the LE and BEI values for the BACE-1 target are lower, which implies that values such as BEI and LE need to be applied in a target-specific context.[3]

The other approach taken by Murray *et al.* is that of virtual screening. Virtual screening is not new to the BACE field and has been applied many times to reveal new compounds of interest.[23] In addition to the fragment library already described at Astex, the Astex Therapeutics Library of Available Substances (ATLAS) and their corporate collection of compounds was virtually screened. They applied the following property restrictions in virtual screening of these two collections of compounds: molecular weights <300, CLogP <3 and the number of HBDs <4. In addition, the list of chemical suppliers was restricted to those with which they had prior experience. Their 3D virtual screen, which consisted of docking the compounds which satisfied the aforementioned criteria into the BACE-1 active site, was guided using pharmacophoric constraints. That is, they specifically searched for compounds which contained the key chemical features identified in the compounds depicted in Figure 9.8. Hydrogen bonding interactions with Asp228 and Asp32 were preferentially selected in the virtual screening studies. It is known that BACE-1 is an inherently flexible enzyme and can exist in many protein conformations. Therefore, in their virtual screening protocol, the compounds were docked using a proprietary version of GOLD[26, 27] and scored and ranked using both GoldScore[28] and ChemScore[29] in multiple protein poses. Using the results from virtual screening as a guide to select 65 compounds for crystallographic studies, they found that eight compounds (12%) soaked into the BACE-1 active site. One of these compounds, depicted in Figure 9.9, has an IC_{50} of 310 μM with corresponding LE (BEI) value of 0.32 (17.5). Two additional fragments were identified which make contacts with the catalytic aspartic acids: a piperidine and a hydroxyethylmorpholine. Murray *et al.*'s work highlights the ability of virtual screening to identify fragments which bind in the BACE-1 active site, albeit with low computed LE and BEI values. Although scoring functions are not ideal, they are still able to rank molecules with reasonable reliability. The work of Murray *et al.* and others lends support to the concept of combining virtual screening of fragments prior to analysis of binding by X-ray crystallography.

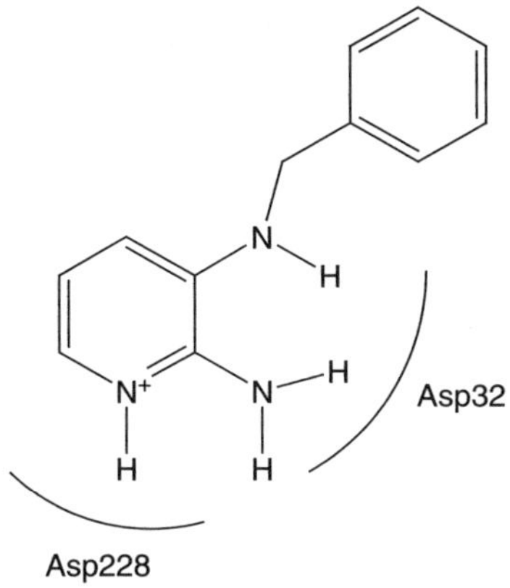

Figure 9.9 *310 μM BACE-1 compound obtained by Murray* et al.[23] *through virtual screening.*

Polgar and Keserü have demonstrated that the incorporation of pharmacophoric constraints can increase the rate of retrieving actives from a 3D virtual screen by limiting the possible conformations and docking poses to the most interesting ones.[30] Using BACE-1 as the target and optimizing the protonation states of the catalytic aspartic acids, they included pharmacophoric constraints derived from a Vertex patent[31] and were further able to improve their enrichment factor from 36 to 41.[30]

In another report, Hajduk and Greer described the successful application of structure-based design in the drug discovery process.[32] Their work might lead one to believe that having the structure of the target protein *should* increase one's probability of success; however, in practice, computational chemists do not find that to be true.[33] In this recent virtual screening study, the authors contend that if one is solely measuring success by the number of active molecules found (or 'hits'), then a simple 2D methodology is best.

9.5 Elaboration of a Fragment via Synthesis

Baxter *et al.* have nicely demonstrated that one can take a 0.9 μM hit, discovered from HTS, to an 11 nM K_i compound.[34] An X-ray crystal structure of the HTS lead helped demonstrate the unique binding pose in the active site which only occupied the non prime side of BACE-1 in a hairpin orientation. The crystal structure revealed that the prime side was not occupied and the incorporation of a cyclohexyl ring resulted in a potent 11 nM compound.[33]

Virtual fragment screening has also been performed at Merck for BACE-1. Two examples of fragment-based screening are described below. The first involves the virtual screening of ∼700 amines in the P1′ pocket, resulting in the identification of new BACE-1 inhibitors. The second example is a retrospective analysis of the virtual screening of the P3 pocket with >9000 amines. In this case, the objective is to identify a fragment with known activity out of the background of the ∼9000 other primary amines as a test of the Merck's proprietary web tools described below.

In the first example of virtual screening of fragments to enhance binding affinity, a set of P1′ substituents were selected prospectively for the BACE-1 inhibitor scaffold depicted at the top of Figure 9.10. Note the scaffold includes an α-methylbenzylamide P3 substituent which is the target of the retrospective analysis in the second example. While the loop conformation near the P3 pocket is shown below to be important for scoring P3 substituents, all fragments in the present example were scored in a single conformation of the BACE-1 active site, consistent with the crystal structure of this compound series. The 'real' and 'virtual' chemistries employed in this reagent scan are also shown in Figure 9.10.

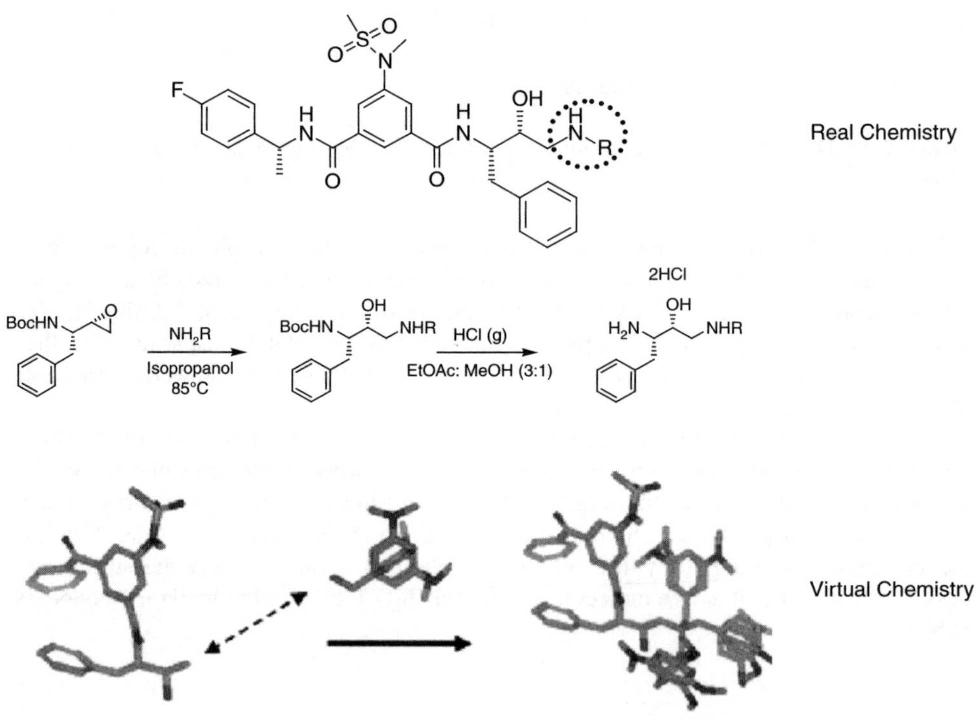

Figure 9.10 P1′ amine substituents were selected prospectively for a BACE-1 inhibitor scaffold. Note the dashed circle (above) and dashed line (below) indicating the point of attachment. The real (top) and virtual (bottom) chemistries are depicted beneath the scaffold.

A set of 701 suitable amine reagents was selected from the ACD.[35] After generation of up to 25 conformers per fragment, a total of 13 089 conformers of 701 virtual BACE-1 inhibitors (i.e. scaffold with annealed amine reagent conformer) were energy minimized in the BACE-1 active site using the Merck Molecular Force Field (MMFF).[36] The best conformer/pose for each reagent was selected using an energy-based scoring metric[37] that emphasizes poses which combine a low conformational energy with a favorable enzyme–inhibitor interaction energy. An arbitrary interaction energy cut-off of –45 kcal mol^{-1} was employed to reduce the number of amines for consideration. Even so, more than 500 amine reagents were selected. The active site space sampled by the amine reagents is illustrated in Figure 9.11.

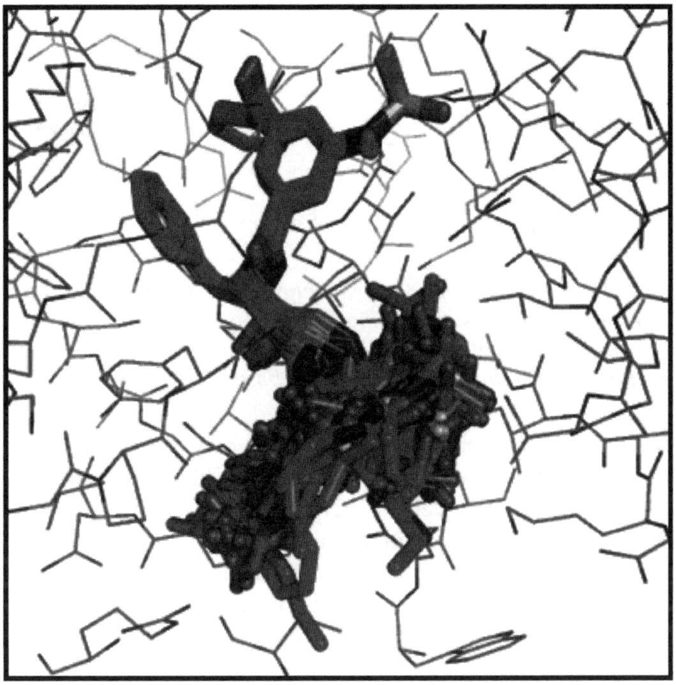

Figure 9.11 *Sampling the S1′ pocket by conformational search over amine fragments.*

Based on these results, it was hypothesized that the S1′ pocket of BACE-1 is rather large, open and promiscuous. To examine this hypothesis, a test set of BACE-1 inhibitors was employed that included (a) previously synthesized compounds containing this scaffold and a variety of P1′ amine reagents, (b) new compounds with amine reagents selected independent of the virtual reagent scan and scoring metric and (c) new compounds that contained amine reagents which scored well in the virtual reagent scan. After synthesis and assay of (b) and (c) inhibitors, a plot of observed activity versus energy score was obtained, as shown in Figure 9.12.

It is clear that both the virtual reagent scan and the independent selection of reagents by the chemist led to active BACE-1 inhibitors. Thus, virtual selection of reagents succeeded in raising interesting reagents to the top of the list, although in some cases their activity appears to have been under-predicted (see the two ether-containing amines highlighted in

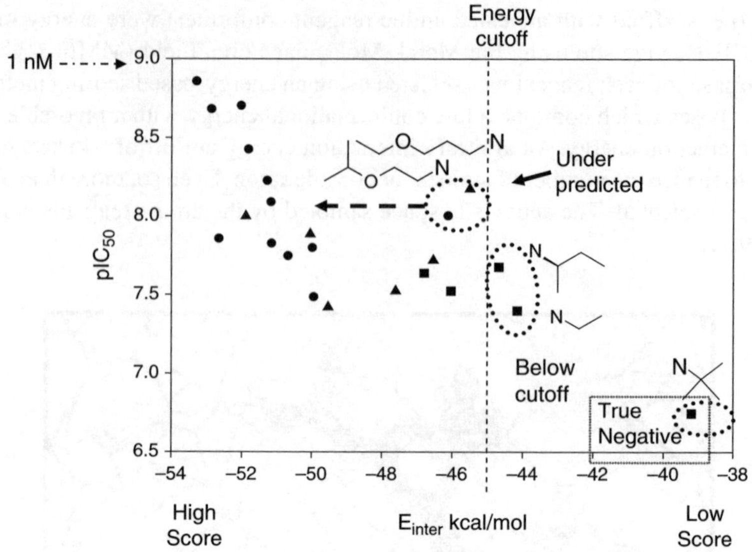

Figure 9.12 *Plot of observed activity versus energy score for compounds sampling the S1′ pocket of BACE-1. (•) Compounds containing amine reagents which were incorporated prior to the study; (■) compounds containing amine reagents selected independent of the scoring; (▲) compounds containing amine reagents selected based on scoring.*

Figure 9.12). In addition, our arbitrary cut-off for the energy score may have missed some interesting inhibitors, e.g. the ethyl- and isobutylamines which are labeled as 'Below cutoff' in Figure 9.12. However, the score versus activity trend is clear, including a 'true negative' control, the *tert*-butylamine reagent, which was predicted (and subsequently observed) to yield an inhibitor with very low BACE-1 activity. In addition to demonstrating the utility of an '*in situ*' virtual reagent selection, this study served to validate the hypothesis that the BACE-1 S1′ pocket is large, open and promiscuous, as a variety of reagents led to potent BACE-1 inhibitors.

While BACE has been a favorite target for modelers as described above, the active site of BACE has been discovered to have a high degree of flexibility, making virtual screening particularly challenging. As seen in Figure 9.13, there are significant differences in several loop conformations. It has been shown that the choice of P2/P3 substituents can affect the the so-called 10s loop. Wild-type BACE exhibits 10s loop up, with an S10–T232 distance >8 Å.[38] If the P3 substituent is large, then the 10s loop will remain up. If the P3 substituent is small and the P2/P3 linker interacts with the T232 hydroxyl, then the 10s loop will remain up and activity will be compromised. However, if the P3 substituent is small and the P2/P3 linker does not interact with the T232 hydroxyl, then the 10s loop will be down, creating a more effective binding pocket for the small substituent.[39]

Capturing this protein flexibility computationally is required to more accurately score virtual molecules to prioritize synthesis of compounds. Figure 9.14 depicts the correlation between binding energy calculated using MMFF[36] and measured pIC_{50} for a diverse set of ligands. The correlation is very high ($R^2 = 0.89$) if the most appropriate 10s loop position is chosen; however, if one uses only 10s up (down) the R^2 is reduced to 0.32 (0.01).[33]

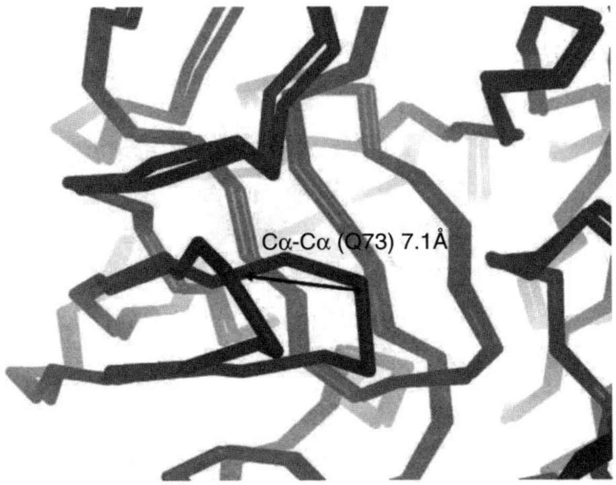

Figure 9.13 *Highlighting the large 10s loop movement in the BACE-1 active site using crystal structures 1W50 and 1W51.*

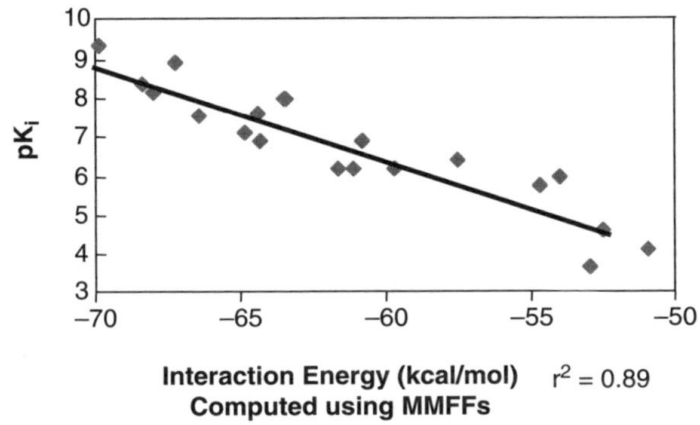

Figure 9.14 *Correlation between binding energy calculated using MMFFs*[36] *and measured Ki for a diverse set of ligands in P3 pocket.*

These results were used to screen a virtual database of primary amine fragments to enhance binding affinity in the P3 pocket using a related scaffold as shown in Figure 9.15. The virtual screening was quickly accomplished through Merck's in-house web-based tools described in Section 9.6 and was in fact a test of these tools.

In the present BACE-1 example, over 9000 amine fragments were combined with the scaffold in Figure 9.15. Up to 25 conformations for each fragment were sampled with the resulting ~25 000 molecules individually energy minimized and scored in two conformations of the BACE-1 active site, one with the 10s loop up and the other with the 10s loop down. In this retrospective analysis, MFCD00041323, the known active component, ranked 51 out of the >9000 candidates with a K_i of 42.4 nM.[40] In this case the best fit was with the 10s loop up, which is consistent with the crystal structure shown in Figure 9.16, pdb code 2IRZ.

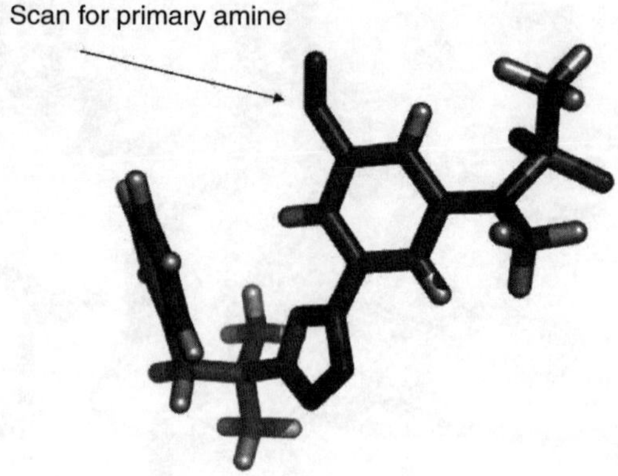

Figure 9.15 *Scaffold for scanning amine fragments for P3 pocket.*

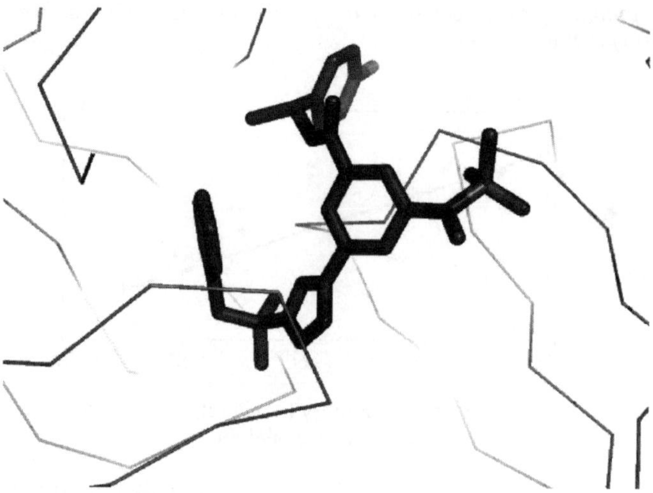

Figure 9.16 *Best fit for targeted active with the 10s loop up is consistent with the crystal structure shown (pdb code 2IRZ).*

The in-house web-tool is described in the next section, which enables modelers and chemists to perform such targeted library designs, especially with respect to scanning of small fragments for enhanced ligand binding affinity.

9.6 Virtual Library Tool Kit

A number of library design tools have been developed by Merck's molecular modeling group to facilitate the design of compound libraries for various purposes, such as small compound libraries in support of specific medicinal chemistry projects or large compound sets to augment Merck's internal sample collection. An important part of the library design process is the evaluation of the chemical properties of the available fragments or reagents and their impact on the fully synthesized library. In addition to chemical property distributions, novelty and diversity may be considerations when spanning chemical space or developing molecules to elucidate structure–activity relationships (SARs).

With the goal of making library design tools widely available to medicinal chemists throughout Merck's Research Laboratories, a web-based library design platform was developed. The web-based method allows users from anywhere on the Merck intranet to access the latest versions of the code without downloading and installing specialized software and allows the developers to restructure and optimize code in response to user feedback and system load.

These web-based library design tools are collectively known as the Virtual Library Tool Kit (VLTK), comprising four major components: (i) Reagent Selector (RS) (ii) Synthon Analysis (SA), (iii) Library Enumeration (LibE) and (iv) Library Analysis (LA). These have been discussed in detail elsewhere.[41–43] Considerable amounts of research and resources were devoted to developing appropriate functional queries and selecting an optimal set of reagents, resulting in tools which distinguish RS and SA from earlier efforts. Seemingly trivial queries such as 'find all primary amines' are found to yield different results when using MDL's ISIS[44] and Daylight's SMARTS.[45, 46] A comprehensive discussion of the salient points in forming typical medicinal chemistry queries is available.[41, 42] A pictorial methodology for selecting reagents based on synthon properties was also developed to facilitate fragment selection.[42, 43] Synthons are usually defined as the fragments of the reagents remaining in the final products. However, for our purposes it is also important to define a point of reference for determining how the synthon will impact the properties of the final products. This point of reference will be the atom(s) of the synthon participating in the newly formed bond(s). In the virtual fragment scan for primary amines in the second BACE-1 example above, the primary amine of the fragment is the reference atom. By combining our unique visual approach for synthon properties with synthon structural clustering, one may quickly select a set of fragments with desired properties, e.g. diversity, molecular weight, number of hydrogen bond acceptors and donors, etc., from the universe of potential candidates. An additional complication arises for multi-dimensional libraries where the properties of the final compounds result from two or more synthetic groups. In this case, the universe of each synthetic group may be chosen so that the final distributions of properties are within acceptable limits of the user-specified constraints. This is accomplished in VLTK through the implementation of the GLARE algorithm recently published by Truchon and Bayly.[47]

The library design tools simplify the process of searching the available reagent databases for fragments, selecting the most interesting subset, performing the virtual library enumeration and evaluating the final virtual library properties. If the library also has a specific target, as in the present example, the VLTK tools also assist conformation generation, docking and scoring.

Fragment selection is accomplished with the RS component of the VLTK. While the selection of fragments can be a time-consuming process, prone to error and missteps, RS allows the user quickly to select fragments with a 'point and click' approach. Intrinsic filtering tools provide the user with fragments sets that have a high likelihood of being available at a reasonable cost in the quantity and purity needed. This implementation of VLTK's user-directed fragment selector is unique in its focus on the structure and properties of the synthons, which come together in a single graphical display (see Figure 9.17) found in the SA component of VLTK. Synthon analysis presents a comprehensive view of the property space covered by the fragments simultaneously with structural similarity. The fragments may be chosen based on chemical property coverage rather than solely on some measure of dissimilarity or cluster membership. The interactive nature of the SA interface ensures that the user is able to assess quickly the chemical property space covered by a fragment, compare it with the coverage of previous selections and contrast it with the chemical diversity of the available universe of fragments. Indicators for the vendor, price and availability for each fragment are also provided.

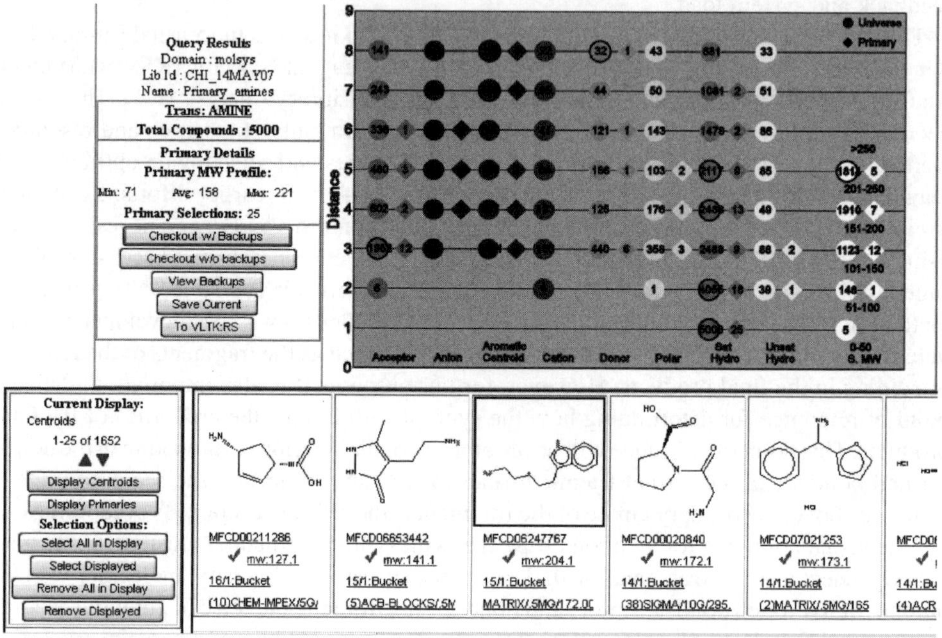

Figure 9.17 *A snapshot of Synthons Analysis for the primary amine query.*

In Figure 9.17, a snapshot of SA for the primary amine query is given. This input to SA is the output from the RS tool where both the ACD and an internal proprietary database of fragments were searched.[35] As can be seen, a list of 5000 low molecular weight primary amines are returned as a direct result of the constraints employed in RS. The 5000 molecules were clustered into 1652 sets. The structure of a representative fragment from each cluster is depicted in the film strip at the bottom of Figure 9.17, sorted by decreasing cluster size. With each structure is given the unique molecular ID, molecular weight, size of the cluster and vendor information. In addition to the cluster size, the number of members in the selected list, termed primary selections, is indicated next to the cluster size. By way of illustration, the representative molecules for the 25 largest clusters have been selected as primary fragments. However, for virtual screening all the fragments would be used. To clarify the data in the film strip, the ACD compound MFCD00211266 has a synthon molecular weight of 127.1 Da, is a primary selection, and is in a cluster (bucket) of 16 members. There are 10 vendors that offer this molecule, with Chem-Impex offering it in 5 g quantities.[48] The chemical properties are displayed in the upper right graphical window, where the number of acceptors, anions, aromatic centroids, cations, donors, polar, saturated hydrocarbons and unsaturated hydrocarbons are shown. The chemical properties of an example molecule containing one acceptor, one donor, two saturated hydrocarbons and three aromatic centroids is shown in Figure 9.18 for illustration. Each fragment in the database is characterized in a similar manner with the results depicted in Figure 9.17. In this figure, the vertical axis of the upper right graph, refers to the number of bonds the property resides from the reference atom, the primary nitrogen in the present example of amine fragments. The last pair of columns on this graphical object displays the distribution of synthon molecular weights. For each property, there are two columns, displayed as circles and diamonds, which represent the properties of the universe of 5000 molecules and the 25 primary selections, respectively. The numbers enclosed by the circle or diamond indicate the number of molecules with the associated property/bond count. Each circle and diamond is also dynamically linked to the film strip, so by clicking on any of these graphical objects, the film strip will be populated with the molecules contributing to its count. The reverse is also true: by clicking on a molecule in the film strip, as is the case for the third molecule, MFCD06247767, the properties in the graphic window corresponding to this molecule are outlined in a heavier line. In this way, the structural and chemical space of all fragments may be quickly navigated. This visual approach coupled with the interactivity of the web page allows users to better optimize the fragment selection process.

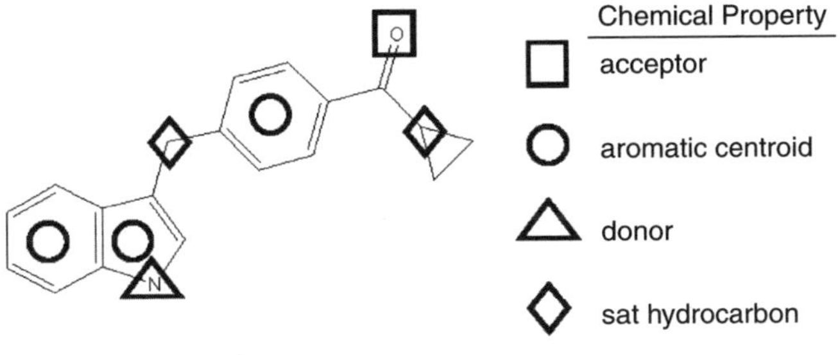

Figure 9.18 *The chemical properties of an example synthon containing one acceptor, one donor, two saturated hydrocarbons and three aromatic centroids.*

Once the chemistry, scaffold and fragments have been determined, the Library Enumeration (LibE) component of VLTK will fully enumerate the library in either 2D or 3D.[43] In many cases, 2D enumeration will be sufficient for analysis since many of the physical properties and even QSAR models often rely only on connection tables. However, in the present example a complete 3D enumeration is required for subsequent docking and scoring in the BACE-1 active site. To preserve the position of the scaffold shown in Figure 9.15, with respect to the active site, the 3D coordinates are written directly into a molecular structure file that is then read into LibE. All fragments are added to the scaffold without changing the scaffold position. Molecules of the fully enumerated molecule will have the correct position and orientation for initiating docking and scoring.

A few physical property distributions of the fully enumerated library are shown in Figure 9.19. The molecular weight, hydrogen bond acceptors (HBAs), hydrogen bond donors (HBDs) and AlogP98[10] are depicted along with a representation of the MDDR[49] and Merck's internal compound collection for reference. In the web application the dotted bars, representing the proposed library, are dynamically linked to the fragments which contribute to molecules with the specific property, sorted by frequency of occurrence. In this way, users can quickly identify which fragments should be replaced or dismissed based on their effect on final property distributions. The virtual library for BACE-1 has a significantly higher molecular weight and HBA and HBD counts and lower calculated AlogP98 than typical drug-like molecules. These properties reflect the size of the BACE-1 active site and hint at difficulties for designing brain-penetrant compounds.

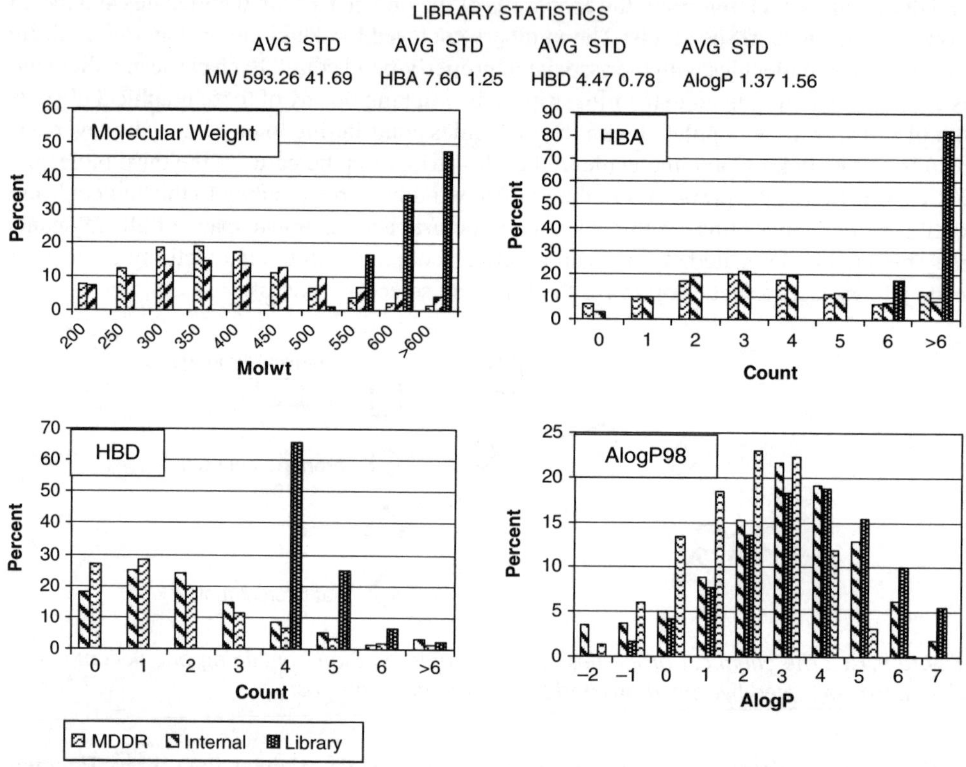

Figure 9.19 Physical property distributions of the fully enumerated virtual library. The average (AVG) and standard deviation (STD) are also provided for each of the four property distributions.

9.7 Conclusion

Although companies are providing fragment-based screening services using a number of empirical techniques, virtual screening of a database of fragments comes with its own unique set of problems. In addition to all the problems that accompany virtual screening of drug-like molecules, e.g. conformational flexibility of ligand/protein and appropriate scoring functions, small fragments exhibit relatively weaker affinity and less specificity. Commercially available fragment libraries are constructed with generally similar objectives such as compliant with Astex's rule of three,[10] yet they exhibit significant differences in their property distributions. Perhaps more important for fragment library design is the intended purpose, since solubility requirements can vary greatly for crystallography, NMR and surface plasmon resonance detection methods. Similarly for virtual screening, fragment databases may be designed with a particular focus. In the examples above, a fragment database of readily available reagents was transformed *in silico* into synthons, reflecting how they would be used synthetically to construct larger targeted drug-like molecules. By utilizing a docked fragment or scaffold in an active site, the difficulty in virtual screening over available fragments to enhance binding affinity is significantly reduced by permitted chemistry and also protein–ligand interactions. Virtual screening of fragment libraries is most successful when the active site is well characterized and the binding modes of the fragments are constrained. By way of example, both retrospective and prospective virtual scans of amine fragments identified active BACE-1 compounds that utilized favorable interactions in the P1′ and P3 pockets, respectively. Another critical component for fragment scanning is the availability of good scoring functions, parameterized for each pocket separately. Although virtual screening of fragments remains a challenging area of research, success can be achieved through application of both steric constraints in the active site and synthetic suitability of fragments.

References

[1] Rees D.C., Congreve M., Murray C.W., Carr R., Fragment-based lead discovery. *Nat. Rev. Drug Discov.* 2004, **3**, 660–672.

[2] Sherman W., Using fragments to couple ligand- and structure-based approaches. Presented at the ACS Meeting, Boston, MA, 18–23 August 2007.

[3] Abad-Zapatero C., Metz J., Ligand efficiency indices as guideposts for drug discovery. *Drug Discov. Today* 2005, **10**, 464–469.

[4] Hopkins A.L., Groom C.R., Alex A., Ligand efficiency: a useful metric for lead selection. *Drug Discov. Today* 2004, **9**, 430–431.

[5] Warren G.L., Andrews C.W., Capelli A.-M., Clarke B., LaLonde J., Lambert M.H., Lindvall M., Nevins N., Semus S.F., Senger S., Tedesco G., Wall I.D., Woolven J.M., Peishoff, C.E., Head M.S., A critical assessment of docking programs and scoring functions. *J. Med. Chem.* 2006, **49**, 5912–5931.

[6] Sherman W., Schrödinger (personal communication).

[7] Tari L.W., Jennings A.J., McRee D.E., Use of high-throughput crystallography and in silico methods for structure-based drug design. In *Industrial Proteomics: Applications for Biotechnology and Pharmaceuticals*, ed. Figeys D., John Wiley & Sons, Inc., Hoboken, NJ, 2005, pp. 107–129.

[8] Verheij H.J., Leadlikeness and structural diversity of synthetic screening libraries. *Mol. Diversity* 2006, **10**, 377–388.

[9] Maybridge, Trevillet, Tintagel, Cornwall, http://www.maybridge.com/Images/pdfs/Ro3frag.pdf.

[10] Ghose A.K., Viswanadhan V.N., Wendoloski J.J., Prediction of hydrophobic (lipophilic) properties of small organic molecules using fragmental methods: an analysis of ALOGP and CLOGP methods. *J. Phys. Chem. A* 1998, **102**, 3762–3772.

[11] Clark D.E., Rapid calculation of polar molecular surface area and its application to the prediction of transport phenomena. 1. Prediction of intestinal absorption. *J. Pharm. Sci.* 1999, **88**, 807–814; Clark D.E., Rapid calculation of polar molecular surface area and its application to the prediction of transport phenomena. 2. Prediction of blood–brain barrier penetration. *J. Pharm. Sci* 1999, **88**, 815–821.

[12] Carr R.A.E., Congreve M., Murray C.W., Rees D., Fragment-based lead discovery: leads by design. *Drug Discov. Today* 2005, **10**, 987–992.

[13] Klopman G., Wang S., Balthasar D.M., Estimation of aqueous solubility of organic molecules by the group contribution approach. Application to the study of biodegradation. *J. Chem. Inf. Comput. Sci.* 1992, **32**, 474–482.

[14] Card G.L., Blasdel L., England B.P., Zhang C., Suzuki Y., Gillette S., Fong D., Ibrahim P.N., Artis D.R., Bollag G., Milburn M.V., Kim S.-H., Schlessinger J., Zhang K.Y.J., A family of phosphodiesterase inhibitors discovered by cocrystallography and scaffold-based drug design. *Nat. Biotechnol.* 2005, **23**, 201–207.

[15] Baurin N., Aboul-Ela F., Barril X., Davis B., Drysdale M., Dymock B., Finch H., Fromont C., Richardson C., Simmonite H., Hubbard R.E., Design and characterization of libraries of molecular fragments for use in NMR screening against protein targets. *J. Chem. Inf. Comput. Sci.* 2004, **44**, 2157–2166.

[16] Shuker S.B., Hajduk P.J., Meadows R.P., Fesik S.W., Discovering high-affinity ligands for proteins: SAR by NMR. *Science* 1996, **274**, 1531–1534.

[17] Evotec, Hamburg, http://www.evotec.com/en/csi/index.aspx.

[18] Braisted A., Oslob J., Delano W., Hyde J., McDowell R., Waal N., Yu C., Arkin M., Raimundo B., Discovery of a potent small molecule IL-2 inhibitor through fragment assembly. *J. Am. Chem. Soc.* 2003, **125**, 3714–3715.

[19] Annis D.A., Athanasopoulos J., Curran P.J., Felsch J.S., Kalghatgi K., Lee W.H., Nash H.M., Orminati J.P.A., Rosner K.E., Shipps G.W., Thaddupathy G.R.A., Tyler A.N., Vilenchik L., Wagner C.R., Wintner E.A., An affinity selection–mass spectrometry method for the identification of small molecule ligands from self-encoded combinatorial libraries. Discovery of a novel antagonist of *E. coli* dihydrofolate reductase *Int. J. Mass Spectrom.* 2004, **238**, 77–83.

[20] Glaser, V., When smaller is better. *Bio-IT World*, July 2007.

[21] Cambridge Healthtech Institutes Drug Discovery Chemistry 2007, La Jolla, CA, 13–16 May 2007.

[22] Baurin N., Mozziconacci J.-C., Arnoult E., Chavatte P., Marot C., Morin-Allory L., 2D QSAR consensus prediction for high-throughput virtual screening. An application to COX-2 inhibition modeling and screening of the NCI database. *J. Chem. Inf. Comput. Sci.*, 2004, **44**, 276–285.

[23] Murray C.W., Callaghan O., Chessari G., Cleasby A., Congreve M., Frederickson M., Hartshorn M.J., McMenamin R., Patel S., Wallis N., Application of fragment screening by X-ray crystallography to β-secretase. *J. Med. Chem.* 2007, **50**, 1116–1123; Congreve M., Aharony D., Albert J., Callaghan O., Campbell J., Carr R.A.E., Chessari G., Cowan S., Edwards P.D., Frederickson M., McMenamin R., Murray C.W., Patel S., Wallis N., Application of fragment screening by X-ray crystallography to the discovery of aminopyridines as inhibitors of β-secretase. *J. Med. Chem.* 2007, **50**, 1124–1132.

[24] Hills I.D., Vacca J.P., Progress toward a practical BACE-1 inhibitor. *Curr. Opin. Drug Discov. Dev.* 2007, **10**, 383–391.

[25] Hartshorn M.J., Murray C.W., Cleasby A., Frederickson M., Tickle I.J., Jhoti H., Fragment-based lead discovery using X-ray crystallography. *J. Med. Chem.* 2005, **48**, 403–413.

[26] Jones G., Willett P., Glen R.C., *et al.* Molecular recognition of receptor sites using a genetic algorithm with a description of desolvation. *J. Mol. Biol.* 1995, **245**, 43–53.

[27] Jones G., Willett P., Glen R., *et al.* Development and validation of a genetic algorithm for flexible docking. *J. Mol. Biol.* 1997, **267**, 727–748.
[28] Eldridge M.D., Murray C.W., Auton T.R., *et al.* Empirical scoring functions. 1. The development of a fast empirical scoring function to estimate the binding affinity of ligands in receptor complexes. *J. Comput.-Aided Mol. Des.* 1997, **11**, 425–445.
[29] Verdonk M.L., Cole J.C., Hartshorn M., *et al.* Improved protein–ligand docking using GOLD. *Proteins* 2003, **52**, 609–623.
[30] Polgar T., Keserü G.M., Ensemble docking into flexible active sites. Critical evaluation of FlexE against JNK-3 and β-secretase. *J. Chem. Inf. Model.* 2006, **46**, 1795–1805.
[31] Vertex, Patent Application WO02/088101, 2002.
[32] Hajduk P.J., Greer J., A decade of fragment-based drug design: strategic advances and lessons learned. *Nat. Rev. Drug Discov.* 2007, **6**, 211–219.
[33] McGaughey G.B., Sheridan R.P., Bayly C.I., Culberson J.C., Kreatsoulas C., Lindsley S., Maiorov V., Truchon J.-F., Cornell W., Comparison of topological, shape and docking methods in virtual screening. *J. Chem. Inf. Model.* 2007, **47**, 1504–1519.
[34] Baxter E.W., *et al.*, 2-Amino-3,4-dihydroquinazolines as inhibitors of BACE-1 (β-site APP cleaving enzyme), use of structure based design to convert a micromolar hit into a nanomolar lead. *J. Med. Chem.* 2007, **50**, 4261–4264.
[35] *MDL Available Chemicals Directory (MDL ACD).* MDL Information Systems, San Leandro, CA.
[36] Halgren T.A., Merck Molecular Force Field. I. Basis, form, scope, parameterization and performance of MMFF94. *J. Comput. Chem.* 1996, **17**, 490–519.
[37] Holloway M.K., McGaughey G.B., Coburn C.A., Stachel S.J., Jones K.G., Stanton E.L., Gregro A.R., Lai M.-T., Crouthamel M.-C., Pietrak B.L., Munshi S.K., Evaluating scoring functions for docking and designing β-secretase inhibitors. *Bioorg. Med. Chem. Lett.* 2006, **17**, 823–827.
[38] Hong L., Koelsch G., Lin X., Wu S., Terzyan S., Ghosh A.K., Zhang X.C., Tang J., Structure of the protease domain of memapsin 2 (β-secretase) complexed with inhibitor. *Science* 2000, **290**, 150–153.
[39] Stachel S.J., Coburn C.A., Steele T.G., Crouthamel M.-C., Pietrak B.L., Lai M.-T., Holloway M.K., Munshi S.K., Graham S.L., Vacca J.P., Conformationally biased P3 amide replacements of β-secretase inhibitors. *Bioorg. Med. Chem. Lett.* 2006, **16**, 641–644.
[40] Rajapakse H.A., Nantermet P.G., Selnick H.G., Munshi S., McGaughey G.B., Lindsley S.R., Young M.B., Lai M.-T., Espeseth A.S., Shi X.-P., Colussi D., Pietrak B., Crouthamel M.-C., Tugusheva K., Huang Q., Xu M., Simon A.J., Kuo L., Hazuda D.J., Graham S., Vacca J.P., Discovery of oxadiazoyl tertiary carbinamine inhibitors of β-secretase (BACE-1). *J. Med. Chem.* 2006, **49**, 7270–7273.
[41] Kraker B, Chakravorty S.J., Mosley R., Culberson J.C., Feuston B.P., Sheridan R.P., Conway J.F., Forbes J.K., Kearsley S.K., Virtual library tool kit: Comparison of functional group queries with Daylight and ISIS software. Unpublished work.
[42] Mosley R.T., Culberson J.C., Kraker B., Feuston B.P., Sheridan R.P., Conway J.F., Forbes J.K., Chakravorty S.J., Kearsley S.K., Reagent selector: using synthon analysis to visualize reagent properties and assist in basis set selection. *J. Chem. Inf. Model.* 2005, **45**, 1439–1446.
[43] Feuston B.P., Chakravorty S.J., Conway J.F., Culberson J.C., Forbes J.K., Kraker B., Lennon P.A., Lindsley C., McGaughey G.B., Mosley R., Sheridan R.P., Valenciano M., Kearsley S.K., Web enabling technology for the design, enumeration, optimization and tracking of compound libraries. *Curr. Top. Med. Chem.* 2005, **5**, 773–783.
[44] MDL Information Systems, San Leandro, CA, www.mdli.com.
[45] Weininger D., SMILES, a chemical language and information-system. 1. Introduction to methodology and encoding rules. *J. Chem. Inf. Comput. Sci.* 1988, **28**, 31–36.

[46] Weininger D., James C.A., *Daylight CIS Systems Administration*, Daylight CIS, Irvine, CA, 1993.

[47] Truchon J.F., Bayly C.I., GLARE: a new approach for filtering large reagent lists in combinatorial library design using product properties. *J. Chem. Inf. Model.* 2006, **46**, 1536–1548.

[48] Chem-Impex International, Wood Dale, IL,www.chemimpex.com.

[49] *MDL Drug Data Report, Version 2005.1*, MDL Information Systems, San Leandro, CA, 2005.

10

Capture Methods for Fragment-based Discovery

Stig K. Hansen and Daniel A. Erlanson

10.1 Introduction

During the past decade, fragment-based lead discovery techniques have become more widely used and are now often applied in conjunction with traditional high-throughput biochemical screening (HTS). While HTS remains a very productive method for finding chemical ligands, fragment-based lead discovery, in which leads are built progressively by expanding or combining small pharmacophores, offers unique opportunities to discover novel and unusual chemical starting points.[1, 2] Fragment-based approaches can identify binding modes and interactions that may not be accessible to larger preassembled molecules that often contain multiple pharmacophores. In addition, fragment-based discovery often leads to the synthesis of compounds that are non-obvious and structurally more diverse, thereby complementing the chemotypes that may be found in HTS libraries. The main challenge with this discovery approach lies in identifying and assembling the fragments. Here we review the concept of covalent and dative fragment capture, where a bond is used to stabilize the interaction between a weak fragment and a target protein, and describe how these techniques can facilitate fragment-based lead discovery.

Fragment-Based Drug Discovery: A Practical Approach Edited by Edward R. Zartler and Michael J. Shapiro

10.2 Principle of Capture Methods

The concept of fragment capture can be traced to the decades-old technique of affinity labeling[3, 4] that entails modifying a ligand with a reactive functionality to label selectively the target protein. Although historically this was often done to elucidate the biology of protein–small molecule interactions, more recently the concept has been applied to lead discovery.[5–7] Covalent capture methods have been particularly useful in the application of fragment-based lead discovery by facilitating the detection of small drug fragments that often have low binding affinity due to limited interactions with the target protein. Identifying ligands with high micromolar to low millimolar binding affinities using biochemical or binding assays can be complicated by issues such as protein denaturation, binding at multiple sites on the protein and compound aggregation.[8, 9] Covalent capture methods can overcome these complications and facilitate fragment discovery by forming a covalent bond between the fragment and the protein.

The principle of covalent capture methods is illustrated in Figure 10.1. A target protein with a native or engineered reactive functional group X is reacted with a collection of small fragments, each of which contains a complementary reactive functional group Y. If the small molecule binds to the protein in the vicinity of X, then the two functional groups can react to form a covalent bond and the resulting covalent complex can be characterized by methods such as mass spectrometry. The reaction between X and Y can be either reversible (such as thiol–disulfide exchange or imine formation) or irreversible (such as epoxide opening or halide displacement). The reversible reaction can be controlled thermodynamically, which offers the advantage that fragments can be readily selected based on binding affinity. In the case of the irreversible reaction, the control will be largely kinetic and the reactivity must be low enough that selection is not dominated by fragment reactivity. Key advantages of covalent capture methods are the detection of weak ligands and, assuming the position of

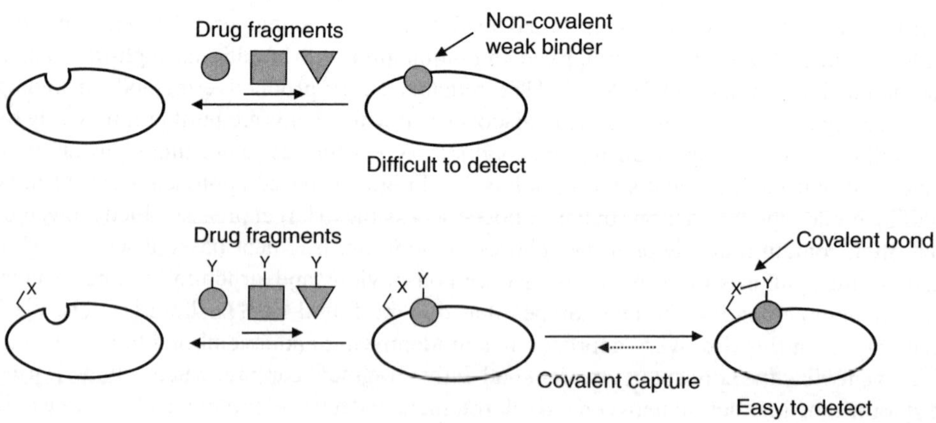

Figure 10.1 *Principle of covalent capture methods. Drug fragments typically have weak binding affinity and can therefore be difficult to detect. By introducing two reactive groups, X and Y, a fragment that binds in the vicinity of X can be captured covalently by the protein target and easily identified by mass spectrometry.*

X is known, information about binding location. This can be exploited to direct the screen to a particular site on a protein. This location is typically the active site, but the method can also be used to investigate allosteric sites or protein–protein binding interfaces. In the following, we will describe various practical approaches to covalent capture and provide examples of their application in studying protein–ligand interactions and drug discovery.

10.3 Reversible Capture Methods

10.3.1 Tethering[10]

Tethering is a capture method that is based on thiol–disulfide exchange, where a free thiol presented by a cysteine residue on a protein surface can react to form a disulfide with a disulfide-containing small fragment.[11, 12] In this method, X is a cysteine residue and Y is a disulfide (Figures 10.1 and 10.2A) and if Y binds close to X, a disulfide bond will form between X and Y. The resulting protein–small molecule conjugate can be detected by mass spectrometry (MS) and the identity of the conjugated fragment can be confirmed based on the mass shift relative to unmodified protein. For screening purposes, fragments of different molecular masses can be pooled and captured hits can be identified based on their unique mass shifts. Tethering is well suited for site-directed ligand discovery, as new cysteine residues can be engineered into the target protein to direct the screen towards a site of interest. Importantly, the sensitivity of the Tethering reaction can be adjusted by addition of a reducing agent, thereby compensating for variability in reactivity

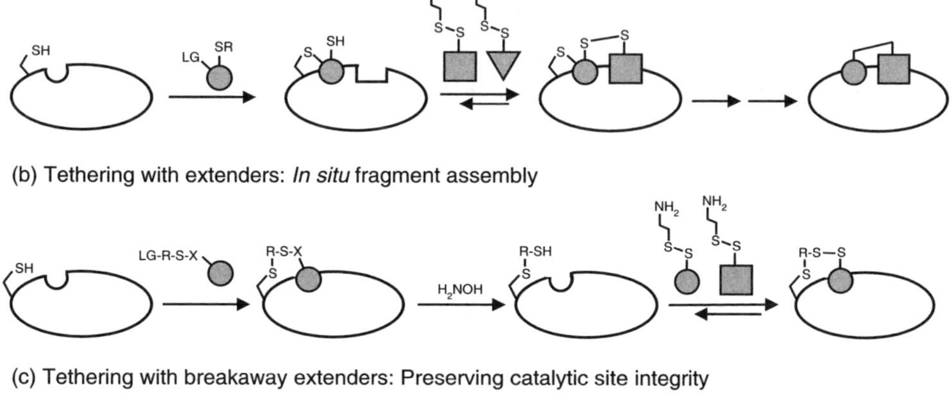

Figure 10.2 *Different versions of Tethering that have been used for covalent capture.*

among different cysteine residues. This method was originally developed and validated with the discovery of nanomolar inhibitors of the enzyme thymidylate synthase[11] and it has since been applied to many other targets, including protein–protein interactions, catalytic sites, inactive conformations of enzymes or even regions outside a catalytic site. Some of these applications are described in the following section.

10.3.2 Tethering at Protein–Protein Interfaces

The binding of interleukin-2 (IL-2) to its receptor constitutes an early event in T-cell activation and has been pursued as a target for novel immuno-modulators.[13] A small-molecule antagonist of IL-2 (Ro26-4550) was first reported by researchers at Hoffman-LaRoche, who identified a compound that binds to IL-2 with low micromolar potency and blocks the binding to the receptor.[14] This small-molecule binding site was used as a starting point for designing a series of Tethering screens for IL-2. Researchers at Sunesis solved the crystal structures of IL-2 in the absence or presence of Ro26-4550, revealing an elongated binding site composed of a highly adaptive hydrophobic subsite and a rigid polar subsite.[15] This structural information was then used to guide the engineering of a series of cysteine residues flanking the Ro26-4550-binding site. Tethering was then used to interrogate the small-molecule binding sites and revealed a number of fragments that can be divided into two classes. Hydrophobic ligands constituted the majority of hits and were identified from a variety of positions around the adaptive subsite. A smaller set of hits bound around a polar subsite, many of which shared similarity with the polar guanidine pharmacophore in Ro26-4550. This is consistent with the notion that hydrophobic interactions are more promiscuous, whereas highly directional polar interactions require precise positioning and produce fewer hits.

Several of the ligands selected around the adaptive subsite of IL-2 were small aromatic carboxylic acids, a pharmacophore that had not previously been explored for this site.[16] Using molecular modeling to guide compound design, aromatic carboxylic acids were fused to a distant relative of Ro26-4550.[17] A small library of 20 compounds yielded eight compounds with sub-micromolar affinity and the most potent compound inhibited the binding of IL-2 to IL-2Rα with an IC_{50} of 60 nM, 50-fold more strongly than the parent compound (Figure 10.3A). The structure of this compound bound to IL-2 was subsequently determined, revealing further movements around the adaptive subsite to accommodate the aromatic acid.[18] Together, these results illustrate the utility of site-directed ligand discovery to identify ligands to a highly flexible protein site that is not amenable to molecular modeling, and the ability to use these ligands to advance medicinal chemistry.

10.3.3 Probing a Ligand Binding Site Using Disulfide Capture and Tethering

G-protein-coupled receptors (GPCRs) constitute a large class of transmembrane receptors that mediate signaling by a plethora of extracellular ligands. GPCRs are unique in their sensitivity to even subtle changes in ligand structure, allowing an agonist to be readily converted into an antagonist. The basis for this triggering mechanism is only beginning to become understood at the molecular level. The complement factor 5a GPCR (C5aR) belongs to this category and its peptide ligands can switch from agonism to antagonism with single amino

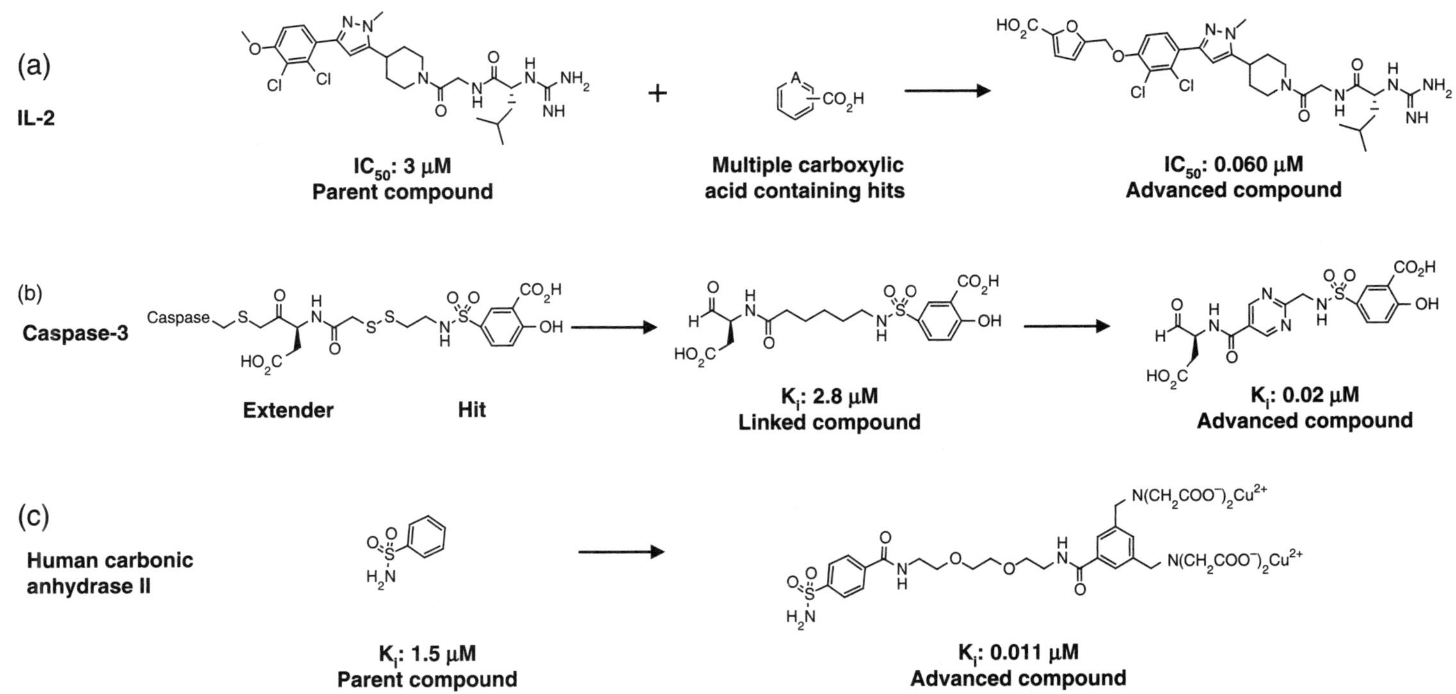

Figure 10.3 *Examples of starting points and progressed compounds using various capture methods.*

acid substitutions. Disulfide capture has recently been used to evaluate the binding of these ligands to the C5a receptor.[19] Cysteine residues were inserted into peptide ligands and in various positions of the C5a receptor and tested for activity. Binding and receptor activation studies showed that the disulfide-captured ligands maintained the agonism/antagonism of their noncovalent parents and confirmed the location of the activation site to within a helical triad previously identified through site-directed mutagenesis. Since disulfide capture compensates for reduction in binding affinity, the size and complexity of ligands can be greatly reduced, allowing a high-resolution mapping of key agonistic/antagonistic motifs and complementing mapping by traditional mutational analysis. Four cysteine insertions in the C5a receptor were subsequently screened by Tethering to identify small-molecule ligands.[20] A subset of these ligands showed agonism or antagonism when tethered to either of two cysteines within the seven-helix bundle. Mutation of Ile116, which is located within 10 Å of the Tethering site, had a significant effect on receptor activation for some ligands. Interestingly, replacing Ile116 with Ala increased the degree of agonism for many ligands, whereas replacing it with a larger Trp residue led to a decrease in agonism and also the number ligands that could serve as agonists. This study illustrates the use of Tethering as a capture method to characterize functionally small ligands prior to selecting the preferred fragments for optimization.

10.3.4 Tethering with Extenders

The examples listed above illustrate the power of Tethering to identify readily a variety of ligands, but, as with other fragment discovery methods, linking and/or expanding these ligands to increase affinity often requires extensive crystallography, molecular modeling or trial and error. A second-generation version of Tethering, Tethering with extenders, provides a powerful solution to this problem by exploiting a known fragment as the basis for capturing a new fragment, essentially allowing the protein to guide the assembly of the two fragments.[21] In this method (Figure 10.2B), the protein is modified with a known ligand referred to as an 'extender,' which is a small fragment that contains both a reactive functionality and a thiol: the reactive functionality covalently labels the protein and the thiol is used for Tethering to capture a fragment that binds nearby. The resulting molecule contains two fragments connected with a disulfide linker that is subsequently replaced with a variety of other linkers and tested for activity in a bioassay. In effect, Tethering with extenders is an *in situ* assembly process that greatly accelerates and simplifies hit identification and permits rapid expansion from a simple fragment to a more elaborate molecule. Moreover, the extender itself can be derived from Tethering.

Sunesis researchers used this approach to discover low micromolar inhibitors of the pro-apoptotic cysteine protease caspase-3. Simple rigidification of the flexible linker connecting the ligand and extender improved potency by more than an order of magnitude, to $K_i = 200$ nM,[21] as did decoration of the linker.[22] Further medicinal chemistry achieved another order of magnitude improvement in potency[23] (Figure 10.3B). The same approach was applied to the related protease caspase-1 (interleukin-1 converting enzyme or ICE) and a series of low nanomolar inhibitors was discovered.[24]

A variant of Tethering with extenders, termed 'breakaway Tethering,' has been developed for enzymes with catalytic sites that do not tolerate the insertion of a cysteine residue due to disruption of structural or functional integrity. Protein tyrosine phosphatases (PTPs) have relatively compact catalytic sites that catalyze the removal of phosphate groups from tyrosine and are involved in many aspects of cell signaling. PTP-1B, for example, is a highly attractive target for anti-diabetic and anti-obesity therapy.[25] In an effort to probe the intact active site of PTP-1B, a cysteine residue was introduced well outside the active site and coupled to a 'breakaway extender', designed based on the insulin receptor peptide substrate. This 'breakaway' extender contains an active site ligand, a 'breakaway' thioester linkage and an irreversible alkylating group designed to react with the newly introduced cysteine (Figure 10.2C). Once the protein has been reacted with the alkylating group, the thioester linkage can be cleaved to release the active site ligand and expose a thiol group for Tethering that is positioned just above the catalytic site. The benefit of using a substrate mimetic 'breakaway extender' is twofold: it provides a recognition sequence to orient the extender properly with respect to a distant cysteine, and it also protects the highly reactive cysteine located in the active site. When this method was applied to PTP-1B, a number of small, negatively charged ligands were identified, some of which had not been previously reported as phosphotyrosine mimetics.[26]

10.3.5 Targeting the Active Site of PTP-1B with Reversible Electrophiles

The catalytic cysteine-215 of PTP-1B is highly reactive toward oxidation and electrophiles and as such it is a tempting target for fragment discovery. Ockey and Gadek assembled a set of 19 reversible electrophiles, such as aldehydes, nitriles and boronic acids.[27] They then used electrospray ionization mass spectrometry to look for one-to-one complexes; and three of the compounds were found to form covalent complexes. The dissociation constants ranged from 25 to 150 μM and one of the compounds was also able to inhibit PTP-1B with an IC_{50} of 60 μM.

10.3.6 Naturally Occurring Reactive Fragments

An interesting natural example of covalent capture of fragment-sized molecules has recently been reported. The ion-channel protein TRPA1 responds to a variety of stimuli, including spices and foods such as cinnamon, garlic and horseradish. Two groups independently demonstrated that this response involves covalent modification of three cysteine residues by electrophilic compounds contained in these foods.[28, 29] The active ingredients (cinnamaldehyde in cinnamon, diallyl disulfide in garlic and allyl isothiocyanate in mustard oil) are chemically distinct but react with cysteine residues (Figure 10.4A). They are also very small molecules and likely derive most of their effects from their reactivity rather than from noncovalent binding interactions. It would be an interesting challenge to try to improve the binding energy while decreasing the reactivity.

10.3.7 Discovery of Peptide Ligands Using Covalent Capture

In addition to finding small organic molecules that bind to a protein, covalent capture methods can identify peptides that interact with proteins. Kohda and colleagues used this approach to study the mitochondrial protein Tom20, an import receptor that recognizes an epitope on proteins targeted for the mitochondria.[30] Previous work had characterized this epitope as a five-residue peptide that assumes an amphiphilic helical conformation, and coarse sequence preferences had been worked out. However, Tom20 has both low affinity

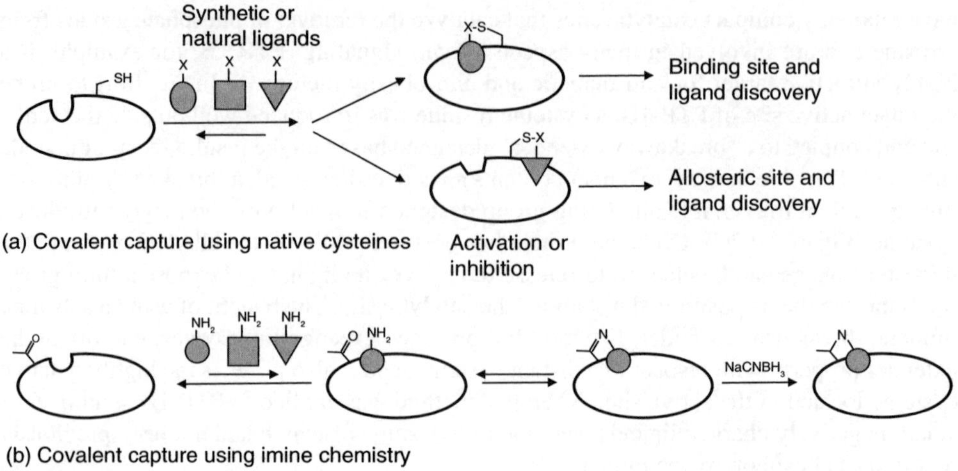

Figure 10.4 *Examples of covalent capture methods. (A) Covalent modification of native cysteines has been shown to modulate ion-channel activity and in the case of enzymes lead to allosteric inhibition. These findings can be important for structural-functional characterization or as starting points for drug discovery. (B) Reversible covalent capture using imine chemistry.*

and specificity for proteins and refining the sequence specificity through traditional methods proved intractable. Kohda and colleagues were able to apply covalent capture techniques to solve the problem. They used a single cysteine residue in the cytosolic domain of Tom20 to capture 19-residue peptides containing a C-terminal cysteine residue.[30] They made the peptide more amenable to noncovalent interactions by inserting a glycine residue as a spacer between the cysteine and the recognition elements of the peptide.

They screened seven libraries, each containing 19 peptides of different masses, and found the sequences that bound most tightly to Tom20. Matrix-assisted laser desorption/ionization time-of-flight mass spectrometry (MALDI-TOF MS) served as a readout that allowed the ionized peptides to be observed directly. Using this approach, Kohda and colleagues discovered that the recognition site spans six residues, not the previously established five, and were able to refine sequence preferences further.

10.3.8 Covalent Capture Using Imines

Linkers other than disulfides can also be used for reversible site-directed ligand discovery (Figure 10.4B). Imine bonds, which form reversibly in water, have been used for performing dynamic combinatorial chemistry on an enzyme.[31–33] In particular, Abell and colleagues took advantage of the fact that the protein aspartate decarboxylase (ADC) contains an N-terminal pyruvoyl group at the active site; the ketone of this reactive functionality can form imines with amine-containing molecules.[34] They reacted 55 primary amines with ADC along with the reductant sodium cyanoborohydride. This trapped the metastable imine as the stable secondary amine and the resulting complexes could be analyzed using MALDI-TOF MS. Nine of the 55 amines formed adducts with ADC at 2 mM. Two of these were decarboxylated, indicating that they were functioning as substrates. The binding SAR was consistent with the crystal structure of ADC.[35]

10.3.9 Dative Capture Using Metal Chelation

The same concept of covalent capture can also be applied to noncovalent interactions, such as metal–ligand bonds. In an interesting application of this concept, Mallik, Srivastava and coworkers have attached copper-chelating moieties to small ligands, such as aryl sulfonamides, for carbonic anhydrase.[36] The resulting molecules can then present copper to histidine residues located just outside the active site. In one example, this technique resulted in taking a 1.5 μM fragment to an 11 nM inhibitor[37] (Figure 10.3C). Crystallographic characterization of some of the inhibitors revealed that, in at least one case, the copper atom does interact with the targeted histidine residue while the sulfonamide interacts with the catalytic zinc in the active site.[38]

Metal chelation has also been applied in studies of the structural-functional basis for seven-trans-membrane receptor activation. Using the β_2-adrenergic receptor as a model system, Elling *et al.* introduced several point mutations to form a metal binding site between trans-membrane helices III, VI and VII and showed that the receptor could be activated by metal ions alone or metal ions chelated by phenanthroline or bipyridine.[39] Based on a homology model built over the rhodopsin receptor crystal structure, helices III, VI and VII would have to move inwards to form a metal chelation site, suggesting that movement of these helices is critical for receptor activation.

10.4 Irreversible Capture Methods

The reversible capture method for site-directed ligand discovery is more commonly used than irreversible approaches. While reversible approaches select ligands on the basis of inherent binding affinity, irreversible chemistries may select compounds with no inherent affinity because species that react faster will out-compete slower reacting species with higher affinity. Consider a ligand that contains two electrophiles, each of which reacts with a protein-bound nucleophile at a different rate. If the reactivity difference between the two electrophiles is sufficiently large, the protein is likely to react with the 'hotter' electrophile regardless of fragment orientation or affinity.

Over a decade ago, work on the enzyme aldolase reductase elegantly demonstrated this point. The noncovalent inhibitor alrestatin was modified to contain various electrophiles: α-chloroacetamide, α-bromoacetamide or α-iodoacetamide. Noncovalent interactions between inhibitors and protein would not have changed, but molecules behaved differently based on the electrophile: the weakest showed reversible inhibition, whereas the iodoacetamide displayed almost complete irreversible inhibition.[40] These results are an important warning: if a reaction is too facile, irreversible reactions can obscure true binding affinities.

Nonetheless, some research indicates that irreversible chemistries can be used for site-directed ligand discovery. Meares and colleagues constructed ligands containing metal chelates and reactive functionalities by engineering cysteine residues into an antibody that tightly binds metal chelates.[41] Using reactive functionalities such as haloacetamides and acrylamide, they were able to capture ligands even in very complex mixtures such as cell media. In a separate example, Belshaw and colleagues converted a potent reversible inhibitor into an irreversible one.[42] They added acrylamide moieties to analogs of cyclosporin A, which allowed cysteine-modified cyclophilin A to capture these moieties. As expected, the acrylamide-containing cyclosporin A irreversibly modified the engineered cyclophilin A but not wild-type protein lacking the additional cysteines.

Importantly, less reactive ligands may be more selective in complex systems where highly reactive ligands may react with a variety of proteins. Sames and colleagues showed this by attaching more than a dozen different electrophiles to benzenesulfonamide, a nanomolar inhibitor of the enzyme carbonic anhydrase.[43] Probes containing electrophiles such a vinyl sulfones bound to many different proteins, but probes containing a less-reactive epoxide reacted only with carbonic anhydrase, forming a one-to-one adduct with the enzyme. Furthermore, the reaction did not occur in the presence of other ligands that bind in the active site, demonstrating that the labeling was specific in crude cell extracts. Thus, if epoxides were used in a fragment library, epoxide opening could, like disulfides, be useful for site-directed ligand discovery.

Even disulfide bonds are not always reversible: use of a highly reactive methanethiosulfonate (MTS) reagent under nonreducing conditions gives complete disulfide formation with cysteine-containing proteins. Rosenberry and colleagues introduced a cysteine residue just outside the active site of acetylcholinesterase and functionalized it with several different MTS-derivatized ligands.[44] The length of the linker and the nature of the inhibitor resulted in different activity profiles for the modified enzyme and molecular modeling suggested that the different ligands bind to different areas of the large active site.

The reactivity of natural cysteines has recently been exploited to discover an allosteric site in the phosphatase PTP1B. While evaluating a series of cysteine-reactive probes for their potential to selectively label the active site cysteine, Hansen *et al.* discovered that the small molecule 4-(aminosulfonyl)-7-fluoro-2,1,3-benzoxadiazole (ABDF) reacted selectively and quantitatively with a single cysteine residue.[45] Surprisingly, although the V_{max} value of the enzyme was reduced, the K_m value was not substantially affected and the enzyme still retained some activity, suggesting that the modification could not be at the active-site cysteine. Proteomics experiments subsequently revealed that the modification is in fact on cysteine-121, which is located approximately 8 Å from the active-site cysteine. Although crystallization of the modified protein was not successful, it is clear that ABDF selectively targets a previously cryptic allosteric site (Figure 10.4A).

Finally, several groups have reported irreversible chemistries that work selectively in the presence of proteins or even whole cells. These include the Huisgen cycloaddition recently improved by Sharpless and others,[46–48] chemical ligation,[49] the formation of oximes or hydrazones and the Staudinger ligation.[50]

Until recently, irreversible chemistries have not been applied to the systematic discovery of novel ligands. Instead, such methods have proven to be more useful for ligand-directed protein discovery, in which ligands specific to a certain protein or class of proteins are used to identify related proteins from a crude cell extract.[4, 51] Irreversible site-directed ligand discovery forms a continuum with affinity labeling and activity-based protein profiling, in which a reactive (often substrate-like) molecule binds covalently to a protein of interest.[52, 53] These techniques are typically used to probe enzyme mechanism or function or to identify a specific protein or class of proteins from a complicated mixture such as a cell lysate. However, more recently, Cravatt and co-workers have started to use these techniques to identify ligands also, including previously uncharacterized ligands.[54]

10.5 Opportunities with Fragment Capture Methods

The fragment capture methods described above provide some of the benefits of other fragment-based screening methods, but go beyond these in two important ways. First, covalent capture allows a site-directed screening approach based on homology models and in the absence of direct structural information, and therefore is applicable to proteins that are not amenable to NMR or crystallography. Second, covalent capture methods permit target-directed fragment assembly, whereby the target protein guides the covalent assembly of fragments with compatible binding modes (this method is described in Section 10.3.4).

The ability to explore chemical diversity space effectively remains a formidable challenge in small-molecule drug discovery. It has been estimated that the number of possible small molecules suitable for drug discovery exceeds 10^{60}, but only a fraction of these, at most a few million, can possibly be synthesized and screened at one time.[55] A combinatorial fragment-based approach has the potential to probe a larger fraction of the chemical diversity space. In fragment-based lead discovery, small drug fragments are screened against a target protein and only fragments with binding affinity for the target are actually expanded or linked with other fragments to generate new starting points for lead discovery. Using this approach, a collection of 10^4 fragments can theoretically probe a chemical diversity space of 10^8 molecules, still a small fraction of total diversity space, but a significant increase relative to traditional HTS (Figure 10.5). Most importantly, a chemical diversity space of 10^8 molecules can be explored by screening and synthesizing as few as 10^4 molecules, representing tremendous savings of cost and time.

Although a combinatorial fragment-based approach provides some great advantages, implementation of effective screening methods has presented a new set of challenges. One

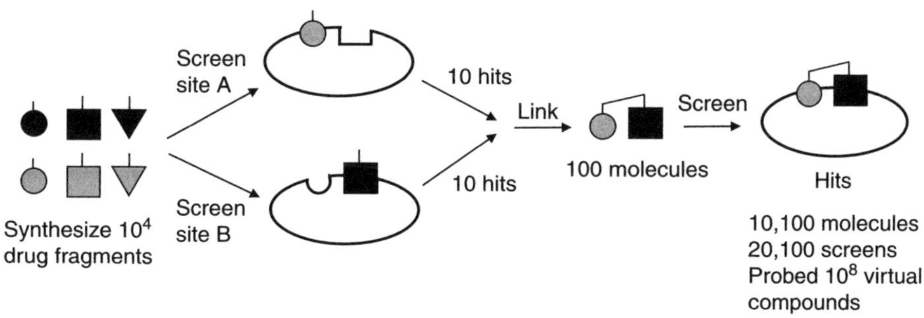

Figure 10.5 Fragment-based capture methods permit a combinatorial approach to lead discovery; 10^4 fragments can be combined to create 10^8 two-fragment molecules. Rather than synthesizing and screening 10^8 compounds, the 10^4 fragments can be screened individually and only the fragments that hit will be converted into two-fragment molecules. Compared with HTS, the fragment-based approach can be accomplished with far less synthetic and screening effort.

of the major limitations of covalent capture methods is the availability of a reactive residue or the ability to engineer a reactive residue at the site of interest. The use of covalent capture methods also requires an investment in custom fragment libraries with reactive functional groups. The covalent linkage itself may provide some limitations with respect to the binding orientations and interactions that can be surveyed from the attachment point, although this limitation can be overcome by using multiple attachment points around the binding site of interest.

10.6 Conclusions

Covalent capture methods have been primarily used to characterize protein–ligand interactions and to discover novel ligands that bind to specific regions on a protein target. Irreversible chemistries have been widely used to identify and characterize protein–ligand interactions and for affinity labeling studies. By contrast, reversible capture methods have primarily been used as screening techniques for protein–ligand and protein–small molecule interactions for drug discovery. Tethering with extenders is probably the most sophisticated of these methodologies by enabling *in situ* assembly of individual fragments in addition to fragment selection. The ability to bypass the fragment linking process solves one of the greatest challenges in combinatorial fragment assembly and opens the gates to probe a much wider chemical diversity space with a minimal synthetic effort.

The past decade has seen significant advances in applying covalent capture technology to fragment-based drug discovery, which is itself an emerging technology that has already had a significant impact on the drug discovery process in many companies. Ultimately, the success of these technologies will be judged by the impact they will have on delivering successful clinical products.

10.7 Acknowledgments

We would like to thank Robert S. McDowell, Monya Baker and Michael Romanowski for critical reading of the manuscript.

References

[1] Hajduk, P. J. and Greer, J. (2007). A decade of fragment-based drug design: strategic advances and lessons learned. *Nat. Rev. Drug. Discov.* **6**, 211–219.

[2] Jahnke, W. and Erlanson, D. A., eds (2006). *Fragment-based Approaches in Drug Discovery*. Methods and Principles in Medicinal Chemistry, Vol. 34. Wiley-VCH Verlag GmbH, Weinheim.

[3] Wold, F. (1977). Affinity labeling – an overview. *Methods Enzymol.* **46**, 3–14.

[4] Campbell, D. A. and Szardenings, A. K. (2003). Functional profiling of the proteome with affinity labels. *Curr. Opin. Chem. Biol.* **7**, 296–303.

[5] Makara, G. M. and Athanasopoulos, J. (2005). Improving success rates for lead generation using affinity binding technologies. *Curr. Opin. Biotechnol.* **16**, 666–673.

[6] Way, J. (2000). Covalent modification as a strategy to block protein–protein interactions with small-molecule drugs. *Curr. Opin. Chem. Biol.* **4**, 40–46.
[7] Rademann, J. (2004). Organic protein chemistry: drug discovery through the chemical modification of proteins. *Angew. Chem. Int. Ed.* **43**, 4554–4556.
[8] Rishton, G. M. (2003). Nonleadlikeness and leadlikeness in biochemical screening. *Drug Discov. Today* **8**, 86–96.
[9] McGovern, S. L., Helfand, B. T., Feng, B. and Shoichet, B. K. (2003). A specific mechanism of nonspecific inhibition. *J. Med. Chem.* **46**, 4265–4272.
[10] Tethering is a registered service mark of Sunesis Pharmaceuticals, Inc., for its fragment-based drug discovery.
[11] Erlanson, D. A., Braisted, A. C., Raphael, D. R., Randal, M., Stroud, R. M., Gordon, E. M. and Wells, J. A. (2000). Site-directed ligand discovery. *Proc. Natl. Acad. Sci. USA* **97**, 9367–9372.
[12] Erlanson, D. A., Wells, J. A. and Braisted, A. C. (2004). Tethering: fragment-based drug discovery. *Annu. Rev. Biophys. Biomol. Struct.* **33**, 199–223.
[13] Nelson, B. H. and Willerford, D. M. (1998). Biology of the interleukin-2 receptor. *Adv. Immunol.* **70**, 1–81.
[14] Tilley, J. W., Chen, L., Fry, D. C., Emerson, S. D., Powers, G. D., Biondi, D., Varnell, T., Trilles, R., Guthrie, R., Mennona, F., Kaplan, G., LeMahieu, R. A., Carson, M., Han, R.-J., Liu, C.-M., Palermo, R. and Ju, G. (1997). Identification of a small molecule inhibitor of the IL-2/IL-2Ra receptor interaction which binds to IL-2. *J. Am. Chem. Soc.* **119**, 7589–7590.
[15] Arkin, M. R., Randal, M., DeLano, W. L., Hyde, J., Luong, T. N., Oslob, J. D., Raphael, D. R., Taylor, L., Wang, J., McDowell, R. S., Wells, J. A. and Braisted, A. C. (2003). Binding of small molecules to an adaptive protein–protein interface. *Proc. Natl. Acad. Sci. USA* **100**, 1603–1608.
[16] Braisted, A. C., Oslob, J. D., Delano, W. L., Hyde, J., McDowell, R. S., Waal, N., Yu, C., Arkin, M. R. and Raimundo, B. C. (2003). Discovery of a potent small molecule IL-2 inhibitor through fragment assembly. *J. Am. Chem. Soc.* **125**, 3714–3715.
[17] Raimundo, B. C., Oslob, J. D., Braisted, A. C., Hyde, J., McDowell, R. S., Randal, M., Waal, N. D., Wilkinson, J., Yu, C. H. and Arkin, M. R. (2004). Integrating fragment assembly and biophysical methods in the chemical advancement of small-molecule antagonists of IL-2: an approach for inhibiting protein–protein interactions. *J. Med. Chem.* **47**, 3111–3130.
[18] Thanos, C. D., Randal, M. and Wells, J. A. (2003). Potent small-molecule binding to a dynamic hot spot on IL-2. *J. Am. Chem. Soc.* **125**, 15280–15281.
[19] Buck, E., Bourne, H. and Wells, J. (2005). Site-specific disulfide capture of agonist and antgonist peptides on the C5a receptor. *J. Biol. Chem.* **280**, 4009–4012.
[20] Buck, E. and Wells, J. A. (2005). Disulfide trapping to localize small-molecule agonists and antagonists for a G protein-coupled receptor. *Proc. Natl. Acad. Sci. USA* **102**, 2719–2724.
[21] Erlanson, D. A., Lam, J. W., Wiesmann, C., Luong, T. N., Simmons, R. L., DeLano, W. L., Choong, I. C., Burdett, M. T., Flanagan, W. M., Lee, D., Gordon, E. M. and O'Brien, T. (2003). *In situ* assembly of enzyme inhibitors using extended tethering. *Nat. Biotechnol.* **21**, 308–314.
[22] Allen, D. A., Pham, P., Choong, I. C., Fahr, B., Burdett, M. T., Lew, W., DeLano, W. L., Gordon, E. M., Lam, J. W., O'Brien, T. and Lee, D. (2003). Identification of potent and novel small-molecule inhibitors of caspase-3. *Bioorg. Med. Chem. Lett.* **13**, 3651–3655.
[23] Choong, I. C., Lew, W., Lee, D., Pham, P., Burdett, M. T., Lam, J. W., Wiesmann, C., Luong, T. N., Fahr, B., DeLano, W. L., McDowell, R. S., Allen, D. A., Erlanson, D. A., Gordon, E. M. and O'Brien, T. (2002). Identification of potent and selective small-molecule inhibitors of caspase-3 through the use of extended tethering and structure-based drug design. *J. Med. Chem.* **45**, 5005–5022.
[24] Fahr, B. T., O'Brien, T., Pham, P., Waal, N. D., Baskaran, S., Raimundo, B. C., Lam, J. W., Sopko, M. M., Purkey, H. E. and Romanowski, M. J. (2006). Tethering identifies fragment that yields potent inhibitors of human caspase-1. *Bioorg. Med. Chem. Lett.* **16**, 559–62.

[25] Johnson, T. O., Ermolieff, J. and Jirousek, M. R. (2002). Protein tyrosine phosphatase 1B inhibitors for diabetes. *Nat. Rev. Drug. Discov.* **1**, 696–709.

[26] Erlanson, D. A., McDowell, R. S., He, M. M., Randal, M., Simmons, R. L., Kung, J., Waight, A. and Hansen, S. K. (2003). Discovery of a new phosphotyrosine mimetic for PTP1B using breakaway tethering. *J. Am. Chem. Soc.* **125**, 5602–5603.

[27] Ockey, D. A. and Gadek, T. R. (2004). Discovery of novel PTP1B inhibitors. *Bioorg. Med. Chem. Lett.* **14**, 389–391.

[28] Hinman, A., Chuang, H. H., Bautista, D. M. and Julius, D. (2006). TRP channel activation by reversible covalent modification. *Proc. Natl. Acad. Sci. USA* **103**, 19564–19568.

[29] Macpherson, L. J., Dubin, A. E., Evans, M. J., Marr, F., Schultz, P. G., Cravatt, B. F. and Patapoutian, A. (2007). Noxious compounds activate TRPA1 ion channels through covalent modification of cysteines. *Nature* **445**, 541–545.

[30] Obita, T., Muto, T., Endo, T. and Kohda, D. (2003). Peptide library approach with a disulfide tether to refine the Tom20 recognition motif in mitochondrial presequences. *J. Mol. Biol.* **328**, 495–504.

[31] Huc, I. and Lehn, J.-M. (1997). Virtual combinatorial libraries: dynamic generation of molecular and supramolecular diversity by self-assembly. *Proc. Natl. Acad. Sci. USA* **94**, 2106–2110.

[32] Hochgurtel, M., Kroth, H., Piecha, D., Hofmann, M. W., Nicolau, C., Krause, S., Schaaf, O., Sonnenmoser, G. and Eliseev, A. V. (2002). Target-induced formation of neuraminidase inhibitors from *in vitro* virtual combinatorial libraries. *Proc. Natl. Acad. Sci. USA* **99**, 3382–3387.

[33] Hochgurtel, M., Biesinger, R., Kroth, H., Piecha, D., Hofmann, M. W., Krause, S., Schaaf, O., Nicolau, C. and Eliseev, A. V. (2003). Ketones as building blocks for dynamic combinatorial libraries: highly active neuraminidase inhibitors generated via selection pressure of the biological target. *J. Med. Chem.* **46**, 356–358.

[34] Webb, M. E., Stephens, E., Smith, A. G. and Abell, C. (2003). Rapid screening by MALDI-TOF mass spectrometry to probe binding specificity at enzyme active sites. *Chem. Commun.*, 2416–2417.

[35] Albert, A., Dhanaraj, V., Genschel, U., Khan, G., Ramjee, M. K., Pulido, R., Sibanda, B. L., von Delft, F., Witty, M., Blundell, T. L., Smith, A. G. and Abell, C. (1998). Crystal structure of aspartate decarboxylase at 2.2 Å resolution provides evidence for an ester in protein self-processing. *Nat. Struct. Biol.* **5**, 289–293.

[36] Roy, B. C., Hegge, R., Rosendahl, T., Jia, X., Lareau, R., Mallik, S. and Srivastava, D. K. (2003). Conjugation of poor inhibitors with surface binding groups: a strategy to improve inhibition. *Chem. Commun.* 2328–2329.

[37] Roy, B. C., Banerjee, A. L., Swanson, M., Jia, X. G., Haldar, M. K., Mallik, S. and Srivastava, D. K. (2004). Two-prong inhibitors for human carbonic anhydrase II. *J. Am. Chem. Soc.* **126**, 13206–13207.

[38] Jude, K. M., Banerjee, A. L., Haldar, M. K., Manokaran, S., Roy, B., Mallik, S., Srivastava, D. K. and Christianson, D. W. (2006). Ultrahigh resolution crystal structures of human carbonic anhydrases I and II complexed with 'two-prong' inhibitors reveal the molecular basis of high affinity. *J. Am. Chem. Soc.* **128**, 3011–3018.

[39] Elling, C. E., Frimurer, T. M., Gerlach, L., Jorgensen, R., Holst, B. and Schwartz, T. W. (2006). Metal ion site engineering indicates a global toggle switch model for seven-transmembrane receptor activation. *J. Biol. Chem.* **281**, 17337–17346.

[40] Smar, M. W., Ares, J. J., Nakayama, T., Itabe, H., Kador, P. F. and Miller, D. D. (1992). Selective irreversible inhibitors of aldose reductase. *J. Med. Chem.* **35**, 1117–1120.

[41] Chmura, A. J., Orton, M. S. and Meares, C. F. (2001). Antibodies with infinite affinity. *Proc. Natl. Acad. Sci. USA* **98**, 8480–8484.

[42] Levitsky, K., Ciolli, C. J. and Belshaw, P. J. (2003). Selective inhibition of engineered receptors via proximity-accelerated alkylation. *Org. Lett.* **5**, 693–696.

[43] Chen, G., Heim, A., Riether, D., Yee, D., Milgrom, Y., Gawinowicz, M. A. and Sames, D. (2003). Reactivity of functional groups on the protein surface: development of epoxide probes for protein labeling. *J. Am. Chem. Soc.* **125**, 8130–8133.

[44] Johnson, J. L., Cusack, B., Hughes, T. F., McCullough, E. H., Fauq, A., Romanovskis, P., Spatola, A. F. and Rosenberry, T. L. (2003). Inhibitors tethered near the acetylcholinesterase active site serve as molecular rulers of the peripheral and acylation sites. *J. Biol. Chem.* **278**, 38948–38955.

[45] Hansen, S. K., Cancilla, M. T., Shiau, T. P., Kung, J., Chen, T. and Erlanson, D. A. (2005). Allosteric inhibition of PTP1B activity by selective modification of a non-active site cysteine residue. *Biochemistry* **44**, 7704–7712.

[46] Lewis, W. G., Green, L. G., Grynszpan, F., Radic, Z., Carlier, P. R., Taylor, P., Finn, M. G. and Sharpless, K. B. (2002). Click chemistry *in situ*: acetylcholinesterase as a reaction vessel for the selective assembly of a femtomolar inhibitor from an array of building blocks. *Angew. Chem. Int. Ed.* **41**, 1053–1057.

[47] Kolb, H. C. and Sharpless, K. B. (2003). The growing impact of click chemistry on drug discovery. *Drug Discov. Today* **8**, 1128–1137.

[48] Baskin, J. M., Prescher, J. A., Laughlin, S. T., Agard, N. J., Chang, P. V., Miller, I. A., Lo, A., Codelli, J. A. and Bertozzi, C. R. (2007). Copper-free click chemistry for dynamic *in vivo* imaging. *Proc. Natl. Acad. Sci. USA* **104**, 16793–16797.

[49] Muir, T. (2003). Semisynthesis of proteins by expressed protein ligation. *Annu. Rev. Biochem.* **72**, 249–289.

[50] Cook, B. N. and Bertozzi, C. R. (2002). Chemical approaches to the investigation of cellular systems. *Bioorg. Med. Chem.* **10**, 829–840.

[51] Jeffery, D. A. and Bogyo, M. (2003). Chemical proteomics and its application to drug discovery. *Curr. Opin. Biotechnol.* **14**, 87–95.

[52] Adam, G. C., Sorensen, E. J. and Cravatt, B. F. (2002). Chemical strategies for functional proteomics. *Mol. Cell. Proteomics* **1**, 781–790.

[53] Perret, P., Laube, B., Schemm, R., Betz, H., Goeldner, M. and Foucaud, B. (2002). Affinity labeling of cysteine-mutants evidences contact residues in modeled receptor binding sites. *J. Recept. Signal Transduct. Res.* **22**, 345–356.

[54] Li, W., Blankman, J. L. and Cravatt, B. F. (2007). A functional proteomic strategy to discover inhibitors for uncharacterized hydrolases. *J. Am. Chem. Soc.* **129**, 9594–9595.

[55] Bohacek, R. S., McMartin, C. and Guida, W. C. (1996). The art and practice of structure-based drug design: a molecular modeling perspective. *Med. Res. Rev.* **16**, 3–50.

11

Identification of High-affinity β-Secretase Inhibitors Using Fragment-based Lead Generation

Jeffrey S. Albert and Philip D. Edwards

11.1 Introduction: BACE and Its Role in Alzheimer's Disease

A leading hypothesis for the pathophysiology of Alzheimer's disease centers on the abnormal accumulation of plaques in the brain that are composed from the 40/42 amino acid polypeptide called β-amyloid peptide ($A\beta_{40/42}$). According to the amyloid cascade hypothesis, β-amyloid is produced in the brain as the product of two proteolytic processing events; β-amyloid precursor protein (APP) is cleaved first by β-secretase (BACE) and then by γ-secretase (Figure 11.1). Thus formed, the β-amyloid peptide accumulates and deposits as insoluble plaques that are either causative or otherwise associated with progression of Alzheimer's disease.[1] According to this model, BACE is a particularly attractive target for Alzheimer's disease therapy because inhibition should reduce β-amyloid production. The safety of BACE inhibitors is supported by the observation that BACE-knockout mice appear normal except for their inability to produce the β-amyloid peptide.[2–6] Five research groups independently identified BACE in 1999.[7–11] Since then, intense effort has been invested in discovering inhibitors for this enzyme. This area has been extensively reviewed.[1, 12–16]

Fragment-Based Drug Discovery: A Practical Approach Edited by Edward R. Zartler and Michael J. Shapiro

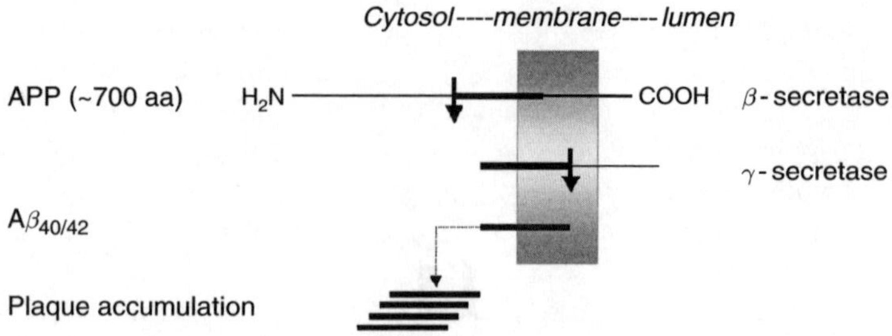

Figure 11.1 *The amyloid hypothesis in Alzheimer's disease. The transmembrane protein amyloid precursor protein (APP) is cleaved first by β-secretase (BACE), then by γ-secretase. The resulting peptidic fragment, $A\beta_{40/42}$, is liberated, leading to plaque accumulation. These plaques are associated with neuronal degeneration.*

11.2 Overview: Fragment-based Lead Generation

At the outset of our program at AstraZeneca, the development of several high-affinity peptide-based inhibitors had been reported together with their binding mechanism as determined by co-crystallography.[17] However, no small-molecule inhibitors had yet been disclosed. We initially used traditional lead generation methods to identify starting points as BACE inhibitors. Our first approach was based on high-throughput screening (HTS) of our entire corporate collection at 10 μM concentration and a subset at 30 μM concentration. We obtained numerous hits in the affinity range IC_{50} 0.5–5 μM, but follow-up efforts showed that these were forming aggregates or were otherwise not active site directed. Our second approach involved the design of peptidomimetics based on published co-crystal structures of BACE or the related aspartyl protease inhibitors renin and HIV protease. All hits that emerged were discounted either because they could not be mechanistically confirmed as competitive inhibitors or because they lacked the properties to make them suitable as CNS drugs.

As an alternative, we adopted fragment-based approaches for lead generation. These methods rely on identifying small compounds (fragments) that bind to the target with fewer, but stronger binding interactions than those typically associated with drugs.[18] This approach is fundamentally different from traditional methods (such as HTS) where hit compounds are typically larger and bind to the target over a larger interaction area, but using overall weaker binding points.

Among the principal advantages of fragment-based lead generation methods are the following. First, because the fragments are small (generally 150–250 Da), structure space can be efficiently sampled with smaller screening sets.[19–21] Therefore, instead of screening millions of typical drug-like compounds, only thousands of small compounds may need to be screened. This makes it possible to use lower throughput screening methodologies. A second advantage is that because of the small size of any fragment-based hit, even relatively weak binding affinity must result from efficient interactions with the target. A 500 Da hit with an affinity of 1 μM may be more challenging to optimize than a 150 Da hit with 1 mM affinity because the larger, higher affinity hit probably binds with a wider network of weaker individual interactions that must be maintained or improved during the optimization process. In the case of the smaller, weaker hit, binding is likely due to fewer, but stronger, individual interactions. As a result, optimization may effectively be achieved by maintaining these and adding (or improving) additional interactions.[22] Addition of such interaction points (and concomitant increases in molecular weight) can be better accommodated while retaining drug-like properties if the initial hit has lower molecular weight. Additional information on fragment-based methods can be found among the several recent and excellent reviews.[18, 22–26]

Among the principal disadvantages of fragment-based lead generation methods are the following. First, by starting from weaker affinity hits (100 μM to 5 mM), considerable structural modifications are likely to be needed during optimization to reach typical drug affinity (< 100 nM). Such optimization can be greatly accelerated with structural knowledge (e.g. NMR or crystallography). A second disadvantage is that typical biological screens may not work reliably at very high ligand concentrations and thus cannot detect weak binders. This frequently necessitates the development of alternative analytical methods. As discussed below, we employed three screening and analysis methods: NMR spectroscopy, surface plasmon resonance (SPR) spectroscopy and crystallography.

11.3 Identification of Fragment-based Leads by NMR Screening

Whereas typical HTS hits have affinity < 1 μM, the smaller and simpler hits emerging from fragment screening typically have affinities in the range 100 μM–5 mM. Because of the high ligand concentration, we could not use conventional enzymatic assays for screening. NMR spectroscopy is particularly well suited for the detection of such starting points because of its sensitivity, throughput and experimental robustness.[27] We designed a screening set of approximately 5000 compounds (150–350 Da) using a combination of diversity-based and target-based considerations. The diversity-based subset included a range of compound classes containing features that are associated with good CNS drug-like features. The target-based subset was influenced by knowledge of the BACE binding site. For example, the BACE active site contains two aspartic acids and a nearby lipophilic region.[17] Thus our screening library was biased toward lipophilic, weakly basic amines and lipophilic alcohols since both are precedented as binders to related aspartic acid proteases.[15]

Ligand binding can be detected by a number of NMR spectroscopic methods.[28, 29] We employed[30] the water-LOGSY[31, 32] approach; this is a 1D experiment that detects magnetization changes to a small molecule if it binds to its macromolecular target. Multiple compounds can be screened at the same time, the experiment uses relatively small amounts of enzyme and signal assignment is not necessary. At high ligand concentrations the false positive rate can be substantial due to non-selective binding or ligand aggregation. Therefore, it is very important to have a means to confirm binding selectivity. Unlike 2D methods, it is not possible to assess directly the ligand binding mode because the biological target is essentially invisible. However, if a high-affinity ligand (for example a peptide) is available it can be used for mechanistic validation. Once binding is detected for a compound in the screening set, the high-affinity ligand can be added. Displacement of the screening compound by that ligand can be observed in the NMR experiment. This provides good support that the binding sites are co-localized and that binding is selective. Frequently, it is observed that the small molecule is not displaced by the high-affinity reference compound. In these cases the project team may choose to de-prioritize such hits or initiate other efforts to elucidate the biding mode. Our typical protocol is illustrated in Figure 11.2.

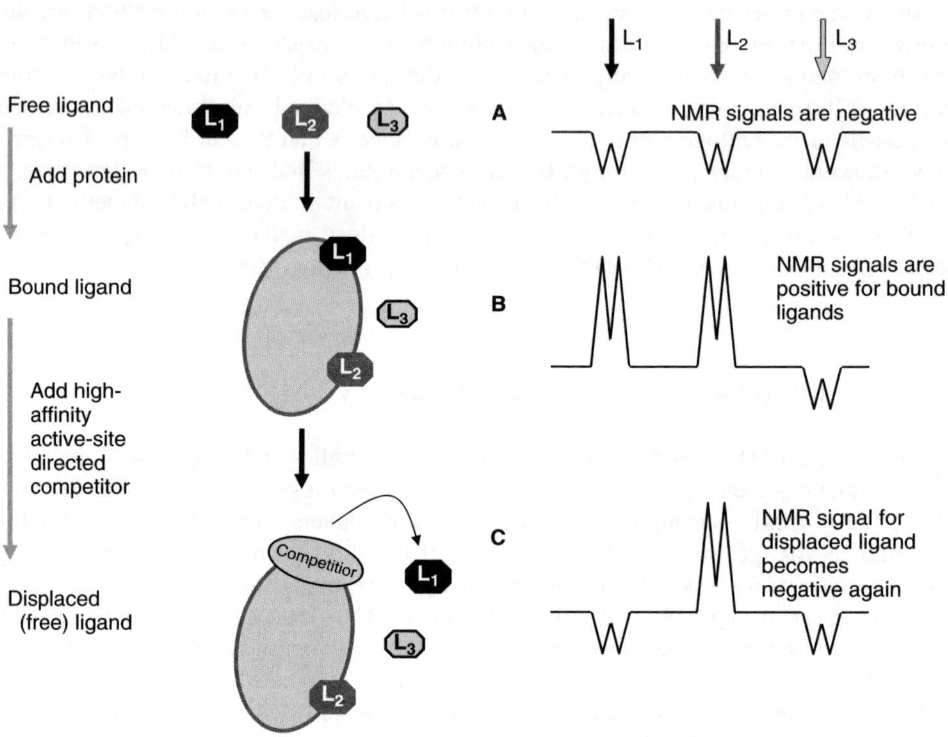

Figure 11.2 *Illustration of water-LOGSY NMR screening protocol. (A) Three representative ligands (L_1, L_2, and L_3) are combined in solution and show negative NMR signals for each resonance in their spectra. (B) The target protein (large oval) is added to the mixture. Any ligand that binds (in this case, L_1 and L_2) shows a positive NMR signal. Those that do not bind (L_3) show no change in the NMR signal. (C) An active site-directed, high-affinity competitor is added to the mixture. Any weaker ligand that occupied the active site is displaced back into free solution and its NMR shift becomes negative again. Any ligand that is not displaced by the competitor continues to have a positive shifted NMR signal. According to this example, it would be suggested (but not proven) that L_1 is a competitive binder, the L_2 is a non-competitive binder and L_3 does not bind.*

To increase throughput, compounds were screened in groups of five in the presence of BACE. If binding was detected for any of the compounds, each member in the set was retested individually. The retesting process involved (1) analyzing the NMR spectrum of the ligand alone in solution, (2) analyzing in the presence of BACE and (3) adding the high-affinity active site directed peptidic inhibitor OM99-2[33] to the mixture to assess if it displaces the ligand. Many hits from the initial screen were eliminated during the confirmation step either because they were not displaced by addition of OM99-2 or because the ligand was self-aggregating (aggregation causes NMR signal inversion under these experimental conditions and thus appears as a false positive). Representatives among the confirmed hits are shown in Figure 11.3. Although there was considerable diversity among the hits, typical features included a hydrogen bonding group tethered to a hydrophobic region.

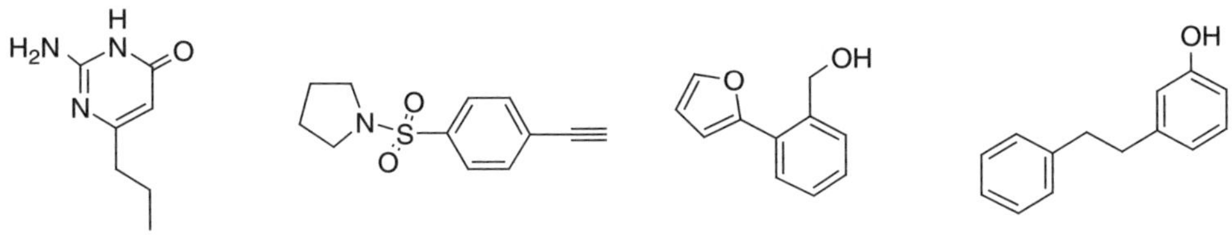

1
IC_{50} 2500 μM
LE = 0.32

Figure 11.3 *Representatives among the hits in the NMR screen. The affinity of* **1** *was determined by SPR spectroscopy (below).*

Because of the number and diversity of the hits from NMR screening, it was critical to establish an efficient protocol to focus our efforts. Ideally, hits would be prioritized according to detailed understanding of the binding mechanism using, for example, NMR structural elucidation or crystallography. As our program progressed, we increasingly relied on crystallography for hit evaluation (discussed below). However, during the early phase of our program we did not have access to these methods.

In the absence of detailed structural knowledge of the ligand binding mode (other than the observation that it was displaced by OM99-2), each NMR screening hit was progressed by evaluating a set of structurally and/or chemically related analogs. This served two purposes. First, it helped validate the hit as being representative of a potential series. Second, it helped to define SAR and assess our likelihood of progressing the hit. Typically, we tested 10–25 compounds related to each. In cases where suitable analogs were not available in the collection, we synthesized them. Many initial hits were eliminated in this process.

11.4 Determining Binding Affinities Using Surface Plasmon Resonance Spectroscopy

According to our experimental methods, NMR spectroscopy could detect compounds that bound to BACE, but did not provide a quantitative measure of the binding affinity, which, of course, is needed to compare hits and generate SAR. Standard enzymatic assays for BACE could not be used for our weak-affinity hits because of the high ligand concentration. The water-LOGSY experiment can be configured quantitatively; however, we found it advantageous to use surface plasmon resonance (SPR) spectroscopy[30] instead because of its ability to monitor even weak binding interactions ($IC_{50} \approx 5$ mM) with relatively high throughput.

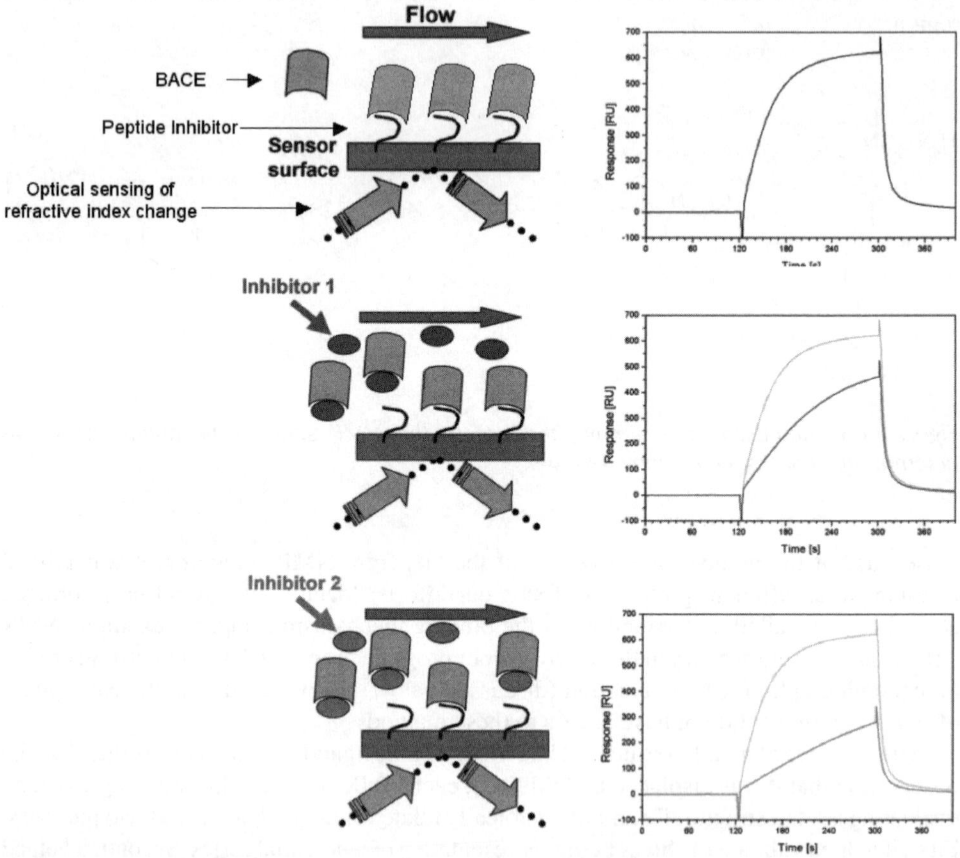

Figure 11.4 *Illustration of the SPR experiment. The binding of BACE to the immobilized small peptidic reference compound is detected by a change in refractive index at the chip surface. Preincubation of BACE with a test compound blocks the interaction with the immobilized reference compound in a manner that is dependent on the concentration of the test compound and its affinity for BACE. In practice, we could rank affinities by comparison of the SPR response at a fixed concentration of the test compound and then quantitatively determine the affinity by analyzing the concentration response dependence. In this example, the inhibitor in the middle panel (Inhibitor 1) has weaker affinity than the inhibitor in the lower panel (Inhibitor 2) because, at equivalent concentrations, Inhibitor 2 more extensively blocks the interaction between BACE and the immobilized peptidic inhibitor.*

To use SPR,[34] a component of the biological interaction (e.g. the biomolecule, screening target or reference compound) is attached to an optical sensor chip. The other component is then flowed in solution past the chip. Binding of the ligand to the surface-bound target causes a change in the effective mass and this is detected by a change in the refractive index of the sensor. This experiment can be configured in many ways, including options for which component is immobilized. In general, response sensitivity is related to the degree of change in effective molecular mass due to ligand binding on the immobilized receptor. Thus, immobilization of a small ligand to the surface and flowing past a much larger biomolecule will result in greater sensitivity than the converse. We employed the peptide-based, active-site directed inhibitor KTEEISEVN-statine-VAEF for immobilization.[35] The typical protocol is illustrated in Figure 11.4; flowing a solution of BACE results in a large response when the biomolecule binds to the immobilized small peptidic inhibitor. If a test compound, at a particular concentration (inhibitor 1 in Figure 11.4), is incubated with BACE, the interaction between BACE and the immobilized reference compound will be blocked in a manner that is dependent on the affinity of the test compound. If a higher affinity test compound is used (e.g. inhibitor 2), the interaction between BACE and the immobilized peptide will be further reduced. In this manner, the binding affinity can be evaluated.[36] To quantitate the binding affinity, the experiment can be repeated at multiple inhibitor concentrations.[35]

Using these approaches, the affinity of **1** was determined as IC_{50} 2500 μM. Next, we selected 20 structurally related compounds from the corporate collection for analysis using the same approach. Representatives(**2–5**) are shown in Figure 11.5. It is noteworthy that the new leading compound **(5)** was screened in the original high throughput assays, but was not detected because of its weak affinity.

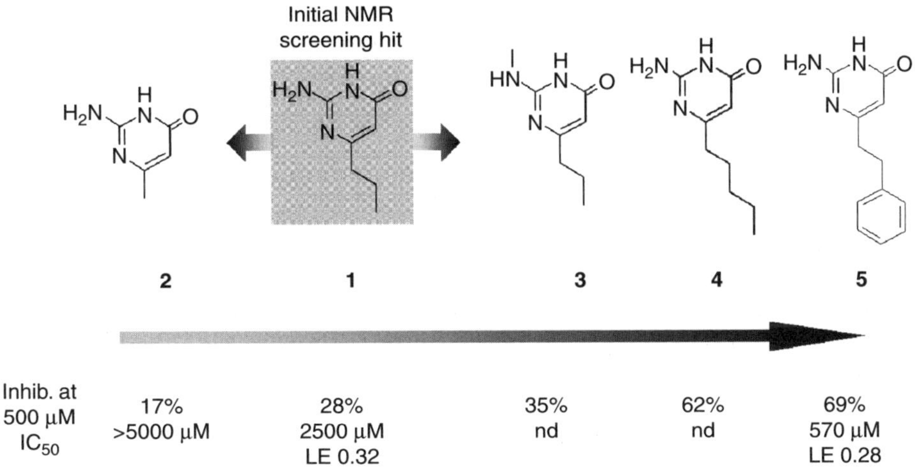

Figure 11.5 *Binding affinity analysis using SPR. Based on the original NMR hit* (**1**)*, we identified structurally related compounds with improved binding. nd, not determined; LE, ligand efficiency, explained in the text.*

11.5 Using Ligand Efficiency for Hit Evaluation and Optimization

As discussed above, hits with low affinity but small molecular size may have advantages over larger compounds with stronger affinity toward the successful optimization and delivery of drug-like clinical candidates. Several scaling factors have been introduced to assess the binding affinity within the context of ligand size and other properties.[37] These include metrics based on polar surface area, molecular weight and number of heavy (non-hydrogen) atoms. While each method has particular advantages, they all seek to provide some measure of how much each atom in a small molecule contributes to a binding interaction with a macromolecule. Among these methods, we have found it useful to consider a ligand's efficiency as the binding energy per heavy atom (with heavy atom defined as all atoms except hydrogen).[38] In general, the free energy of binding (ΔG) was not determined; instead, IC_{50} values were used (Figure 11.6). Of course, molecular weight is highly correlated with the heavy atom count; based on the analysis of 201 marketed CNS drugs that are assumed to passively distribute into the brain, we find that on average each heavy atom contributes about 14 Da to the molecular weight. [The correlation is MW = (14.051) (heavy atom count) + 0.3829, $R^2 = 0.9734$. When considering all orally delivered drugs, the mean molecular weight per heavy atom is only slightly higher, MW = (14.248) (heavy atom count) + 2.5452, $R^2 = 0.9429$.]

$$\mathrm{LE} = -\Delta G/\mathrm{HAC}$$
$$\mathrm{LE} = RT\ln(K_d)/\mathrm{HAC}$$
$$\mathrm{LE} \approx RT\ln(\mathrm{IC}_{50})/\mathrm{HAC}$$
$$\mathrm{LE} \approx -0.592\ln(\mathrm{IC}_{50})/\mathrm{HAC}$$

Figure 11.6 *Expressions for ligand efficiency (LE). HAC = heavy atom count, K = 298 K, R = 1.987 × 10^{-3} kcal K^{-1} mol^{-1}.*

To improve the likelihood of the successful delivery of orally bioavailable CNS drugs, research programs will aim towards molecular weights of less than around 500 Da (<36 heavy atoms) and affinity/potency for the target of around 10 nM. In Figure 11.7, a compound with these properties is represented by point A. Further increases in the potency or decreases in molecular weight are both desirable and define the shaded region. The ligand efficiency of the compound represented by point A is approximately 0.30. By fixing this value as the target ligand efficiency, the correlation between affinity and heavy atom count is shown as the middle line in Figure 11.7. Any compounds that reside to the left of this line have higher ligand efficiency and can be considered to be tending toward desirable drug-like features. Any compounds that reside to the right of the line have lower ligand efficiency and may be considered to have atoms contributing to a smaller extent to the productive binding interaction. Of course, the delivery of successful drug candidates requires balancing among many properties. In this discussion of drug-likeness, ligand efficiency monitors only the interaction of the ligand with its target and is therefore well suited for the evaluation of early-stage compounds. Other chemical features will influence the optimization of physical

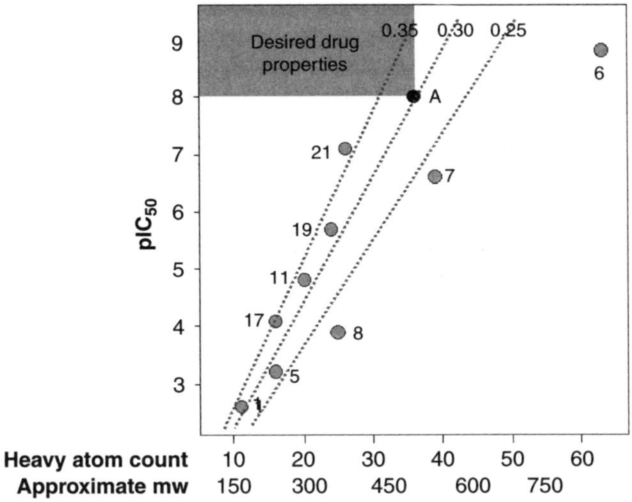

Figure 11.7 *Affinity versus heavy atom count for selected compounds. Lines illustrate profiles at fixed ligand efficiency (0.35, 0.30, 0.25) over the range of affinity and heavy atom count. Point* **A** *designates an idealized compound with 36 heavy atoms (approximate MW 500 Da) and affinity 10 nM). Numbers correspond to compounds described in the text.*

properties, DMPK properties, selectivity, etc. All these features together contribute to the final assessment of drug-likeness.

These considerations are further illustrated as follows. The peptidic inhibitor OM99-2 (**6**, Figure 11.8) binds with very high affinity (1.6 nM) to BACE,[17] but because of its high molecular weight each atom appears to make only a relatively small contribution to the overall binding. This compound has a ligand efficiency of only 0.19; this is indicated in Figure 11.7. Compound **7** was discovered by the Elan group[39] and demonstrated by Astex

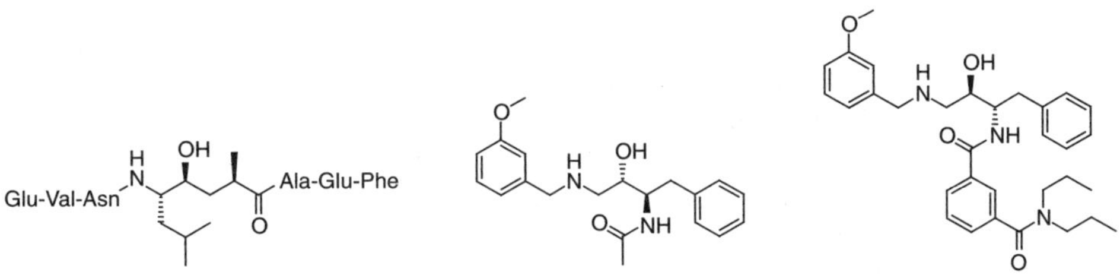

6

OM99–2
$K_i = 1.6$ nM[a]
MW = 893, HAC = 63
LE 0.19

7

IC_{50} = 138 μM[b]
MW 342; HAC 25
LE = 0.21

8

IC_{50} = 230 nM[b]
MW 531, HAC 39
LE = 0.23

Figure 11.8 *Structure of OM99-2 and related peptidic inhibitors.* [a]*Reported by Hong* et al.[17] [b]*Determined by SPR spectroscopy.*

to bind to BACE in a similar manner to **6** using crystallography.[40] We analyzed this compound and analog **8**. In both cases, the binding affinity decreased in a manner that roughly correlated with the decreasing molecular weight, resulting in overall similar ligand efficiency among all three compounds. [The typical reproducibility of most biological assays (2–3-fold for IC_{50}) translates to a reliability for ligand efficiency values of about $\pm 10\,\%$ of the stated value. This is adequate for the assessment of overall trends, although individual small changes should not be over-interpreted.] This suggests that each atom is contributing a small amount of binding energy and the result is that to achieve high binding, a fairly high molecular weight would be necessary. Overall, this would present additional challenges and risk to the successful delivery of a CNS-penetrant drug from this structural class.

It is noteworthy that the ligand efficiency remains low and does not change very much among these three examples, despite an almost 10^5-fold improvement in affinity. As reported by Hajduk,[41] it has been found in several cases that the ligand efficiency remains relatively constant within structural families. Hajduk started with a set of high-affinity drug-like compounds across several projects and did a retrospective analysis to compare the properties of those final compounds with iteratively truncated analogs. It was found that the binding affinity remained remarkably linear with molecular weight throughout the 'deconstruction' process in each family of ligands. This finding suggests that once a chemical lead has been identified, it is possible to estimate the molecular weight of a final compound that will have the desired affinity. Consequently, ligand efficiency should be valuable in helping to rank and prioritize hits among different structural classes and affinity ranges.

Using NMR screening against BACE at 5 mM ligand concentration, we identified numerous hits with affinities ranging between 1 μM and 5 mM. Among those hits, **1** had nearly the weakest affinity (IC_{50} 2500 μM), but the highest ligand efficiency (0.32). Despite the weak affinity, we chose this as a lead for optimization according to the rationale above.

11.6 Optimization of the Isocytosine Series

As described above, screening of compounds that were structurally related to the initial hit compound **1** led to the identification of **5**. Despite its weak binding affinity, compound **5** could be co-crystallized with BACE. Although the resolution was low, the structural information was used with molecular modeling to generate the binding model shown in Figure 11.9. According to this model, the isocytosine nitrogens at positions N1 and C2 interacted directly with the BACE catalytic aspartates and the phenyl ring was directed toward the largely open S_3 pocket.

To fill the S_2' pocket better and improve the CNS drug-like potential (by removing a hydrogen bond donor), we methylated the N3 nitrogen. This afforded compound **9** (Figure 11.10), which had threefold improved affinity.[42] Next, we modified the aryl region to expand further into the S_3 pocket by replacing the phenyl with indole. Addition of this polar functionality was also desirable in order to improve solubility. This led to compound **10**. Combining the *N*-methylation and the indole replacement led to **11**, which had a 35-fold improvement in affinity relative to **5** and slightly improved ligand efficiency. Further survey of aryl analogs **12–14** (Table 11.1) led to the identification of **14** as the first inhibitor in this

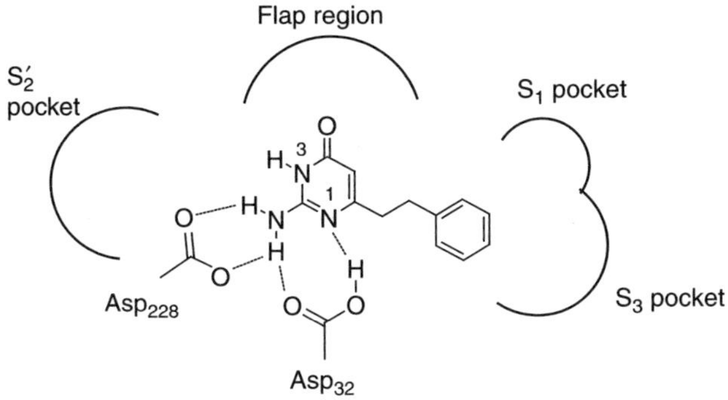

Figure 11.9 *Binding model for* **5** *bound to BACE. This model was generated from a low-resolution crystal structure and computational analysis.*

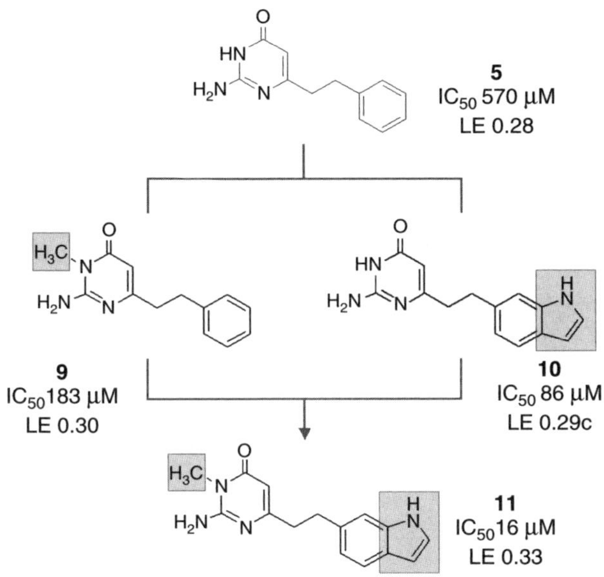

Figure 11.10 *Starting from* **5**, N3 *methylation and indole incorporation led to the identification of* **11**.

series with affinity less than 10 μM. A high-resolution crystal structure was obtained for the **14**–BACE complex[42] (Figure 11.11) and it precisely confirmed the binding model shown in Figure 11.9. In addition to the hydrogen bonding contacts with the catalytic aspartates (Asp32 and Asp228), there is a hydrogen bonding contact between the isocytosine carbonyl and the main-chain amide NH from Gln73. Residues 72 and 73 comprise the flap region, which closes down over the ligand in this case. The bis-aryl region fills the S_1/S_3 pocket and opens a deep pocket at the bottom of S_3 to accommodate the methoxy substituent.

Table 11.1 Optimization of the isocytosine aryl region.

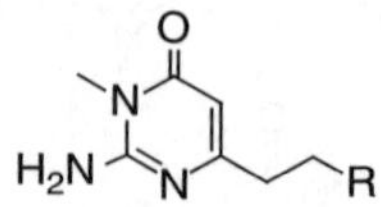

Compound	R	IC_{50} (μM)	LE
12	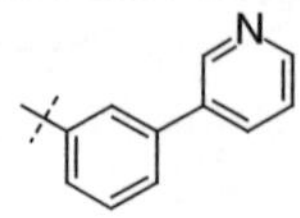	24	0.27
13	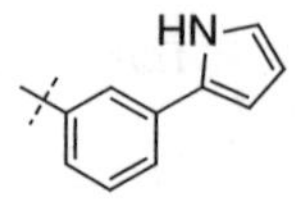	51	0.27
14	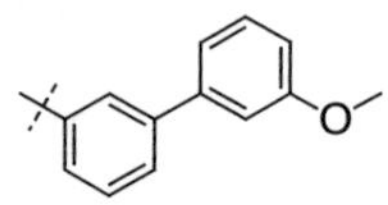	6	0.28

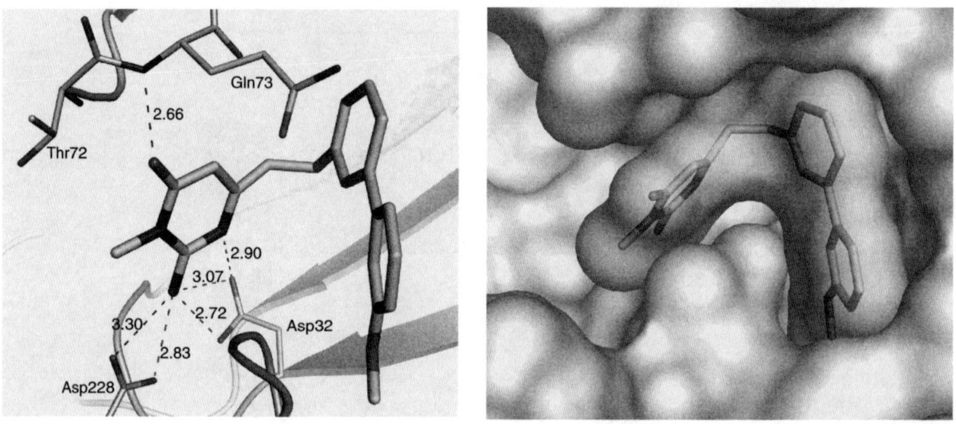

Figure 11.11 *Two views of the crystal structure of* **14***–BACE complex. In the right-hand image, residues 72–73 cover the ligand and have been removed for clarity.*

11.7 Identification of the Dihydroisocytosine Series

Following the identification of **1**, a survey of structurally related compounds from the company collection led to the identification of compound **15** (Figure 11.12). Binding of this compound to BACE could be detected by NMR spectroscopy, but could not be quantitated by SPR spectroscopy. Although the affinity was very weak, the compound was of interest because the nonaromatic nature of the heterocycle made the overall structure significantly different from that of the isocytosines. Despite the weak affinity, it could be crystallized

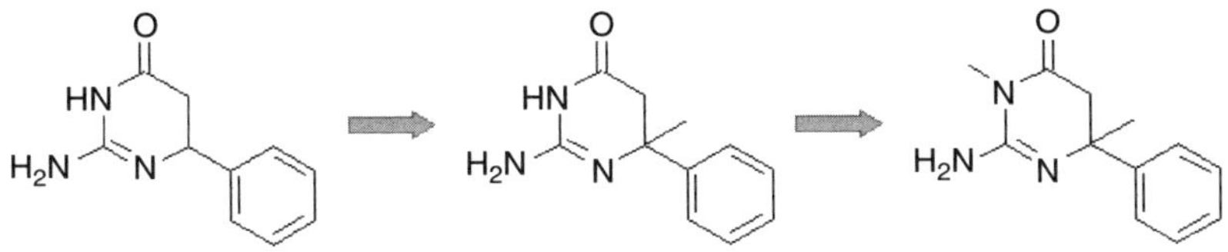

15 IC_{50} >1000 μM

16 138 μM LE = 0.35

17 79 μM LE = 0.35

Figure 11.12 *Compound* **15** *was identified by similarity searching. Methylation of the heterocycle C6 position* **(16)** *and* N3 **(17)** *led to further improved analogs.*

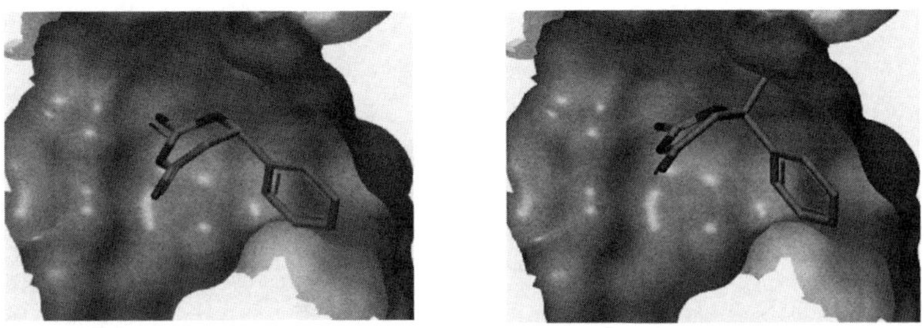

15 **16**

Figure 11.13 *Comparison of crystal structures of BACE with* **15** *and* **16**. *Addition of the C6 methyl group in* **16** *reduces the presumed energetic penalty for the aryl substituent to adopt the pseudo-axial geometry seen in the bound conformation and also makes van der Waals contact with the enzyme.*

with the enzyme (Figure 11.13). The resolution of the structure was relatively low, but sufficient to show that the heterocycle adopted an envelope conformation with the pendant phenyl group in a presumably disfavored pseudo-axial position.[42] Using semiempirical methods with a continuous solvent model, calculations indicated that this pseudo-axial orientation was 1.4 kcal mol^{-1} higher in energy than the pseudo-equatorial orientation. We speculated that replacement of the methine hydrogen with methyl **(16)** would reduce this energetic penalty and thereby improve binding affinity. Indeed, calculations on **16** showed that although the pseudo-axial orientation was still disfavored, the energy difference between the two forms was reduced to only 0.1 kcal mol^{-1}. We prepared **16** and showed that the affinity improved substantially. Crystallographic analysis of this compound showed that the binding orientation was exactly maintained and that the methyl group made good van der Waals contact with the back surface of the binding pocket.

As described above, it was known that N3 methylation was favored for the isocytosine series. The analogous compound in the dihydroisocytosine series **(17)** was then prepared and this resulted in a further improvement in affinity (IC_{50} 79 μM) and maintenance of high ligand efficiency.

11.8 Identification and Optimization of Extended Dihydroisocytosines

We sought to merge the SAR between the isocytosine and the dihydroisocytosine series. Compound **18** contains the isocytosine core from **17** but has the extended liker as in **5** (Figure 11.14). The binding affinity of **18** was stronger than that for **5** or **17** and it maintained high ligand efficiency. A focused library of approximately 15 compounds was prepared based on **18** to explore various phenyl ring substitutions. The best compound to emerge was the biphenyl analog **19**. A second focused library of about 15 compounds was constructed to evaluate substitution at the peripheral phenyl ring and this led to the identification of **20**. After chiral resolution, we identified **21** as the highest affinity inhibitor yet seen. The absolute configuration of **21**, and also of all other stereochemically defined compounds, was determined by crystallography (Figure 11.15).[42]

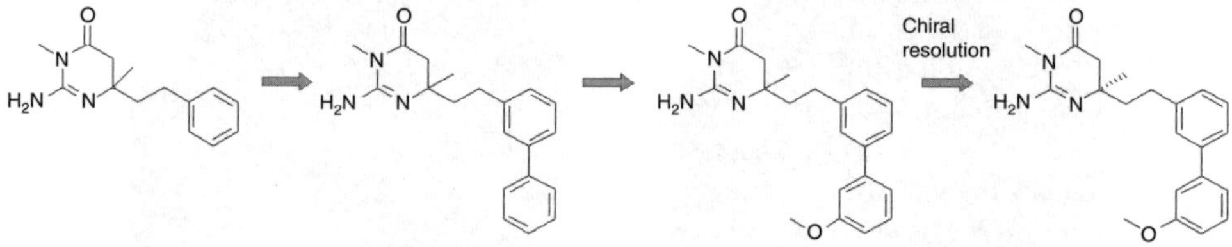

18 IC_{50} 34 μm LE = 0.34

19 IC_{50} 2 μM LE = 0.32

20 IC_{50} 0.2 μM LE = 0.35

21 IC_{50} 80 nM LE = 0.37

Figure 11.14 *Combining elements from* **5** *and* **17** *led to* **18**. *Two cycles of focused library generation led to the identification of* **20**. *Chiral resolution of this led to* **21** *as the highest affinity compound yet identified in our program.*

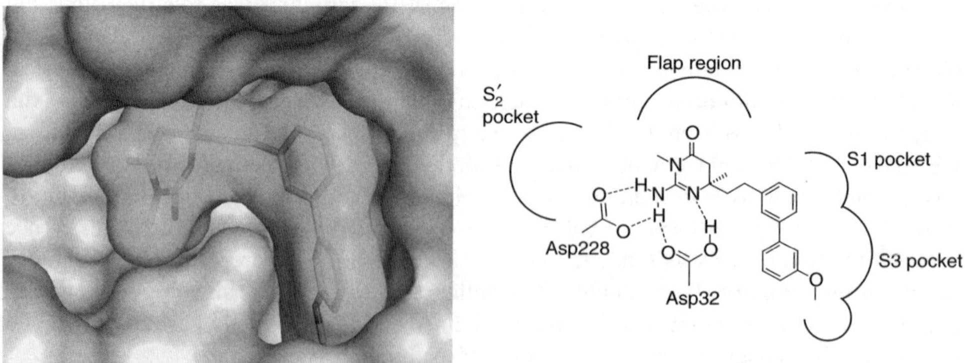

Figure 11.15 *Left: X-ray structure of* **21** *complexed to BACE. Right: schematic illustration of the crystallographic analysis of the* **9**–*BACE complex showing key binding interactions.*

Crystallographic analysis of the **21**–BACE complex showed that the binding mechanism remained consistent. The nitrogens at positions N1 and C2 of the dihydroisocytosine bind to the catalytic aspartates of BACE. The aryl region binds within S3 using interactions that are essentially the same as for the isocytosines. The methoxy substituent binds deeply in the S_3 pocket at a new site termed the S3 sub-pocket, where it may undergo a hydrogen bonding interaction with the Ser229 hydroxyl.

11.9 Summary and Conclusions

When HTS and other conventional methods for lead generation failed to identify useful starting points as BACE inhibitors, we adopted fragment-based approaches. A library of approximately 5000 low molecular weight compounds was screened at high concentration (5 mM) by NMR spectroscopy. Among the numerous hits obtained, many appeared to be false positives and we eliminated all those that were not displaced from their binding site by addition of the high-affinity peptidic ligand OM99-2. To measure binding affinities, we employed a SPR-based assay because conventional enzymatic assays could not be carried out under the high ligand concentrations needed to evaluate such weak binders. One of the key components in evaluating the resulting hits was consideration of the ligand efficiency. Rather than prioritize based on strong affinity, we prioritized based on high ligand efficiency. As a consequence, many higher affinity hits were discounted if the overall efficiency was low. We focused on hit **1** which had very weak affinity (2500 μM) but high ligand efficiency (0.32). Crystallization with BACE allowed us to evaluate directly the mechanism of binding. With this information, we methylated the N3 position and modified the aryl region; together, these changes led to the identification of **14**, which showed 400-fold improved binding.

Once a validated hit has been identified, there exists the obvious potential to find additional structurally related hits. Screening of compounds similar to **1** led to the identification of dihydroisocytosine **15**. Despite its weak affinity (IC_{50} >1000 μM), we were able to use crystallographic information to make improvements rapidly, leading to **17** (IC_{50} 79 μM). The SAR knowledge gained in this process was applied to the original isocytosine hit to explore a related series of extended isocytosines. Structural knowledge, combined with the evaluation of two focused libraries, allowed us to efficiently identify the high-affinity lead compound **21** (IC_{50} 80 nM).

Overall, fragment-based methods succeeded in the identification of useful starting points as BACE inhibitors where other methods failed. By screening for small compounds and allowing for weak binders, the hits that emerge tend to represent the minimal binding units. Among the key consequences are the following. First, by restricting to smaller compounds and (consequently) a smaller volume of structure space, many fewer compounds need to be screened. Whereas our high-throughput screens included >500 000 compounds and delivered no useful hits, our fragment screen included only 5000 compounds and delivered multiple useful hits. A corollary to this is that by iteratively evolving from a minimal binding unit, there exists greater potential to identify truly novel chemical series. For typical HTS (test concentration around 10 μM), hit identification is dependent on already having the appropriate compounds in the collection. This may be more true for some target classes and less so for others according to the target types and chemical classes of historical experience for each company or organization. Furthermore, company compound collections may be increasingly populated by compounds that are available to other companies; consequently, any hits carry additional competitive or intellectual property risks. Such risks are lowered for fragment-based hits because even common hits may be considerably evolved in novel ways during the optimization phase.

A second consequence for fragment-based methods is that by restricting to hits with high ligand efficiency, we improved the likelihood of evolving towards suitable drug candidates with a target potency of ~10 nM while keeping the molecular weight below 500 Da . The optimization of typical drug-like hits from HTS (typically molecular weight 450–500 and target potency 500–1000 nM) may require further increase in molecular weight or efforts to truncate and redesign. Compound **21** has 26 heavy atoms (molecular weight 351 Da) and a ligand efficiency of 0.37. Assuming that the ligand efficiency can be maintained, the target potency (10 nM) should be obtainable for a compound with molecular weight around 420 Da ; this would correspond to addition of as few as three heavy atoms. Fragment-based lead generation methods were successful here because three essential components were in place. First, NMR served as a robust and sensitive hit identification tool and methods were available to identify false positives quickly. Second, we could quantitate binding affinities over a wide range using SPR spectroscopy and thereby build and evaluate structure–activity relationships. Third, the design process was accelerated because we could obtain highly detailed structural information through crystallography on a routine basis, even for compounds with low affinity. Taken together, these approaches allowed us to identify hit **1** and rapidly improve its affinity by 30 000-fold to deliver lead **21** in an efficient manner, requiring only a small number of chemical modifications.

11.10 Acknowledgments

We gratefully acknowledge the contributions from our colleagues and collaborators at Astex Therapeutics, including Gianni Chessari, Miles Congreve, Chris Murray and Sahil Patel. We also thank Lise-Lotte Olsson for computational support and Tonny de Beer, Fredrik Edfeldt, Rutger Folmer and Stefan Geschwinder for scientific input, support and illustrations. Finally, we thank all our other collaborators whose names appear in the cited papers.

References

[1] Dominguez, D. I., De Strooper, B. Novel therapeutic strategies provide the real test for the amyloid hypothesis of Alzheimer's disease. *Trends Pharmacol. Sci.* 2002, **23**, 324–330.

[2] Luo, Y., Bolon, B., Damore, M. A., Fitzpatrick, D., Liu, H., Zhang, J., Yan, Q., Vassar, R., Citron, M. BACE1 (β-secretase) knockout mice do not acquire compensatory gene expression changes or develop neural lesions over time. *Neurobiol. Dis.* 2003, **14**, 81–88.

[3] Cai, H., Wang, Y., McCarthy, D., Wen, H., Borchelt, D. R., Price, D. L., Wong, P. C. BACE1 is the major β-secretase for generation of Aβ peptides by neurons. *Nat. Neurosci.* 2001, **4**, 233–234.

[4] Roberds, S. L., Anderson, J., Basi, G., Bienkowski, M. J., Branstetter, D. G., Chen, K. S., Freedman, S. B., Frigon, N. L., Games, D., Hu, K., Johnson-Wood, K., Kappenman, K. E., Kawabe, T. T., Kola, I., Kuehn, R., Lee, M., Liu, W., Motter, R., Nichols, N. F., Power, M., Robertson, D. W., Schenk, D., Schoor, M., Shopp, G. M., Shuck, M. E., Sinha, S., Svensson, K. A., Tatsuno, G., Tintrup, H., Wijsman, J., Wright, S., McConlogue, L. BACE knockout mice are healthy despite lacking the primary β-secretase activity in brain: implications for Alzheimer's disease therapeutics. *Hum. Mol. Genet.* 2001, **10**, 1317–1324.

[5] Luo, Y., Bolon, B., Kahn, S., Bennett, B. D., Babu-Khan, S., Denis, P., Fan, W., Kha, H., Zhang, J., Gong, Y., Martin, L., Louis, J.-C., Yan, Q., Richards, W. G., Citron, M., Vassar, R. Mice deficient in BACE1, the Alzheimer's β-secretase, have normal phenotype and abolished β-amyloid generation. *Nat. Neurosci.* 2001, **4**, 231–232.

[6] Harrison, S. M., Harper, A. J., Hawkins, J., Duddy, G., Grau, E., Pugh, P. L., Winter, P. H., Shilliam, C. S., Hughes, Z. A., Dawson, L. A., Gonzalez, M. I., Upton, N., Pangalos, M. N., Dingwall, C. BACE1 (β-secretase) transgenic and knockout mice: identification of neurochemical deficits and behavioral changes. *Mol. Cell. Neurosci.* 2003, **24**, 646–655.

[7] Vassar, R., Bennett, B. D., Babu-Khan, S., Kahn, S., Mendiaz, E. A., Denis, P., Teplow, D. B., Ross, S., Amarante, P., Loeloff, R., Luo, Y., Fisher, S., Fuller, J., Edenson, S., Lile, J., Jarosinski, M. A., Biere, A. L., Curran, E., Burgess, T., Louis, J.-C., Collins, F., Treanor, J., Rogers, G., Citron, M. β-Secretase cleavage of Alzheimer's amyloid precursor protein by the transmembrane aspartic protease BACE. *Science* 1999, **286**, 735–741.

[8] Sinha, S., Anderson, J. P., Barbour, R., Basi, G. S., Caccavello, R., Davis, D., Doan, M., Dovey, H. F., Frigon, N., Hong, J., Jacobson-Croak, K., Jewett, N., Keim, P., Knops, J., Lieberburg, I., Power, M., Tan, H., Tatsuno, G., Tung, J., Schenk, D., Seubert, P., Suomensaari, S. M., Wang, S., Walker, D., Zhao, J., McConlogue, L., John, V. Purification and cloning of amyloid precursor protein β-secretase from human brain. *Nature* 1999, **402**, 537–540.

[9] Yan, R., Bienkowski, M. J., Shuck, M. E., Miao, H., Tory, M. C., Pauley, A. M., Brashler, J. R., Stratman, N. C., Mathews, W. R., Buhl, A. E., Carter, D. B., Tomasselli, A. G., Parodi, L. A., Heinrikson, R. L., Gurney, M. E. Membrane-anchored aspartyl protease with Alzheimer's disease β-secretase activity. *Nature* 1999, **402**, 533–537.

[10] Hussain, I., Powell, D., Howlett, D. R., Tew, D. G., Meek, T. D., Chapman, C., Gloger, I. S., Murphy, K. E., Southan, C. D., Ryan, D. M., Smith, T. S., Simmons, D. L., Walsh, F. S., Dingwall, C., Christie, G. Identification of a novel aspartic protease (Asp 2) as β-secretase. *Mol. Cell. Neurosci.* 1999, **14**, 419–427.

[11] Lin, X., Koelsch, G., Wu, S., Downs, D., Dashti, A., Tang, J. Human aspartic protease memapsin 2 cleaves the β-secretase site of β-amyloid precursor protein. *Proc. Natl. Acad. Sci. USA* 2000, **97**, 1456–1460.

[12] Durham, T. B., Shepherd, T. A. Progress toward the discovery and development of efficacious BACE inhibitors. *Curr. Opin. Drug Discov. Dev.* 2006, **9**, 776–791.

[13] Baxter, E. W., Reitz, A. B. BACE inhibitors for the treatment of Alzheimer's disease. *Annu. Rep. Med. Chem.* 2005, **40**, 35–48.

[14] Schmidt, B., Baumann, S., Braun, H. A., Larbig, G. Inhibitors and modulators of β- and γ-secretase. *Curr. Top. Med. Chem.* 2006, **6**, 377–392.

[15] Guo, T., Hobbs, D. W. Development of BACE1 inhibitors for Alzheimer's disease. *Curr. Med. Chem.* 2006, **13**, 1811–1829.

[16] Ghosh, A. K., Kumaragurubaran, N., Tang, J. Recent developments of structure-based β-secretase inhibitors for Alzheimer's disease. *Curr. Top. Med. Chem.* 2005, **5**, 1609–1622.

[17] Hong, L., Koelsch, G., Lin, X., Wu, S., Terzyan, S., Ghosh, A. K., Zhang, X. C., Tang, J. Structure of the protease domain of memapsin 2 (β-secretase) complexed with inhibitor. *Science* 2000, **290**, 150–153.

[18] Leach, A. R., Hann, M. M., Burrows, J. N., Griffen, E. Fragment screening: an introduction. In *Structure-based Drug Discovery*, ed. Hubbard, R. E., Royal Society of Chemistry, Cambridge, 2006, pp. 142–172.

[19] Hann, M. M., Leach, A. R., Harper, G. Molecular complexity and its impact on the probability of finding leads for drug discovery. *J. Chem. Inf. Comput. Sci.* 2001, **41**, 856–864.

[20] Teague, S. J., Davis, A. M., Leeson, P. D., Oprea, T. The design of leadlike combinatorial libraries. *Angew. Chem. Int. Ed.* 1999, **38**, 3743–3748.

[21] Makara, G. M. On sampling of fragment space. *J. Med. Chem.* 2007, **50**, 3214–3221.

[22] Rees, D. C., Congreve, M., Murray, C. W., Carr, R. Fragment-based lead discovery. *Nat. Rev. Drug Discov.* 2004, **3**, 660–672.

[23] Fattori, D. Molecular recognition: the fragment approach in lead generation. *Drug Discov. Today* 2004, **9**, 229–238.

[24] Hajduk, P. J., Greer, J. A decade of fragment-based drug design: strategic advances and lessons learned. *Nat. Rev. Drug Discov.* 2007, **6**, 211–219.

[25] Jahnke, W., Erlanson, D. A. *Fragment-based Approaches in Drug Discovery*, Wiley-VCH Verlag GmbH, Weinheim, 2006.

[26] Jhoti, H., Cleasby, A., Verdonk, M., Williams, G. Fragment-based screening using X-ray crystallography and NMR spectroscopy. *Curr. Opin. Chem. Biol.* 2007, **11**, 485–493.

[27] Zartler, E. R., Mo, H. Practical aspects of NMR-based fragment discovery. *Curr. Top. Med. Chem.* 2007, **7**, 1592–1599.

[28] Lepre, C. A., Moore, J. M., Peng, J. W. Theory and applications of NMR-based screening in pharmaceutical research. *Chem. Rev.* 2004, **104**, 3641–3675.

[29] Coles, M., Heller, M., Kessler, H. NMR-based screening technologies. *Drug. Discov. Today* 2003, **8**, 803–810.

[30] Geschwindner, S., Olsson, L.-L., Albert, J. S., Deinum, J., Edwards, P. D., de Beer, T., Folmer, R. H. A. Discovery of a Novel warhead against β-secretase through fragment-based lead generation. *J. Med. Chem.* 2007, **50**, 5903–5911.

[31] Dalvit, C., Pevarello, P., Tato, M., Veronesi, M., Vulpetti, A., Sundstrom, M. Identification of compounds with binding affinity to proteins via magnetization transfer from bulk water. *J. Biomol. NMR* 2000, **18**, 65–68.

[32] Dalvit, C., Fogliatto, G., Stewart, A., Veronesi, M., Stockman, B. WaterLOGSY as a method for primary NMR screening: practical aspects and range of applicability. *J. Biomol. NMR* 2001, **21**, 349–359.

[33] Ghosh, A. K., Shin, D., Downs, D., Koelsch, G., Lin, X., Ermolieff, J., Tang, J. Design of potent inhibitors for human brain memapsin 2 (β-secretase). *J. Am. Chem. Soc.* 2000, **122**, 3522–3523.

[34] Cooper, M. A. Optical biosensors: where next and how soon? *Drug. Discov. Today* 2006, **11**, 1061–1067.

[35] Sinha, S., Anderson, J. P., Barbour, R., Basi, G. S., Caccavello, R., Davis, D., Doan, M., Dovey, H. F., Frigon, N., Hong, J., Jacobson-Croak, K., Jewett, N., Keim, P., Knops, J., Lieberburg, I., Power, M., Tan, H., Tatsuno, G., Tung, J., Schenk, D., Seubert, P., Suomensaari, S. M., Wang, S., Walker, D., Zhao, J., McConlogue, L., John, V. Purification and cloning of amyloid precursor protein β-secretase from human brain. *Nature* 1999, **402**, 537–540.

[36] Karlsson, R., Kullman-Magnusson, M., Hamalainen, M. D., Remaeus, A., Andersson, K., Borg, P., Gyzander, E., Deinum, J. Biosensor analysis of drug-target interactions: direct and competitive binding assays for investigation of interactions between thrombin and thrombin inhibitors. *Anal. Biochem.* 2000, **278**, 1–13.

[37] Abad-Zapatero, C. Ligand efficiency indices for effective drug discovery. *Expert Opin. Drug Discov.* 2007, **2**, 469–488.

[38] Hopkins, A. L., Groom, C. R., Alex, A. Ligand efficiency: a useful metric for lead selection. *Drug. Discov. Today* 2004, **9**, 430–431.

[39] Maillaird, M., Hom, C., Gailunas, A., Jagodzinska, B., Fang, L. Y., John, V., Freskos, J. N., Pulley, S. R., Beck, J. P., Tenbrink, R. E. Preparation of substituted amines to treat Alzheimer's disease. Patent WO2002002512, 2002.

[40] Patel, S., Vuillard, L., Cleasby, A., Murray, C. W., Yon, J. Apo and inhibitor complex structures of BACE (β-secretase). *J. Mol. Biol.* 2004, **343**, 407–416.

[41] Hajduk, P. J. Fragment-based drug design: how big is too big? *J. Med. Chem.* 2006, **49**, 6972–6976.

[42] Edwards, P. D., Albert, J. S., Sylvester, M., Aharony, D., Andisik, D., Callaghan, O., Campbell, J. B., Carr, R. A., Chessari, G., Congreve, M., Frederickson, M., Folmer, R. H. A., Geschwindner, S., Koether, G., Kolmodin, K., Krumrine, J., Mauger, R. C., Murray, C. W., Olsson, L.-L., Patel, S., Spear, N., Tian, G. Application of fragment-based lead generation to the discovery of novel, cyclic amidine β-secretase inhibitors with nanomolar potency, cellular activity and high ligand efficiency. *J. Med. Chem.* 2007, **50**, 5912–5925.

Index

Note: References to tables are given in bold type; references to figures are given in italic type.

Fragment-Based Drug Discovery: A Practical Approach Edited by Edward R. Zartler and Michael J. Shapiro